Post-genomic Approaches in Cancer and Nano Medicine

RIVER PUBLISHERS SERIES IN RESEARCH AND BUSINESS CHRONICLES: BIOTECHNOLOGY AND MEDICINE

Volume 4

Combining a deep and focused exploration of areas of basic and applied science with their fundamental business issues, the series highlights societal benefits, technical and business hurdles, and economic potentials of emerging and new technologies. In combination, the volumes relevant to a particular focus topic cluster analyses of key aspects of each of the elements of the corresponding value chain.

Aiming primarily at providing detailed snapshots of critical issues in biotechnology and medicine that are reaching a tipping point in financial investment or industrial deployment, the scope of the series encompasses various specialty areas including pharmaceutical sciences and healthcare, industrial biotechnology, and biomaterials. Areas of primary interest comprise immunology, virology, microbiology, molecular biology, stem cells, hematopoiesis, oncology, regenerative medicine, biologics, polymer science, formulation and drug delivery, renewable chemicals, manufacturing, and biorefineries.

Each volume presents comprehensive review and opinion articles covering all fundamental aspect of the focus topic. The editors/authors of each volume are experts in their respective fields and publications are peer-reviewed.

For a list of other books in this series, visit www.riverpublishers.com
http://riverpublishers.com/series.php?msg=Research and Business Chronicles: Biotechnology and Medicine

Post-genomic Approaches in Cancer and Nano Medicine

Kishore R Sakharkar

OmicsVista, Singapore

Meena K Sakharkar

Department of Pharmacy and Nutrition,
University of Saskatchewan, SK, Canada

Ramesh Chandra

Department of Chemistry, Delhi University, India
B. R. Ambedkar Center for Biomedical Research,
University of Delhi, India

Published 2015 by River Publishers
River Publishers
Alsbjergvej 10, 9260 Gistrup, Denmark
www.riverpublishers.com

Distributed exclusively by Routledge
4 Park Square, Milton Park, Abingdon, Oxon OX14 4RN
605 Third Avenue, New York, NY 10158

First published in paperback 2024

Post-genomic Approaches in Cancer and Nano Medicine / by Kishore R Sakharkar, Meena K Sakharkar, Ramesh Chandra.

ISBN: 978-87-93102-86-6 (hbk)
ISBN: 978-87-7004-488-2 (pbk)
ISBN: 978-1-003-33908-3 (ebk)

DOI: 10.1201/9781003339083

Contents

Series Note xv

Preface xvii

Acknowledgements xix

List of Figures xxi

List of Tables xxxi

1 Alternative Splicing and Cancer 1
J.E. Kroll, A.F. Fonseca and S.J.de Souza
1.1 Introduction 1
1.1.1 Overall Features of mRNA Splicing 1
1.1.2 Alternative Splicing 2
1.1.2.1 Types of alternative splicing 3
1.1.2.2 Identification of alternative splicing variants 4
1.1.3 Alternative Splicing and Cancer 6
1.1.3.1 Mutations affecting splicing signals and regulatory elements 6
1.1.3.2 Splicing factors affected in cancer 7
1.1.3.3 Protein families altered by alternative splicing in cancer 8
1.1.3.4 Splicing variants as cancer biomarkers . . 9
1.2 Perspectives 11

2 Non-coding RNAs as Molecular Tools 25
Renu Wadhwa, Yoshio Kato, Ran Gao and Sunil C. Kaul
2.1 Introduction 25
2.2 Ribozyme as Molecular Tool 28
2.2.1 Muscle Differentiation as a Model 28

2.2.2 Cancer Mechanism as a Model 29
2.2.3 Cancer Targets as a Model 34
2.3 siRNA as a Molecular Tool 39
2.3.1 Anticancer siRNAs and Cancer Targets 39
2.3.2 Cancer Drug Target Screening 42
2.4 miRNA as a Molecular Tool 42
2.4.1 Detection of miRNAs 42
2.4.1.1 RT-PCR and Related Technologies 42
2.4.1.2 Hybridization-Based Technologies 46
2.4.1.3 Imaging of miRNAs in Cells 47
2.4.2 miRNA in Cancers 49

3 PPAR Responsive Regulatory Modules in Breast Cancer 61

Meena K Sakharkar, Babita Shashni, Karun Sharma, Ramesh Chandra and Kishore R Sakharkar

3.1 Introduction 61
3.2 PPAR Gamma, Breast Cancer and Energy Metabolism . . . 65
3.3 Glycolysis, Cell pH and NHE1 69
3.4 ROS and Breast Cancer 72
3.5 Natural and Synthetic Ligands of PPARγ 75
3.6 Conclusions 77

4 Animal Model of Cancer and Infection 85

Karun Sharma, Babita Shashni, Meena K Sakharkar, Kishore R Sakharkar and Ramesh Chandra

4.1 Introduction 85
4.2 Mouse Models for Cancer 86
4.3 *In-vivo Methods* 87
4.3.1 Tumor Model by use of Chemical Agents 87
4.3.1.1 DMBA- induced mouse skin papillomas . 88
4.3.1.2 N-methyl, N-nitrosurea (MNU) induced Rat mammary gland carcinogenesis 89
4.3.1.3 DMBA induced rat mammary gland carcinogenesis 89
4.3.1.4 MNU-induced tracheal squamous cell carcinoma in hamster 89
4.3.1.5 Azoxymethane(AOM) induced aberrant crypt Foci in Rat 89

4.3.1.6 1,2 Dimethylhydralazine(DMH) induced colorectal adenocarcinomain rat and mouse 90
4.3.2 Models Involving Cell Lines or Tumor Pieces (Xenograft) 90
4.3.2.1 Cell lines 90
4.3.2.2 Xenograft 91
4.3.3 Transgenic Mice 91
4.3.3.1 Retroviral vector method 92
4.3.3.2 DNA microinjection method 92
4.3.3.3 Nude mouse 94
4.4 Anti-cancer Drugs 94
4.5 Mouse Model of Infection 96
4.5.1 Air Pouch Model 96
4.5.2 Dermatophytosis 97
4.5.3 Endocarditis Model 98
4.5.4 Lung Infection Model 99
4.5.5 Thigh Infection Model 99
4.5.6 Urinary Tract Infection Model 99

5 Potential Application of Natural Compounds for the Prevention and Treatment of Hepatocellular Carcinoma 101
Shikha Satendra Singh, Sakshi Sikka, Gautam Sethi and Alan Prem Kumar
5.1 Introduction 101
5.2 Risk Factors Involved in HCC 103
5.2.1 Viral infections 103
5.2.1.1 Hepatitis B virus (HBV) 103
5.2.1.2 Hepatitis C virus (HCV) 104
5.2.2 Toxins 104
5.2.2.1 Chemical carcinogens 104
5.2.2.2 Other carcinogens 104
5.2.3 Diet and Metabolic factors 105
5.2.4 Genetic Factors 105
5.2.5 Cirrhosis 106
5.3 Types of Liver Cancers 106
5.3.1 Epithelial tumors (Malignant) 106
5.3.1.1 Hepatocholangiocarcinoma (HCC-CC) . . 106
5.3.1.2 Hepatoblastoma 106

5.3.2 Non-Epithelial Tumors (Benign) 106
5.3.2.1 Hepatic angiomyolipoma (AML) 106
5.3.2.2 Hemangioma 107
5.3.3 Non-Epithelial Tumors (Malignant) 108
5.3.3.1 Hepatic Angiosarcoma 108
5.3.3.2 Hepatic epithelioidhemangioendothelioma (EH) 108
5.3.3.3 Embryonal sarcoma 108
5.3.3.4 Hepatic Rhabdomyosarcoma 108
5.3.4 Miscellaneous tumors 109
5.3.4.1 Solitary fibrous tumor 109
5.3.4.2 Hepatic teratomas 109
5.3.4.3 York Sac tumor 109
5.3.4.4 Carcinosarcoma 109
5.3.4.5 Rhabdiod tumor 109
5.3.4.6 Hepatic mesenchymalhamartoma (HMH) 110
5.3.5 Secondary tumors 110
5.4 Dysregulated Signaling Cascades in HCC 110
5.4.1 STAT3 signaling 111
5.4.2 Wnt/β-catenin pathway 113
5.4.3 NFκB pathway 114
5.4.4 PI3K/AKT/mTOR pathway 116
5.4.5 MAPK pathway 119
5.4.6 VEGF pathway 120
5.4.7 Other miscellaneous pathways 121
5.5 Reported Anti-Cancer Effects of Natural Compounds Against Hepatocellular Carcinoma 123
5.5.1 Curcumin 124
5.5.2 Selenium 126
5.5.3 Epigallocatechin-3-gallate (EGCG) 126
5.5.4 Resveratrol 127
5.5.5 Ursolic Acid 129
5.5.6 Pterostilbene 130
5.5.7 Celastrol 131
5.5.8 Honokiol 132
5.5.9 γ- Tocotrienol 132
5.5.10 Butein 133
5.5.11 β- Escin 134

5.5.12 Diosgenin 135
5.5.13 Phyllanthusniruri 136
5.5.14 Oleanolic Acid 136
5.5.15 Mushrooms 137
5.5.16 JuglansMandshurica 138
5.5.17 Quercetin 139
5.5.18 Ginger 139
5.6 Conclusions 142

6 Nanomaterials: A Ray of Hope in Infectious Disease Treatment 167

Rashmi M. Bhande and C.N.Khobragade
6.1 Introduction 167
6.2 Systemic Applications of Nanoparticles 170
6.2.1 Antimicrobial Nanotechnology Based Drug Delivery 171
6.2.1.1 NPs for efficient antimicrobial drug delivery 171
6.2.1.2 Liposome for antimicrobial drug delivery . 171
6.2.1.3 Solid lipid (SL) NPs 172
6.2.1.4 Polymeric NPs 173
6.2.1.5 Dendrimers 174
6.2.1.6 Drug-infused nanoparticles 175
6.2.1.7 Chitosan Nanoparticle 175
6.2.1.8 Silver Nanoparticle 176
6.2.1.9 Copper Nanoparticle 176
6.2.1.10 Titanium Nanoparticle 177
6.2.1.11 Magnesium Nanoparticles 177
6.2.1.12 Zinc Nanoparticle 178
6.2.1.13 Nitric oxide-releasing Nanoparticles . . . 178
6.2.1.14 Immunomodulatory effects of nanotechnology-based drug delivery systems 179
6.2.1.15 Nanotechnology-based vaccines and immunostimulatory adjuvant 179
6.2.1.16 Synthetic polymers 180
6.2.1.17 Nanoemulsions 180
6.2.1.18 Immune-stimulating complexes 180
6.2.1.19 Cytidine-phosphate-guanosine (CpG) motifs 181
6.2.1.20 Fullerenes (C60) and fullerene-derivatives 181
6.2.1.21 Carbon nanotubes (CNTs) 181

6.2.1.22 Surfactant-based nanoemulsions 182
6.3 Nanocarriers 182
6.3.1 Types of nanocarriers 182
6.3.1.1 Liposome 182
6.3.1.2 Polymeric micelles 182
6.3.1.3 Polymer blended nanoparticles 183
6.3.1.4 Fluorescent Nanoparticles 183
6.4 Synergism of Antibiotics with Zinc Oxide Nanoparticles: A Study of Urinary Tract Infections 183
6.4.1 Structural and Morphological Evaluation of Synthesized ZnO NPS 184
6.4.2 Surface Analysis of Synthesized ZnO NPS 185
6.4.3 Optical Analysis 186
6.4.4 TIME–KILL ASSAY 186
6.5 Appications of Nanoparticles 189
6.6 Conclusions 191

7 Nanomedicine for the Treatment of Oxidative Stress Injuries 199

Toru Yoshitomi, Long Binh Vong and Yukio Nagasaki

7.1 Introduction 199
7.2 Preparation and Characterization of Redox Polymers and Nanoparticles 200
7.2.1 Design and Preparation of Redox Polymers and Redox Nanoparticles 200
7.2.2 Safety of RNPs 202
7.2.3 pH-Sensitive Disintegration of RNP^N 203
7.3 Treatment of Ischemia-Reperfusion Injuries with pH-Sensitive Redox Nanoparticles 205
7.3.1 Ischemia-Reperfusion Injury 205
7.3.2 Biodistribution and Morphological Change of RNP^N 206
7.3.3 Therapeutic Effect of RNPs on Renal Ischemia-Reperfusion Injury in Mice 208
7.4 Oral Nanotherapy with pH-Insensitive Redox Nanoparticle for the Treatment of Inflammatory Bowel Disease 209
7.4.1 Inflammatory Bowel Disease (IBD) 209
7.4.2 Specific Accumulation of Orally Administered RNP^O without Uptake in the Blood Stream 211

7.4.3 Therapeutic Effect of RNP^{O} in a Mouse Model of Colitis 213
7.5 Treatment of Other Oxidative Stress Injuries 213
7.6 Conclusion 214

8 Rational Design of Multifunctional Nanoparticles for Targeted Cancer Imaging and Therapy 219
Arun K. Iyer
8.1 Introduction 219
8.1.1 Mechanism of Tumor Selective Delivery: Passive and Active Targeting 221
8.2 Rational Design of Nanoparticles 224
8.2.1 Shape of Nanoparticles 226
8.2.2 Size and Surface Charge of Nanoparticles 227
8.2.3 Surface Functionalization of Nanoparticles 228
8.3 Types of Nanoparticle Systems for Cancer Therapy 229
8.3.1 Polymeric Nanoparticles 229
8.3.2 Polymeric Micelles and Dendrimers 231
8.3.3 Lipid-Based Nanoparticles 233
8.4 Multifunctional Theranostic Nanosystems 235
8.5 Illustrative Examples of Multifunctional Nanosystems . . . 238
8.5.1 Iron Oxide Nanoparticles-Based Theranostic Systems 238
8.5.2 Quantum-Dots-Based Nano-Theranostic Agents . . . 239
8.5.3 Gold Nanoparticles-Based Theranostic Agents . . . 241
8.5.4 Silica Nanoparticles and Carbon Nanotubes-Based Theranostic Agents 243
8.6 Conclusions and Future Directions 244

9 Nanomedicine for the Treatment of Breast Cancer 267
Surendra Nimesh, Nidhi Gupta and Ramesh Chandra
9.1 Introduction 267
9.2 Etiology of Breast Cancer 268
9.2.1 Histology and Diagnosis of Breast cancer 269
9.2.2 Luminal A subgroup 270
9.2.3 Luminal B Subgroup 270
9.2.4 Basal-like Carcinomas, a Subgroup of Triple-negative Breast Cancers 271
9.2.5 Normal Breast 271

9.2.6 Claudin-low 271
9.3 Nanomedicine 272
9.4 Targeted Nanomedicine 273
9.4.1 Passive Targeting 273
9.4.2 Active Targeting 275
9.5 Nanomedicince for the Treatment of Breast Cancer 275
9.5.1 Liposomes 275
9.5.2 Gold Nanoparticles 277
9.5.3 Carbon Nanotubes 278
9.5.4 Human Serum Albumin Nanoparticles 278
9.5.5 Other Nanomaterials 279
9.6 Conclusions 279

10 Nanoparticle-Based Drug Delivery Systems: Associated Toxicological Concerns and Solutions **287**

James Lyons, Aniruddha Bhati, Jaimic Trivedi and Arati Sharma

10.1 Introduction 287
10.1.1 Nanotechnology and Nanoparticles (NP) 288
10.1.2 Nanotechnology and Cancer 289
10.1.2.1 Nanoparticles in diagnostics 289
10.1.2.2 MRI (Magnetic Resonance Imaging) . . . 290
10.1.2.4 Cantilevers 290
10.1.2.5 Bio-barcode 291
10.2 Nanoparticles as Onco-therapeutics 291
10.2.1 Drug Delivery Systems (DDSs) 291
10.2.1.1 Polymeric biodegradable nanoparticles . . 292
10.2.1.2 Metallic nanoparticles 292
10.2.1.3 Ceramic nanoparticles 292
10.2.1.4 Polymeric micelles 293
10.2.1.5 Dendrimers 293
10.2.1.6 Liposomes 294
10.2.2 Nucleic acid carriers 295
10.2.3 Magnetofection 295
10.3 Nanotoxicology 295
10.3.1 Important physiochemical properties, which causes toxicity 296
10.3.1.1 Particle size 296
10.3.1.2 Particle composition and charge 296

10.3.1.3 Particle surface area 296
10.3.2 Nanodrug delivery systems and toxicity 296
10.3.2.1 Liposome toxicity 297
10.3.2.2 Dendrimer-based drug delivery system toxicity 297
10.3.2.3 Metallic nanoparticle toxicity 297
10.3.2.4 Quantum dot toxicity 298
10.4 Assessment of Nanotoxicology 298
10.4.1 *In vitro* cell culture assays 298
10.4.1.1 Biocompatibility assays 299
10.4.1.2 Hemolytic and platelet aggregation tests . 299
10.4.1.3 Reactive Oxygen Species (ROS) and oxidative stress detection assays 299
10.4.1.4 Genotoxicity assays 299
10.4.2 *In vivo* animal assays 299
10.4.2.1 Dose-range determination 299
10.4.2.2 Pharmacokinetics 299
10.4.2.3 Immunotoxicity 300
10.5 Toxicity Modulation to Optimally Exploit Clinical Potential of Nanoparticles as a Therapeutics 300
10.5.1 Size 300
10.5.2 Shape 301
10.5.3 Charge 302
10.5.4 Masking to escape immune rejection 302
10.5.5 Leaching of the constituents entrapped in nanoparticles 302
10.5.6 Adding targeting moieties to nanoparticles 303
10.6 Future Implication 303
10.7 Conclusions 304

11 Biodegradable Carrier Systems for Drug and Vaccine Delivery 315
Anil Mahapatro and Dinesh Singh
11.1 Introduction 315
11.2 Nanoparticle Fabrication 318
11.2.1 Dispersion of preformed polymers 320
11.2.1.1 Polymerization Methods 323
11.2.1.2 Ionic gelation method for hydrophilic polymers 324

11.2.1.3 Biodegradable polymer matrix for nanoparticle fabrication 324
11.2.1.4 Nanoparticle functionalization 325
11.3 Specific Applications of Biodegradable Nps 328
11.3.1 Tumor Targeting 328
11.3.1.1 Nanoparticles for Oral delivery 329
11.3.1.2 Nanoparticles for vaccine adjuvants and gene delivery 330
11.3.1.3 Nanoparticles for drug delivery into the brain 331
11.4 Conclusions 332

About the Editors **341**

About the Contributors **343**

Index **351**

Series Note

The deciphering in 2003 of the nucleotide sequence of the human genome, which followed the determination of the chromosomal sequences of an array of model microorganisms including *Escherichia coli*, *Bacillus subtilis*, and *Saccharomyces cerevisiae*, is a landmark in biology that has paved the way for a revolution in research and development in biotechnology and pharmaceutical sciences. Moreover, several novel discovery tools have since emerged, including dramatically enhanced computers, sequencing instruments with dramatically higher throughout and decreased costs, softwares conferring the ability to generate and manage very large amounts of data, and systems biology tools which have enabled *in silico* experiments and the creation of virtual patients or virtual microbes to both accelerate and increase the scope of pharmaceutical and biotechnological research. Whereas more than 10 years have already elapsed since this major scientific milestone, the translation into novel products, perhaps best exemplified by the current focus of pharmaceutical companies on personalised medicine, has only reached mainstream at the beginning of the present decade.

The impact of post-genomic approaches in cancer and nano-medicine development is the focal point of the present monograph. Starting with a review of underlying bases of cancer and the biology of coding and non-coding RNAs, principles of discovery of novel drugs including advances in animal models for oncology are laid out here. Revisiting the potential of natural compounds for the treatment and prevention of carcinomas, the discussion subsequently explores one of the next innovation S-curves in cancer therapeutics using nanomaterials as a case study. The ultimate purpose of the journey is to accelerate the development of disease-modifying pharmaceuticals, and answer unmet medical needs to enable cancer patients worldwide achieve remission and, ideally, cure.

Alain Vertès, Basel, Switzerland
Pranela Rameshwar Rutgers, USA
Paolo di Nardo, Roma, Italy

Preface

Cancer is a complex disease involving genomic alterations across several molecular mechanisms. Systematic and comprehensive elucidation of the molecular landscape of a wide range of cancers complemented by genome-wide approaches to interrogating the function of cancer genes and the vulnerabilities of tumors will pave the way for understanding the basic molecular mechanisms of cancer and applying this knowledge to transform the practice of cancer medicine. Alternative splicing has critical roles in normal cell function and development and can promote growth and survival in cancer. Aberrant splicing can lead to loss-of-function in tumor suppressors or activation of oncogenes and cancer pathways. Cancer-specific changes in splicing profiles can occur through mutations that are affecting splice sites and splicing control elements, and also by alteration in the expression of proteins that control splicing decisions. Chapter 1 presents a comprehensive review on alternative splicing and how it contributes to tumorigenesis by producing splice isoforms that can stimulate cell proliferation and cell migration or induce resistance to apoptosis and anticancer agents. Chapter 2 discusses the use of non-coding RNAs as molecular tools to understand the molecular mechanism of cell proliferation control during carcinogenesis, differentiation and drug-induced cytotoxicity. Malignant cells exhibit metabolic changes, when compared to their normal counterparts. Chapter 3 delineates the identification and validation of novel targets of nuclear hormone receptor (PPAR-γ) in glycolytic pathway and their role in breast cancer pathophysiology. Animal experiments have contributed significantly to our understanding of mechanisms of disease and mouse has been the model of choice. Chapter 4 describes various mouse/rat models for cancer and infectious diseases. Chapter 5 describes the use of natural compounds for hepatocellular carcinoma.

The advent of nanotechnology promises revolutionizing many fields including oncology, by proposing advanced systems for cancer treatment.

Targeted drug delivery systems are among the most successful examples of nanotechnology. In the past few years, there has been significant momentum in the field of nanomedicine with the development of novel nanoparticles for the diagnosis and treatment of cancer. Their small size, large surface area-to-volume ratio, and surface characteristics enable them to have viable carrier for site specific delivery of vaccines, genes, drugs and other biomolecules in the body. They also have compatibility with different administration routes, which makes them highly attractive in many aspects of oncology and infectious diseases. Chapter 6 through Chapter 11 discuss the use of nanoparticles in cancer therapeutics. In putting together this book, we have tried to bring to table the contributions of various experts towards some key aspects in drug discovery with focus on cancer and naomedicine. As editors of this book, we are grateful to all the contributors who have made this book possible.

Acknowledgements

On behalf of all the authors, we would like to thank all our mentors, colleagues and friends who instilled in us the culture of science. Without support from them, we could not have written this book. The unconditional love and support from our families is gratefully acknowledged.

Finally, we would like to take this opportunity to acknowledge the services of the team of River Publishers and everyone who collaborated in producing this book.

Kishore R. Sakharkar

Meena K. Sakharkar

Ramesh Chandra

List of Figures

Figure 1.1	Most important splicing signals	2
Figure 1.2	Molecular diversity increases as information flows from genes to proteins	3
Figure 1.3	Types of alternative splicing events: (A) exon skipping, (B) 3' alternative border, (C) 5' alternative border, (D) intron retention, (E) dual-specific splice site, (F) mutually exclusive exons and (G) exon skipping of multiple adjacent exons	4
Figure 1.4	RNA-Seq mapping, as implemented by TopHat. (A) Mapping of splice-sites boundaries. (B) Amount of aligned reads (read depth)	5
Figure 1.5	CD44 isoforms associated to cancer	10
Figure 2.1	Schematic representation of action of ribozymes and miRNA	26
Figure 2.2	Schmatic representation of use of randomized ribozyme library for identification of genes involved in muscle differentiation: adopted from Wadhwa *et al.* [26]	30
Figure 2.3	Demonstration of ARF-Per19p interaction in mouse (p19ARF), but not in human (p14ARF) cells. Pex19p specific ribozymes increased the activity of p19ARF only as shown in G. Adopted from Wadhwa *et al.* [39]	33
Figure 2.4	Use of randomized for identification of genes involved in killing od cancer cells by Ashwagandha leaf extract (i-Extract)and its pure phytochemical, Withanone. Adopted from Widodo *et al.* [46] . . .	37
Figure 2.5	Use of mortalin staining as a reporter for induction of senescence in cancer cells and hence the identification of anticancer siRNAs. Adopted from Gao *et al.* [65]	41
Figure 2.6	A representative flowchart of miRNA detection. Three major approaches to detect miRNAs;	

hybridization, RT-PCR and cellular imaging are shown 43
Figure 2.7 Flowchart describing the modification of miRNAs for amplification is shown. In order to amplify miRNA pools from the total RNA, miRNAs can be tagged with adaptors based on their chemical characteristics as shown 44
Figure 2.8 Schematic representation of molecular spotter probe. The probe is designed to be quenched in the presence of precursor RNA. It emits fluorescence only upon hybridization to the mature RNA 47
Figure 2.9 Schematic representation of the dual-color sensor vector system. Reporter system enables the real-time, quantitative detection of miRNA in single cells using dual color fluorescent proteins 48
Figure 2.10 Demonstration of targeting of p53 and $p21^{WAF1}$ by miR-296. Adopted from Yoon et al [88] 52
Figure 3.1 PPARγ activation mechanism. Upon ligand activation PPARγ heterodimerizes with Retinoid X Receptor (RXR) in nucleus and binds to PPRE and/or PACM motifs in the promoter region and modulates the expression of genes downstream. The consensus PPRE site consists of a direct repeat of the sequence AGGTCA separated by a single/double nucleotide, which is designated as DR-1 site/DR-2 site and PACM site consist of 15 bp consensus sequence, TTCATTTGGACATTG. The PACM motifs are reported to be more common than PPREs 62
Figure 3.2 Molecular targets of PPARγ and pathways associated [Adapted from 27] 63
Figure 3.3 Genomic structure of the human PPAR gamma gene (5' end) and PPARγ mRNA splicing forms and protein variants. There are seven isoforms of PPARγ with common exons 1–6 64
Figure 3.4 Transcriptional regulation of PPARγ gene targets. A PPAR protein binds PPRE/ PACM motifs in combination with retinoid X receptors (RXRs) upon

ligand activation. The two paired up proteins then regulate transcription of various genes e.g. PPARγ upon activation is known to down regulate glycolytic genes - PGK1 and PKM2, pH regulator -NHE1, anti-oxidant enzyme - MnSOD in breast cancer cells 64

Figure 3.5 PPARγ activation inhibits many malignancies . . . 65

Figure 3.6 Metabolic targets of PPARγ. Many glycolytic enzymes are over expressed in cancers. Glycolytic enzyme pyruvate kinase-muscle 2 (PKM2) is a key regulator of tumor metabolism which promotes tumor growth and Warburg effect by switching between its dimeric form the active one, which has higher affinity for substrate Phosphoenol pyruvate (PEP) to tetrameric form the inactive form, with lower affinity for substrate PEP and vice-versa. This switching behavior of PKM2 keeps a balance of activation of many pathways including, glycerol, serine/glycine, ether/ester phospholipid pyrimidine biosynthesis (in green) and oxidative metabolism for energy production, thereby promoting tumor growth and tumor cell proliferation 67

Figure 3.7 Regulation of glycolytic genes - PGK1 and PKM2 by PPARγ in human breast cancer cell lines – MDA-MB-231 and MCF-7. The human breast cancer cells were exposed to 10 μM of PPARγ inhibitor GW9662 for 4 h followed by 5 μM and 10 μM of 15d-PGJ2 for 48 h at 37 °C. PPARγ activation by 15d-PGJ2 down regulated the expression of glycolytic enzyme, PGK1 and PKM2 in breast cancer cell lines - MDA-MB-231 (7A and 7C) and MCF-7 (7B and 7D). Inhibiting the activation of PPARγ by the PPARγ inhibitor GW9662, did not affect the expression of PGK1 and PKM2, suggesting the transcriptional regulation of these glycolytic genes by PPARγ [Adapted from 27] 68

Figure 3.8 Apoptosis and PPARγ. Apoptosis was initiated in human breast cancer cells lines MDA-MB-231 and MCF-7 upon 15d-PGJ2 activation of PPARγ.

The breast cancer cells were exposed to 5 μM and 10 μM of 15d-PGJ2 for 48 h at 37 °C. Caspase dependent apoptosis was confirmed in MDA-MB-231 and MCF-7 by expression studies for active caspase 8 (8A and 8B) and chromatin condensation as assessed by nuclear specific dye Hoechst (8C and 8D) [Adapted from 27] 69

Figure 3.9 Disruption of mitochondrial potential in human breast cancer cells by PPARγ ligand. The human breast cancer cells were exposed to 5 μM and 10 μM of 15d-PGJ2 for 48 h at 37 °C. PPARγ ligand, 15d-PGJ2 induced loss of mitochondrial potential in human breast cancer cell lines -MDA-MB-231 and MCF-7 as assessed by potential dependent dye, JC-1. In healthy/non-apoptotic cells, JC-1 exists as a monomer in the cytosol (green) and accumulates as J-aggregates in the active mitochondria, which appear red. In apoptotic cells, due to loss mitochondrial potential relatively lower dye J-aggregates accumulates in the mitochondria than cytoplasm where it is remains as a monomer. Control breast cancer cell lines had higher Red/Green flouresence ratio than 15d-PGJ2 treated test cells, suggesting the loss of mitochondrial potential by PPARγ ligand, 15d-PGJ2 [Adapted from 27] 70

Figure 3.10 PPARγ activation by 15d-PGJ2 represses NHE1 mRNA and protein levels in human breast cancer cell lines - MDA-MB-231 and MCF-7. The breast cancer cells were exposed to 1 μM, 3 μM and 5 μM of 15d-PGJ2 for 24 h at 37 °C. A significant down regulation of NHE1 expression was observed. A similar pattern was followed at transcriptional level, suggesting PPARγ regulation of NHE1 in breast cancer cell lines [Adapted from 44] 71

Figure 3.11 PPARγ novel ligand, Hydroxy hydroquinone (HHQ) induces intracellular Reactive Oxygen Species (ROS) formation in human breast cancer

Figure 3.11 cell lines - MDA-MB-231 and MCF-7. The breast cancer cells were treated with 12.5 μM and 25 μM of HHQ for 48 h at 37 °C. Intracellular ROS formation was found to be significantly increased in HHQ treated cells as compared to control cells in a dose dependent manner was reported. The effective enhancement of ROS production by HHQ correlates to its cytotoxicity nature
[Adapted from 52] 73

Figure 3.12 PPARγ activation by 15d-PGJ2 represses MnSOD mRNA and protein levels in human breast cancer cell lines - MDA-MB-231 and MDA-MB-468. The breast cancer cells were exposed to 3 μM, 5 μM and 10 μM of 15d-PGJ2 for 24 h at 37 °C. A significant repression of MnSOD level was observed in dose-dependent manner [Adapted from 44] 74

Figure 4.1 Retroviral vector method 92

Figure 4.2 DNA microinjection method 93

Figure 4.3 Anti cancer drug targets 95

Figure 4.4 The Cell cycle 96

Figure 4.5 Air pouch model 97

Figure 4.6 Urinary tract infection 100

Figure 5.1 Deregulation of JAK/STAT3 pathway in Hepatocellular carcinoma: Upregulation of IL-6, Mutations in gp130, Methylation of SOCS has been reported as plausible mechanisms for deregulation of JAK/STAT3 pathway. IL6: Interleukin-6; JAK:Janus Kinase-2; STAT3:Signal Transducer and Activation of transcription 3; SOCS: Suppressor of cytokine signaling 112

Figure 5.2 Deregulation of β-catenin in hepatocellular carcinoma: Upregulation of FZD7, mutations in AXINS, production of stable beta-catenin and increase in cell-cell adhesion have been postulated as plausible mechanisms for deregulation of beta-catenin pathway in HCC. FZD7: Frizzled-7 protein; GSK3B: Glycogen Synthase kinase 3 beta; TCF: Transcription factor 114

Figure 5.3 Deregulation of NFkB in hepatocellular carcinoma: Elevated levels of IL-6 and TNFα have been postulated in constitutive activation of NFkB, thereby, deregulation in hepatocellular carcinoma due to upregulation of pro-survival signals. NFkB:Nuclear factor kappa-light chain-enhancer of activated B cells; TNF: Tumor necrosis factor; IL-6: Interleukin-6; IKK: IκB kinase 115

Figure 5.4 Deregulation of PI3K/Akt/mTOR pathway in hepatocellular carcinoma: Elevated levels of Akt phosphorylation, overexpression of phoshomTOR, somatic mutations of PTEN have been postulated to play a plausible role in deregulation of PI3K/Akt/mTOR pathway in hepatocellular carcinoma. PTEN : Phosphatase and tensin homolog; PI3K : Phophoinositide-3-kinase; IGF : Insulin-like growth factor; EGF : Epidermal growth factor; PDGF : Platelet derived growth factor; PDK1 : Phosphoinositide-dependent kinase-1; PIP2: Phosphophatidylinositol 4, 5-biphosphate; PIP3:Phosphophatidylinositol 3, 4, 5-triphosphate; Akt: Protein kinase B 118

Figure 6.1 The XRD spectra of ZnO nanoparticles 184

Figure 6.2 (a) The SEM image of synthesized ZnO NPs. (b) TEM images of ZnO NPs. (c) HR-TEM of ZnO NPs (d) SAED pattern 185

Figure 6.3 XPS spectra of ZnO nanoparticles. (a, b, c represents the scan over wide range and magnified band structure at Zn and O level) 186

Figure 6.4 UV-Visible spectrum of ZnO NPs 187

Figure 6.5 Time –Kill curve of *E.coli* 187

Figure 6.6 Time –Kill curve of K.pneumoniae 188

Figure 6.7 Time –Kill curve of *S.paucimobilis* 189

Figure 6.8 Time –Kill curve of *P.aeruginosa* 189

Figure 7.1 Chemical structures of redox polymers possessing nitroxide radicals, PEG-b-PMNT and PEG-b-PMOT, and a redox nanoparticle (RNP) 201

Figure 7.2 *In vitro* characterization of RNP^{N} and RNP^{O} **(a)** Effect of pH on the light scattering intensities of

RNP^N (closed circle) and RNP^O (open circle). The normalized scattering intensity (%) is expressed as the value relative to that at pH 8.2 **(b,c)** X-band ESR spectra of (b) RNP^N and (c) RNP^O from pH 5.6 to 8.2. Reprinted with permission from [15] and [19] and modification.© American Chemical Society (2009) and Elsevier B.V. (2011), respectively. 204

Figure 7.3 Environmental-signal-enhanced polymer drug therapy using RNP^N for the treatment of oxidative stress injuries 206

Figure 7.4 Time profile of drug concentration in **(a)** blood and **(b)** injured kidney. (white circle, RNP^O; black circle, RNP^N; white square, TEMPOL) **(c)** Therapeutic effect of RNP on renal IR. BUN and Cr levels in the plasma of mice are measured at 24 h after reperfusion following 50 min of ischemia. Drugs are administered at 5 min after reperfusion. Sham veh, sham-operated and vehicle-treated groups; IR veh, vehicle . . . .
treated group; IR RNP^N, RNP^N-treated group; IR (RNP^O, RNP^O-treated group; IR TEMPOL, TEMPOL-treated group. Values are expressed as mean ± SE. *$P < 0.0001$ as compared to IR veh. **$P < 0.005$ as compared to IR veh. ***$P < 0.05$ as compared to IR veh. n = 7, ANOVA) Reprinted with permission from [19] and modification. © Elsevier B.V. (2011). 207

Figure 7.5 Oral nanotherapy with RNP^O reduces the inflammation in UC patients. RNP^O is stable, withstands the harsh conditions of the GIT, and reaches the colon to scavenge ROS, especially at sites of inflammation. 210

Figure 7.6 **(a)** Specific accumulation of RNP^O in the colon. Accumulation of LMW TEMPOL, RNP^O, and polystyrene latex particles in the colon. The data are expressed as mean ± SE, $n = 3$. **(b,c)** Therapeutic effect of RNP^O on DSS-induced colitis in mice. **(a)**(b) Changes in DAI. DAI is the summation of the stool consistency index (0–3), c the fecal bleeding index (0–3), and the weight loss index (0–4). The data are expressed as mean ± SE, *P <0.05, **P < 0.01, and ***P < 0.001 vs. control group; ‡P < 0.05 and P< 0.001 vs. DSS groups, n = 6–7, two-way ANOVA, followed by the Bonferroni post-hoc test. **(c)** The survival rate of mice with 3% (wt/vol) DSS treatment for 15 d. From 5 d, drugs are orally administered daily until 15 d. The number of surviving mice is counted until 15 d, n = 6. Reprinted with permission from [20]. © Elsevier B.V. (2012). 212

Figure 8.1 Concept of EPR effect for tumor targeted drug delivery. (Adapted from Ref. [25] with permissions Elsevier B.V.) 221

Figure 8.2 The schematic shows the passive and active targeting mechanisms of multifunctional image guided nanoparticles and the difference in the vasculature of normal and tumor tissues; drugs and small molecules diffuse freely in and out of the normal and tumor blood vessels due to their small size and thus the effective drug concentration in the tumor drops rapidly with time. However, macromolecular drugs and nanoparticles can passively target tumors due to the leaky vasculature or the EPR effect, however they cannot diffuse back into blood stream due to their large size and impaired lymphatic clearance, leading to enhanced tumor accumulation and retention. Targeting molecules such as antibodies or peptides decorated on the surface of nanoparticles can selectively bind to cell surface receptors/antigens overexpressed by tumor cells and can be taken up by receptor-mediated endocytosis (active targeting). The image guiding molecules and contrast agents conjugated/encapsulated in the nanoparticles can be useful for targeted imaging and (non-invasive) visualization of nanoparticle accumulation/localization, as well as for mechanistic understanding of events and efficacy of drug treatment simultaneously (Reprinted with permissions from Ref. [58] Bentham Science. 223

Figure 8.3 RGD peptide functionalized polycaprolactone-gold microparticle design for colon cancer screening. A. Fabrication steps; **B**. 3D design and; **C**. Scanning Electron Micrograph (SEM) revealing a size of ~1.5 μm for the microparticle. Please see Ref. [59] and [63] for details 224

Figure 8.4 Schematic of Type of Nanoparticles (NP) Used for Cancer Imaging and Therapy 234

Figure 8.5 Publications on Multifunctional Theranostic Nanosystems. Number of annual publications on theranostic (red bars) and multifunctional (blue bars) nanosystems in medical applications, for the years 2000 to 2012. (Search results obtained from Google scholar and Highwire press) 236

Figure 8.6 Schematic Illustration of Nano-theranostic Agent in Action. Multifunctional theranostic nanoparticle (NP) endowed with tumor homing ligands encapsulated with drugs, genes and imaging agents when inject intravenously into the blood stream, can selectively target tumor tissues and cells. Furthermore, the single nanoconstruct can be used for simultaneous cancer detection/imaging, diagnosis and measurement of treatment response, all performed non-invasively, in real-time 237

Figure 9.1 Types of nanoparticles, (1) nanospheres and (2) nanocapsules 273

Figure 9.2 Representation of tumor targeting by nanoparticles via EPR effect 274

Figure 9.3 Different types of liposomes 276

Figure 10.1 Various approaches to address the toxicological concerns assoiciated with nanoparticledbased drug delivery systems 301

Figure 10.2 Schematic Theranostic, a multifunctional particle 304

Figure 11.1 Multifunctional nanoparticles. Multifunctional nanoparticles can combine a specific targeting agent (usually with an antibody or peptide) with nanoparticles for imaging (such as quantum dots or magnetic nanoparticles), a cell-penetrating agent (e.g., the polyArg peptide TAT), a stimulus-selective element for drug release, a stabilizing polymer to ensure biocompatibility polyethylene glycol most frequently), and the therapeutic compound. Development of novel strategies for controlled released of drugs will provide nanoparticles with the capability to deliver two or more therapeutic agents. Adapted from ref [18] Copyright 2009 Wiley interscience 320

Figure 11.2 Schematic representation of various techniques for the preparation of polymer nanoparticles. SCF: supercritical Fluid technology, C/LR: controlled/livingradical. Adapted from ref [5] Copyright 2011 Elsevier 321

Figure 11.3 Schematic representation of the emulsification-evaporation technique. Adapted from ref [19] Copyright 2006
Elsevier 323

Figure 11.4 3PEGylated nanoparticles are able to avoid clearance from the blood stream by repelling protein adsorption, thus prolonging nanoparticle circulation time within the body. Adapted from ref [44] Copyright 2013 Elsevier 326

Figure 11.5 Schematic representation of different drug-targeting approaches. Adapted from ref [9] Copyright 2012 Elsevier 327

List of Tables

Table 2.1 microRNAs associated with human cancers 50

Table 3.1 List of natural and synthetic ligands of PPARγ . . 76

Table 5.1 Types of liver cancer and pre-cancerous forms . . . 107

Table 5.2 List of natural compounds and their mode of action in HCC 140

Table 8.1 Polymer–Drug Conjugates in Clinical Trials 231

Table 10.1 Various classes of nanoparticles 288

Table 10.2 Issues surrounding nanotoxicology 293

Table 10.3 Issues surrounding nanotoxicology 298

Table 11.1 Examples nanotechnology based products that are being tested or approved for commercial use. Adapted from ref [9] Copyright 2012 Elsevier . . . 317

Table 11.2 Nanoparticles for drug delivery and the subsequent therapeutic improvement. Adapted from ref [9] Copyright 2012 Elsevier 319

Table 11.3 Polymeric nanoparticles: general advantages and drawbacks of the preparation methods Adapted from ref [19] Copyright 2006 Elsevier 322

1

Alternative Splicing and Cancer

J. E. Kroll[1,2], A. F. Fonseca[1,2] and S. J. de Souza[2]

[1]Institute of Bioinformatics and Biotechnology, Natal, Brazil
[2]Brain Institute, UFRN, Natal, Brazil

1.1 Introduction

Next-generation sequencing is allowing an exhaustive exploration of the complexity of the human transcriptome through RNA-Seqtechnologies [1, 2]. What would be a dream ten years ago, 20 million reads from the transcriptome of a cellor tissue, is nowadays considered a small RNA-Seq output. That technological advance brings different challenges related to the bioinformatics analysis of sequences derived from these technologiess incethereads are shorter than conventional Expressed Sequence Tags (ESTs) and identification of transcript variants is problematic. Inspite of these problems, RNA-Seq is widely used in a variety of models and experimental conditions.

The field of alternative splicing is booming especially in disease models. The impact of alternative splicing in cancer has been recognized and it is expected that soon research on these topics will have a profound effect in clinical practices around the globe.

In this chapter, we present a broad introduction to the field of alternative splicing giving a special emphasis in its impact in cancer research.

1.1.1 Overall Features of mRNA Splicing

The interrupted nature of almost all eukaryotic genes forces them to a post-transcriptional process, called splicing, in which introns are removed and exons arespliced together [3]. This processing is done by the spliceosome, a ribonucleo protein (RNP) composed by five small nuclear RNAs (U1,U2,U4,U5 and U6), and by more than 150 proteins [4–7].

Post-genomic Approaches in Cancer and Nano Medicine, 1–24.

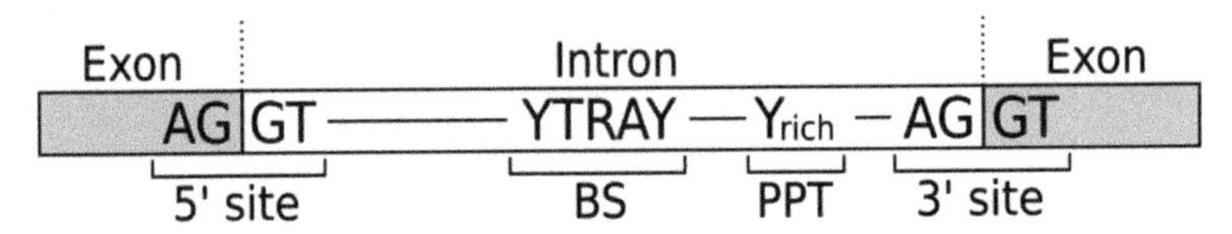

Figure 1.1 Most important splicing signals.

The spliceosome recognizes rexons and introns through signals present in cis along the RNA [8]. The most important signals are the 5'(donor) and 3'(acceptor) splice sites, which are conserved in more than 95% of all exon–exon borders [4, 5]; the branch site (BS); and the polypyrimidinetract (PPT), which is located upstream the 3'splice site [3, 9] (Figure 1.1). In metazoans, those signals are not enough to promote the splicing process, and they may represent less than half of the information needed [10].

The splicing process is also regulated by other cis-acting regulatory elements. They correspond to short sequences (4 upto 18 nucleotides), which usually interact with trans-acting factors, determining whether a splicesite is used or not. These splicing regulatory elements (SREs) are found in exonicandintronic sequences, and can be classified as enhancers (ESE and ISE) or silencers (ESS and ISS). They are needed for the constitutive splicing, as for the regulation of the alternative splicing [11–16].The most known regulatory proteins are these rine/arginine-rich (SR) proteins and heterogeneous nuclear ribonucleoproteins (hnRNP) [17, 18]. SR proteins are known to bind ESEs and promote splicing, while hnRNPs are known to bind to ESSs and ISSs, repressing the cognition of adjacent splicing sites [19].

1.1.2 Alternative Splicing

Alternative splicing is defined, basically, as a process in which identical pre-mRNA molecules are processed in different ways in terms of usage of exons/introns borders. It is a fundamental process in complex organisms [3, 20], and is responsible for creating a large diversity of proteins from a relatively small number of genes [21] (Figure 1.2). It affects a variety of process such as subcellular localization and stability of mRNA and ultimately the efficiency of mRNA transcription.

Almost all human genes, as annotated by the International Human Genome Sequencing Consortium [22] possess more than one mRNA through alternative splicing [23]. Different events resulted from that process can be

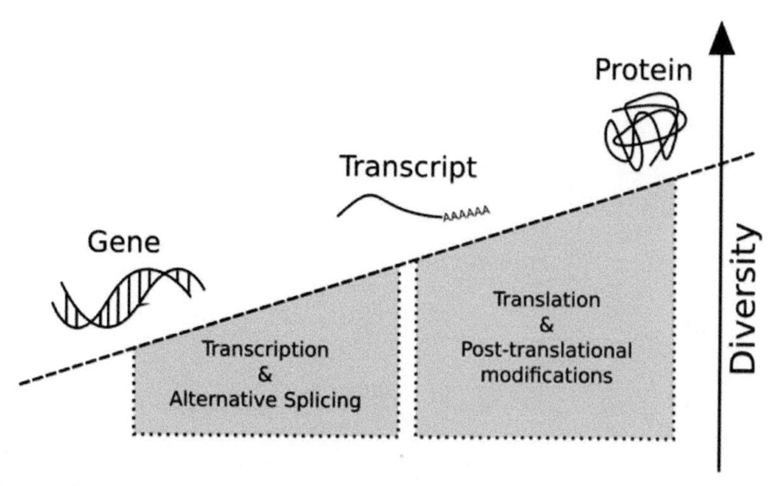

Figure 1.2 Molecular diversity increases as information flows from genes to proteins.

observed in human development [24] and about 18% of all events seems to be tissue-specific [25].

Alternative splicing events are able to shape the activity of many cellular processes [26], and abnormalities in that process can result in different diseases, such as cancer [27–29], is chemia [30] and other kinds of human disorders [31–33].

1.1.2.1 Types of alternative splicing

Five types of alternative splicing events are currently known: exon skipping, alternative 5' and 3' splice sites, intron retention, and dual-specific splice site, which was recently identified and characterized [34]. Exon skipping events are identified when an exon is removed from a transcript, and alternative 5' or 3' splice sites are identified when different 5' or 3' sites are used by an exon, respectively. Intron retention events are identified when an intron is not removed and is still present in the mature mRNA [35], and dual-specific splice sites, the most uncommon type of event, are characterized when the same splice site can eventually work as 5' or 3' [34] (Figure 1.3).

When two or more simple events are observed occurring in the same mRNA, they are characterized as complex alternative splicing event [36]. That kind of event exists due to the complex splicing regulation and high frequency of simple events. The most radical case of a complex event, for example, occurs in the gene Dscam of *Drosophila*. It has a cluster of

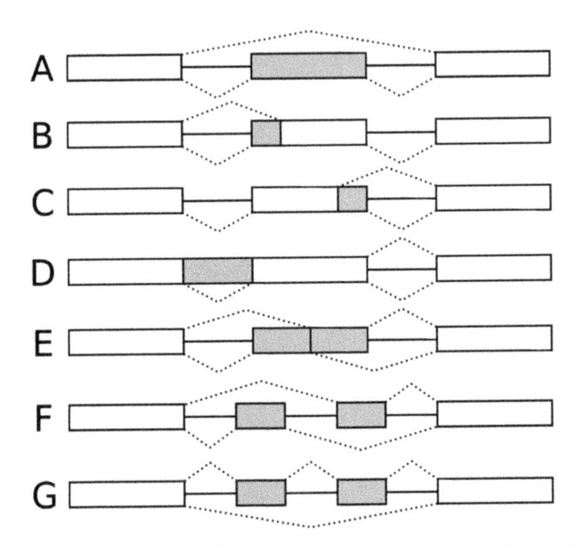

Figure 1.3 Types of alternative splicing events: (A) exon skipping, (B) 3' alternative border, (C) 5' alternative border, (D) intron retention, (E) dual-specific splice site, (F) mutually exclusive exons and (G) exon skipping of multiple adjacent exons.

48 mutually exclusive exons that can, theoretically, generate thousands of alternative splicing variants [37].

One of the most important complex events is the mutually exclusive exons, in which the presence of an exon promotes the skipping of an adjacent exon, and vice versa (Figure 1.3). The most frequent complex event, however, is the skipping of multiple adjacent exons. Alternative promoters and alternative *polyadenylation,* due to alternative splicing events, are also classified as complex events [3, 38–40].

1.1.2.2 Identification of alternative splicing variants

The identification of alternative splicing events is far from easy. In recent years, genome-wide approaches based on expressed sequence tags (EST), microarray and RNA-Seq have become standard methods for the investigation of alternative splicing events. EST databases and microarray technologies were extensively explored by some studies [41–43]. The first initiatives using large datasets were produced from EST sequences, which are found in databases such as dbEST that currently has more than 20 million of human sequences [44]. Traditionally, expressed sequences have been a rich source for alternative splicing identification. Through a basic alignment of an EST against a reference genome, it is possible to detect any type of event, including those classified as complex. Different algorithms have been developed for the alignment of expressed sequences, such as SIM4 and BLAT [45, 46]. These algorithms consider splice-sites boundaries, which are needed to differ a simple indel from an intronic region [47].

An alternative for ESTs is microarray. Highly parallel microarray platforms have been widely used, since it allows the identification and quantification of alternative splicing events for a specific sample [48]. Some arrays are based on differential hybridization techniques, such as exon-level or exon–exon junctions [49], which allow the identification of sample signatures with a good precision [50–52]. However, conventional microarrays are limited, since their probes are designed to target complementary sequences. This means that alternative splicing events can be detected only if they are already known [53].

Differently from conventional microarrays, tiling array hybridization method seems to be a better choice since it can detect novel transcripts [54]. This technology has been explored by several studies to define profiles between samples, to identify alternative splicing events and to discover novel biomarkers [55, 56].

Finally, next generation sequencing platforms provide a high-throughput method for entire transcriptomes (RNA-Seq) [47, 57]. In general, RNA-Seq samples consist of a pool of purified RNA, which is converted to cDNA library and sequenced by NGS technologies. Such methodology has shown unlimited possibilities of analysis, such as gene expression, single nucleotide variation calling, fusion gene detection, absolute quantification and identification of splicing variants [58–60]. Currently, platforms from llumina, Applied BiosystemsSOLiD and Roche 454 have been used for those purposes [47]. Each of those technologies has its peculiarities, which are associated with different strategies and tools of analysis.

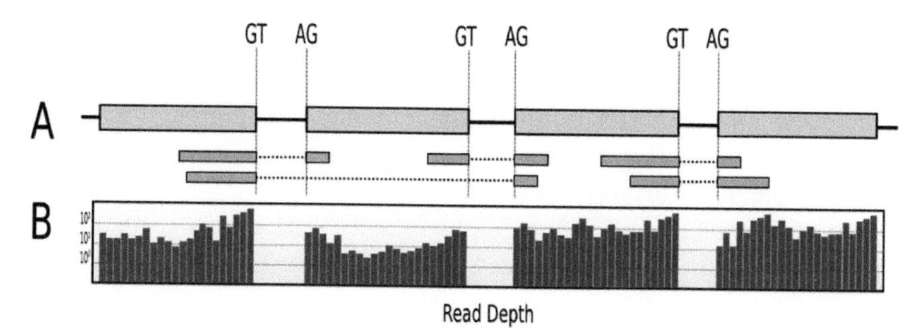

Figure 1.4 RNA-Seq mapping, as implemented by TopHat. (A) Mapping of splice-sites boundaries. (B) Amount of aligned reads (read depth).

Transcriptome analysis using RNA-Seq data require high bioinformatics background, and many bioinformatics tools have been developed, such as TopHat, MMES, Split-Seek and SpliceMap [61–64]. Those algorithms first align RNA-Seq sequence reads against a reference genome. For the unmapped reads, the algorithms dynamically divides these in short fragments and align them to genome based on splicing junctions (Figure 1.4). Finally, the transcriptome is assembled for analysis of the transcripts through the use of algorithms like Scripture and Cufflinks, which are genome-guided methods [65, 66].

1.1.3 Alternative Splicing and Cancer

In cancer, a large number of alternative splicing events were already described [67, 28], and many of them show a relationship with mutations in splicing signals and elements from the spliceosomal complex [68]. Reports suggest, furthermore, that some events can also be the result of the differential expression of SR proteins and hnRNPs [69], as like the result of epigenetic mechanisms that affect the splice-site recognition efficiency [70].

It is usually a hard task to determine whether a cancer-specific alternative splicing event is biologically meaningful [9, 71]. The problem is that many of those events usually introduce premature stop codons within the transcripts, which can have their expression inhibited through different mechanisms [72–74]. However, studies showed that alternative splicing is able to shape the activity of a large set of proteins [27, 67], such as transcription factors, cellular signal transducers and components from the extracellular matrix [28,75].

1.1.3.1 Mutations affecting splicing signals and regulatory elements

Mutations affecting 5' or 3' splicing sites are related to different diseases and types of cancer, and they are a powerful way to promote alternative splicing, since these sites are very conserved and sensible to any variation. The 5' and 3' splicing sites usually show the dinucleotides GT and AG in the intronic sequence side, respectively [76, 77], and mutations can lead to different alternative splicing events, such as exon skipping, activation of a cryptic splice-site, and intron retention [78].

An important example is TP53, a classical cancer-related gene with tumor suppressor activity. Dozens of different mutations within TP53 splicing sites were already reported for different types of cancer. Interestingly, some of those mutations seems to be "neutral" because they do not change the amino-acid, although they are able to affect splicing [79]. Recently, a report showed a *novel*

TP53 splice-site mutation related to osteosarcoma development. Sequence analysis observed a mutation at the donor splice-site of intron 6 (G to A transition) that created a 6 amino acid insertion [80].

Other examples include the genes SMARCB1, MLH1, ATM, BRCA1 and NF2 [81–86], which were identified by different studies. Currently, due to technological advances, genome-wide analyses have been possible, allowing the discovery of many other cancer-related genes and splicing events at once. For example, recently a deep whole-genome sequencing and transcriptome sequencing was performed for 19 lung cancer cell lines and three lung tumor/normal pairs. A total of 106 splice-site mutations associated with alternative splicing events were observed, including mutations in several known cancer-related genes [87].

Some mutations can also affect the PPT signal, which is target of a PTB (polypyrimidine tract-binding protein) that has several roles in RNA processing, such as alternative splicing, mRNA stability, mRNA localization and translation [88]. In colorectal cancer, for example, a mutated PPT in MLH1 is known to cause an exon skipping that truncates the protein and induces cancer development [89].

Although mutations within splicing regulators can show less influence over alternative splicing events, it is still being possible. One example is KLF6, a transcription factor that has a tumor suppressor role. A single nucleotide polymorphism was reported to create a new binding site for the SR protein SRp40, which increased the expression of the KLF6-SV1 isoform [90]. That new isoform lacks a zinc finger domain [91], antagonizing the wild-type KLF6 function in a negative manner. Hence, KLF6-SV1 over-expression accelerates prostate cancer progression and metastasis, and its polymorphism association with prostate cancer suggests an increased cancer risk [78].

1.1.3.2 Splicing factors affected in cancer

SR proteins are one of the most important splicing factors, and they can function both as splicing enhancers or silencers by interfering with the spliceosome assembly [92, 93]. Small changes in their expression levels can deregulate alternative splicing events, affecting cellular behavior [94–97]. Reports have shown that SR protein levels change during tumorigenesis, and that SR proteins levels are lower in tumors than in the normal counterparts [94].

Alterations in SR proteins expression in tumorigenesis clearly affect alternative splicing. In mouse mammary adenocarcinoma, the altered expression of SR proteins change the splicing pattern of CD44, which is induced to express metastatic mRNA variants [98, 99]. Some SR proteins, such as PTBP, showed increased levels during ovarian tumorigenesis, while other factors, such as SRSF1 and U2AF65, did not show variations in their expression levels [100]. There is a clear association between the levels of specific SR proteins and cancer. The mechanisms behind their regulation remain to be elucidated [101].

1.1.3.3 Protein families altered by alternative splicing in cancer

Although most multi-exonic genes undergo alternative splicing, its effect in some protein families seems to have a deeper impact in the physiology of cells and tissues. These protein families are usually related to regulatory phenomena. For example, DNMT3b, a DNA methyltransferase, lacks an exon resulting in a truncated protein, which is related to hypomethylation on genomic pericentromeric satellite regions in liver tumors [102]. Another example is PASG, a homolog of SNF2 that encodes a chromatin-remodeling protein. A specific alternative 5' splicing site in PASG seems prevalent in different samples of acute leukemia [103]. The dysfunction of those proteins directly affects the splice-site recognition efficiency by the spliceosome complex [70].

Transcription factors are a frequent target of alternative splicing. Alternative splicing of NRSF, a transcription-silencing factor, is known to generate a truncated protein in lung cancer [104]. Other examples are related to hormones receptors, such as AIB1, which is a nuclear hormone receptor co-activator. In tumor samples, a large amount of AIB1 variants lacking exon 3 is observed. They are known to promote estrogen receptor-mediated transcription, which is related to the development of breast cancer [105]. Variant forms of AR (androgen receptor) were also observed for breast cancer [106].

Many alternative splicing events can be observed for proteins from the cell surface, which are expressed by genes such as CD44 and by genes from the family of fibroblast growth factor receptors (FGFRs). Among genes from other protein families, there is SVH that is overexpressed in liver cancer and may play a role in cell growth and survival. From a few different variants, only SVH-B is observed in tumor, and its inhibition can cause apoptosis [107]. Other proteins from cell surface have adhesion function. One example is MUC1, which has an essential role in forming protective mucous barriers on epithelial surfaces and is involved in metastasis. In thyroid cancer cells, MUC1 shows a specific cryptic exon [108].

An interesting class of proteins is related to cell signaling, especially the soluble ones that can transmit oncogenic signals. One example is NF1, which has a tumor suppressor function by interfering through the ras signal transduction pathway. In medulloblatomas and primitive neuroectodermal tumors, weaker NF1 suppressor variants are predominant, differently from observed for normal brain tissue [109]. Another example is SYK, a tyrosine kinase related to the supression of metastasis. In breast cancer, a specific SYK variant was observed to fail to prevent metastasis [110].

Finally, extracellular proteins are usually secreted in the extracellular matrix and have essential roles in metastatic growth. uPG (urokinase-type plasminogen activator), for example, is a secreted protein, and some of its variants are involved in the degradation of extracellular matrix proteins [111]. Similar to uPG, WISP1 gene shows a variant that lacks an exon that caused invasion of cells through collagen in gastric carcinoma [112].

1.1.3.4 Splicing variants as cancer biomarkers

Biomarkers of cancer-specific alternative splicing variants are of great importance due to their specificity, which make them excellent targets for diagnosis and/or ther apeutic treatments. Monoclonalanti bodies for cancer-specific targets have been used since the 70's against solid tumors [113]. In 2001, about 700 monoclonal antibodies against cancer were estimated to be in clinical tests, sponsored by more than 200 biotechnology laboratories [114]. Currently, due to the rapid scientific and technological development, those numbers shall surpass dozens of thousands.

The genes from the surfaceome project [115] are known to codify proteins from the cellular surface, where they perform an important role in the inter cellular communication and can be easily spotted by antibodies, for example. Many of those proteins are related to the cancer development, in which many cases of specific alternative splicing events were already identified. Among the genes of cellular surface usually affected, there are the well-known CD44 and the fibroblast growth factor receptors (FGFR), both previously discussed.

CD44 shows the most studied alternative splicing events for cancer. It codifies more than 20 protein isoforms known due to the varied incorporation of 10 alternative exons close to the extra cellular domains. Its normal and prevalent forms do not have alternative exons, and the isoforms associated to cancer are characterized by the presence of the v4–7or v8–10 exons (Figure 1.5) [28]. Those isoforms are expressed in different kinds of cancer in several animal species, and are, hence, important therapeutic targets. Antibodies against CD44 isoforms already showed to be able to reduce the tumor skill to develop metastasis and resistance against different therapeutics [116], although they showed limitations against solid tumors [113]. Clinically, through the presence

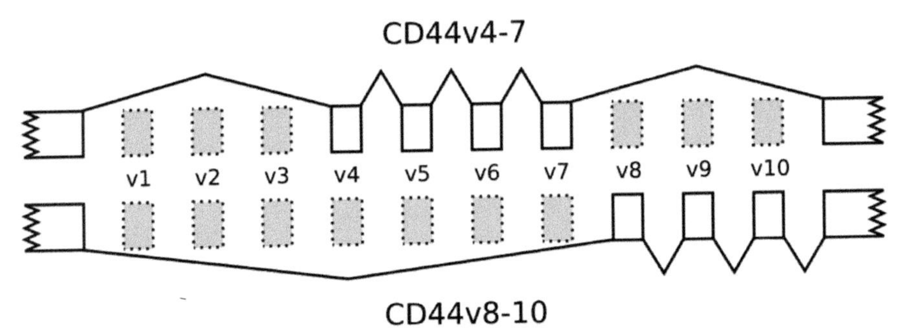

Figure 1.5 CD44 isoforms associated to cancer.

of some CD44 variants in blood, it is possible to differentiate among different tumor status. For example, the down-regulation of CD44 v6 variant is related to tumor malignancy and favorable prognosis in prostate cancer, while benign prostate tissues overexpress CD 44v7–9 variant [117, 118].

FGFRs have been detected in normal and malignant cells and have a crucial role in cell differentiation and development. All genes from that group undergo alternative splicing events, which are needed to regulate the binding specificity between FGFRs and fibroblast growth factors(FGFs) [119]. An interesting feature is that soluble FGFRs, which lack the trans membrane domains due to aberrant alternative splicing events, have been identified in blood from breast cancer patients [120]. Some examples, such as FGFR1 beta isoform, FGFR3 and FGFR2 showed an important role in cancer development [121], colorectal tumorigenesis [122] and prostate cancer [123], respectively.

Another interesting family of proteins is the GPCR, which is a super family of trans membrane receptors that have a large distribution and capacity to identify a large number of ligands. Hence, the activation or inhibition of GPCRs signals can affect several pathophysiological processes [124]. It is known that about 50% of the genes from the GPCRs family do not have introns within codifying regions, and the other genes can undergo alternative splicing and generate different protein isoforms, which can differ in their properties of signalization and regulation [124]. Those events have been considered of being of great importance, since they were already identified for different kinds of cancer, including melanoma, breast, ovary, prostate, liver and gastro intestinal cancer [67, 125, 126].

The detection of biomarkers based on alternative splicing variant scan be done throughs simple standard techniques. Microarrays, for example, have proved to be a robust method [127], and it is able to resolve alternative splicing signatures of different tumor subclasses [128, 129]. Alternative splicing variants can also be detected through real-time quantitative PCR by using specific primers, or through the use of antibodies that target variant proteins. Targeting protein variants with antibodies is the most common strategy for therapeutics. However, there are methods able to interfere in alternative splicing mechanisms. For example, antisense-based methods have been developed to silence specific alternative splicing events through RNA antisense, RNA interference, and hybrid protein nucleic acids [28].

Short anti senseoligonucleotides(AONs) complementary to regions from target pre-mRNAs are being used to interfere in disease processes. It can degrade the target mRNA, as like promote exon inclusion or skipping by blocking splice-sites, enhancers or suppressors. The inducing of exon skipping has been successfully applied to restore the reading frame of mutated genes in different diseases. A major advantage of AONs is the ability to target specific variants. One limitation, however, is its efficiency in delivering AONs to specific tissues for a specific duration [78].

1.2 Perspectives

The availability of genome and transcriptome sequences from thousands of human tumors will allow the identification of many genetic and epigenetic alterations causally involved in cancer. Together with clinical data, this information will represent a priceless resource for the development of new diagnostic and therapeutic strategies. We envisage that splicing variants will emerge as valuable targets for such strategies.

References

[1] Mutz, K. O., A. Heilkenbrinker, M. Lönne, J. G. Walter, F. Stahl., "Transcriptome analysis using next-generation sequencing," Current Opinion in Biotechnology, Vol. 24, N. 1, 2012, pp. 22–30.

[2] Martin, J. A., and Z. Wang., "Next-generation transcriptome assembly," Nature Review Genetics, Vol. 12, N. 10, 2011, pp. 671–82.

[3] Black, D. L., "Mechanisms Of Alternative Pre-messenger RNA Splicing," Annual Review Of Biochemistry, Vol. 72, 2003, pp. 291–336.

[4] Deckert, J., K. Hartmuth, D. Boehringer, N. Behzadnia, C. L. Will, B. Kastner, H. Stark, H. Urlaub and R. Lührmann, "Protein Composition And Electron Microscopy Structure Of Affinity-purified Human Spliceosomal B Complexes Isolated Under Physiological Conditions," Molecular And Cellular Biology, Vol. 26, No. 14, 2006, pp. 5528–43.

[5] Hartmuth, K., H. Urlaub, H. P. Vornlocher, C. L. Will, M. Gentzel, M. Wilm and R. Lührmann, "Protein Composition Of Human Pre-spliceosomes Isolated By A Tobramycin Affinity-selection Method," Proceedings Of The National Academy Of Sciences Of The United States Of America, Vol. 99, No. 26, 2002, pp. 16719–24.

[6] Jurica, M. S. and M. J. Moore, "Pre-mRNA Splicing: Awash In A Sea Of Proteins," Molecular Cell, Vol. 12, No. 1, 2003, pp. 5–14.

[7] Zhou, Z., L. J. Licklider, S. P. Gygi and R. Reed, "Comprehensive Proteomic Analysis Of The Human Spliceosome," Nature, Vol. 419, No. 6903, 2002, pp. 182–5.

[8] Berget, S. M., "Exon recognition in vertebrate splicing," The Journal Of Biological Chemistry, Vol. 270, 1995, pp. 2411–2414.

[9] Graveley, B. R., "Alternative Splicing: Increasing Diversity In The Proteomic World," Trends In Genetics, Vol. 17, No. 2, 2001, pp. 100–7.

[10] Lim, L. P. and C. B. Burge, "A Computational Analysis Of Sequence Features Involved In Recognition Of Short Introns," Proceedings Of The National Academy Of Sciences Of The United States Of America, Vol. 98, No. 20, 2001, pp. 11193–8.

[11] Cáceres, J. F. and A. R. Kornblihtt, "Alternative Splicing: Multiple Control Mechanisms And Involvement In Human Disease," Trends In Genetics, Vol. 18, No. 4, 2002, pp. 186–93.

[12] Woodley, L. and J. Valcárcel, "Regulation Of Alternative Pre-mRNA Splicing," Briefings In Functional Genomics Proteomics, Vol. 1, No. 3, 2002, pp. 266–77.

[13] Blencowe, B. J., "Exonic Splicing Enhancers: Mechanism Of Action, Diversity And Role In Human Genetic Diseases," Trends In Biochemical Sciences, Vol. 25, No. 3, 2000, pp. 106–10.

[14] Graveley, B. R., "Sorting Out The Complexity Of SR Protein Functions," RNA, Vol. 6, No. 9, 2000, pp. 1197–211.

[15] Cartegni, L., S. L. Chew and A. R. Krainer, "Listening To Silence And Understanding Nonsense: Exonic Mutations That Affect Splicing," Nature Reviews Genetics, Vol. 3, No. 4, 2002, pp. 285–98.

[16] Fairbrother, W. G., R. F. Yeh, P. A. Sharp and C. B. Burge, "Predictive Identification Of Exonic Splicing Enhancers In Human Genes," Science, Vol. 297, No. 5583, 2002, pp. 1007–13.

[17] Sanford, J. R., J. Ellis and J. F. Cáceres, "Multiple Roles Of Arginine/serine-rich Splicing Factors In RNA Processing," Biochemical Society Transactions, Vol. 33, No. 3, 2005, pp. 443–6.

[18] Singh, R. and J. Valcárcel, "Building Specificity With Nonspecific RNA-binding Proteins," Nature Structural Molecular Biology, Vol. 12, No. 8, 2005, pp. 645–53.

[19] Cartegni, L., S. L. Chew and A. R. Krainer, "Listening To Silence And Understanding Nonsense: Exonic Mutations That Affect Splicing," Nature Reviews Genetics, Vol. 3, No. 4, 2002, pp. 285–98.

[20] Maniatis, T. and B. Tasic, "Alternative Pre-mRNA Splicing And Proteome Expansion In Metazoans," Nature, Vol. 418, No. 6894, 2002, pp. 236–43.

[21] Cork, D. M., T. W. Lennard, A. J. Tyson-Capper, D. M., "Progesterone Receptor (Pr) Variants Exist In Breast Cancer Cells Characterised As Pr Negative," Tumour Biology: The Journal Of The International Society For Oncodevelopmental Biology And Medicine, Vol. 33, No. 6, 2012, pp. 2329–40.

[22] International Human Genome Sequencing Consortium, "Finishing the euchromatic sequence of the human genome," Nature, Vol. 431, No. 7011, 2004, pp. 931–45.

[23] Johnson, J. M., J. Castle, P. Garrett-Engele, Z. Kan, P. M. Loerch, C. D. Armour, R. Santos, E. E. Schadt, R. Stoughton and D. D. Shoemaker, "Genome-wide Survey Of Human Alternative Pre-mRNA Splicing With Exon Junction Microarrays," Science, Vol. 302, No. 5653, 2003, pp. 2141–4.

[24] Black, D. L., Grabowski, P. J., "Alternative Pre-mRNA Splicing And Neuronal Function," Progress In Molecular And Subcellular Biology, Vol. 31, 2003, pp. 187–216.

[25] Markovic, D. and D. K. Grammatopoulos, "Focus On The Splicing Of Secretin Gpcrs Transmembrane-domain 7," Trends In Biochemical Sciences, Vol. 34, No. 9, 2009, pp. 443–52.

[26] Hsu, S. N. and K. J. Hertel, "Spliceosomes Walk The Line: Splicing Errors And Their Impact On Cellular Function," RNA Biology, Vol. 6, No. 5,, pp. 526–30.

[27] Wang, G. S. and T. A. Cooper, "Splicing In Disease: Disruption Of The Splicing Code And The Decoding Machinery," Nature Reviews Genetics, Vol. 8, No. 10, 2007, pp. 749–61.

[28] Venables, J. P., "Aberrant And Alternative Splicing In Cancer," Cancer Research, Vol. 64, No. 21, 2004, pp. 7647–54.

[29] Kirschbaum-Slager, N., R. B. Parmigiani, A. A. Camargo and S. J. de Souza, "Identification Of Human Exons Over expressed In Tumors

Through The Use Of Genome And Expressed Sequence Data," Physiological Genomics, Vol. 21, No. 3, 2005, pp. 423–32.

[30] Daoud, R., G. Mies, A. Smialowska, L. Oláh, K. A. Hossmann and S. Stamm, "Ischemia Induces A Translocation Of The Splicing Factor Tra2-beta 1 And Changes Alternative Splicing Patterns In The Brain," The Journal Of Neuroscience: The Official Journal Of The Society For Neuroscience, Vol. 22, No. 14, 2002, pp. 5889–99.

[31] Faustino, N. A. and T. A. Cooper, "Pre-mRNA Splicing And Human Disease," Genes Development, Vol. 17, No. 4, 2003, pp. 419–37.

[32] Garcia-Blanco, M. A., A. P. Baraniak and E. L. Lasda, "Alternative Splicing In Disease And Therapy," Nature Biotechnology, Vol. 22, No. 5, 2004, pp. 535–46.

[33] Pagani, F. and F. E. Baralle, "Genomic Variants In Exons And Introns: Identifying The Splicing Spoilers," Nature Reviews Genetics, Vol. 5, No. 5, 2004, pp. 389–96.

[34] Zhang, C., M. L. Hastings, A. R. Krainer and M. Q. Zhang, "Dual-specificity Splice Sites Function Alternatively As 5' And 3' Splice Sites," Proceedings Of The National Academy Of Sciences Of The United States Of America, Vol. 104, No. 38, 2007, pp. 15028–33.

[35] Kim, E., A. Magen and G. Ast, "Different Levels Of Alternative Splicing Among Eukaryotes," Nucleic Acids Research, Vol. 35, No. 1, 2007, pp. 125–31.

[36] Kroll, J. E., P. A. Galante, D. T. Ohara, F. C. Navarro, L. Ohno-Machado and S. J. de Souza, "Splooce: A New Portal For The Analysis Of Human Splicing Variants," RNA Biology, Vol. 9, No. 11, 2012, pp. 1339–43.

[37] Schmucker, D., J. C. Clemens, H. Shu, C. A. Worby, J. Xiao, M. Muda, J. E. Dixon and S. L. Zipursky, "Drosophila Dscam Is An Axon Guidance Receptor Exhibiting Extraordinary Molecular Diversity," Cell, Vol. 101, No. 6, 2000, pp. 671–84.

[38] Breitbart, R. E., A. Andreadis and B. Nadal-Ginard, "Alternative Splicing: A Ubiquitous Mechanism For The Generation Of Multiple Protein Isoforms From Single Genes," Annual Review Of Biochemistry, Vol. 56, 1987, pp. 467–95.

[39] Letunic, I., R. R. Copley and P. Bork, "Common Exon Duplication In Animals And Its Role In Alternative Splicing," Human Molecular Genetics, Vol. 11, No. 13, 2002, pp. 1561–7.

[40] Beaudoing, E., S. Freier, J. R. Wyatt, J. M. Claverie and D. Gautheret, "Patterns Of Variant Polyadenylation Signal Usage In Human Genes," Genome Research, Vol. 10, No. 7, 2000, pp. 1001–10.

[41] Brett, D., H. Pospisil, J. Valcarcel, J. Reich and P. Bork, "Alternative splicing and genome complexity," Nature Genetics, Vol. 30, No. 1, 2001, pp. 29–30.

[42] Modrek, B. and C. Lee, "A genomic view of alternative splicing," Nature Genetics, Vol. 30, No. 1, 2001, pp. 13–19.

[43] Brett, D., J. Hanke, G. Lehmann, S. Haase, S. Delbruck, S. Krueger, J. Reich and P. Bork, "EST comparison indicates 38% of human mRNAs contain possible alternative splice forms," Febs Letters, Vol. 474, No. 1, 2000, pp. 83–86.

[44] Benson, D. A., M. Cavanaugh, K. Clark, I. Karsch-Mizrachi, D. J. Lipman, J. Ostell, and E. W. Sayers, "Genbank," Nucleic Acids Research, Vol. 41, Database issue, 2013, pp. D36–42.

[45] Florea, L., G. Hartzell, Z. Zhang, G. M. Rubin and W. Miller, "A Computer Program for Aligning a cDNA Sequence with a Genomic DNA Sequence," Genome Research, Vol. 8, No. 9, 1998, pp. 967–974.

[46] Kent, W. J., "BLAT – the BLAST-like alignment tool," Genome Research, Vol. 12, No. 4, 2002, pp. 656–664.

[47] Feng, H., Z. Qin and X. Zhang, "Opportunities and methods for studying alternative splicing in cancer with RNA-Seq," Cancer Lett., Vol. 12, No. 1, 2012, pp. 1–13.

[48] Yeo, G., D. Holste, G. Kreiman and C. B. Burge, "Variation in alternative splicing across human tissues," Genome Biology, Vol. 5, No., 2004, pp. R74.

[49] Blencowe, B. J., "Alternative Splicing: New Insights from Global Analyses," Cell, Vol. 126, No. 1, 2006, pp. 37–47.

[50] Zhang, X. Z., A. H. Yin, X. Y. Zhu, Q. Ding, C. H. Wang and Y. X. Chen, "Using an exon microarray to identify a global profile of gene expression and alternative splicing in K562 cells exposed to sodium valproate," Oncology Reports, Vol. 27, No. 4, 2012, pp. 1258–1265.

[51] Zhang, Z., S. Pa, J. Tchou and R. V. Davuluri, "Isoform-level expression profiles provide better cancer signatures than gene-level expression profiles," Genome Medicine. Vol. 13, No. 4, 2013, pp. 33.

[52] Li, H. R., J. Wang-Rodriguez, T. M. Nair, J. M. Yeakley, Y. S. Kwon, M. Bibikova, C. Zheng, L. Zhou, K. Zhang, T. Downs, X. D. Fu, J. B. Fan, "Two-dimensional transcriptome profiling: identification of messenger RNA isoform signatures in prostate cancer from archived paraffin-embedded cancer specimens," Cancer Research, Vol. 66, No. 8, 2006, pp. 4079–4088.

[53] Malone, J. H. and B. Oliver, "Microarrays, deep sequencing and the true measure of the transcriptome," BMC Biology, Vol. 9, No. 1, 2011, p. 34–43.

[54] Bertone, P., V. Stolc, T. E. Royce, J. S. Rozowsky, A. E. Urban, X. Zhu, J. L. Rinn, W. Tongprasit, M. Samanta, S. Weissman, M. Gerstein, M. Snyder, "Global identification of human transcribed sequences with genome tiling arrays," Science, Vol. 306, No. 5705, 2004, pp. 2242–2246.

[55] Rajan, P., D. J. Elliott, C. N. Robson, H. Y. Leung, "Alternative splicing and biological heterogeneity in prostate cancer," Nature reviews Urology, Vol. 6, No. 8, 2009, pp. 454–460.

[56] Wang, E. T., R. Sandberg, S. Luo, I. Khrebtukova, L. Zhang, C. Mayr, S. F. Kingsmore, G. P. Schroth, C. B. Burge, "Alternative isoform regulation in human tissue transcriptomes," Nature, Vol. 456, No. 7221, 2008, pp. 470–6.

[57] Ozsolak, F. and P. M. Milos, "RNA sequencing: advances, challenges and opportunities," Nature Reviews Genetics, Vol. 12, No. 2, 2011, pp.87–98.

[58] Levin, J. Z., M. F. Berger, X. Adiconis, P. Rogov, A. Melnikov, T. Fennell, C. Nusbaum, L. A. Garraway and. A. Gnirke, "Targeted next-generation sequencing of a cancer transcriptome enhances detection of sequence variants and novel fusion transcripts," Genome Biology, Vol. 10, No. 10, 2009, pp. R115.

[59] Chen, G., R. Li, L. Shi, J. Qi, P. Hu, J. Luo, M. Liu, and T. Shi, "Revealing the missing expressed genes beyond the human reference genome by RNA-Seq," BMC Genomics, Vol. 12, No. 1, 2011, pp. 590–599.

[60] Sultan, M., M. H. Schulz, H. Richard, A. Magen, A. Klingenhoff, M. Scherf, M. Seifert, T. Borodina, A. Soldatov, D. Parkhomchuk, D. Schmidt, S. O'Keeffe, S. Haas, M. Vingron, H. Lehrach and M. L. Yaspo, "A global view of gene activity and alternative splicing by deep sequencing of the human transcriptome," Science, Vol. 321, No. 5891, 2008, pp. 956–990.

[61] Trapnell, C., L. Pachter and S. L. Salzberg, "TopHat: discovering splice junctions with RNA-seq," Bioinformatics, Vol. 25, No. 9, 2009, pp. 1105–1111.

[62] Wang, L., Y. Xi, J. Yu, L. Dong, L. Yen and W. Li, "A Statistical Method for the Detection of Alternative Splicing Using RNA-Seq," PLoS ONE, Vol. 5, No. 1, 2010, pp. e8529.

[63] Ameur, A., A. Wetterbon, L. Feuk and U. Gyllensten, "Global and unbiased detection of splice junctions from RNA-Seq data," Genome Biology, Vol. 11, No. 3, 2010, pp. R34.

[64] Au, K. F., H. Jiang, L. Lin, Y. Xing and W. H. Wong, "Detection of splice junctions from paired-end RNA-seq data by SpliceMap," Nucleic Acids Research, Vol. 38, No. 14, 2010, pp. 4570–4578.

[65] Guttman, M., M. Garber, J. Z. Levin, J. Donaghey, J. Robinson, X. Adiconis, L. Fan, M. J. Koziol, A., Gnirke, C. Nusbaum, J. L. Rinn, E. S. Lander and A. Regev, "Ab initio reconstruction of cell type-specific transcriptomes in mouse reveals the conserved multi-exonic structure of lincRNAs," Nature Biotechnology, Vol. 28, No. 5, 2010, pp. 503–510.

[66] Trapnell, C., B.A Williams, G. Pertea, A. Mortazavi, G. Kwan, M. J. van Baren, S. L. Salzberg, B. J. Wold and L. Pachter, "Transcript assembly and quantification by RNA-Seq reveals unannotated transcripts and isoform switching during cell differentiation," Nature Biotechnology, Vol. 28, No. 5, 2010, p.511–515.

[67] Srebrow, A. and A. R. Kornblihtt, "The Connection Between Splicing And Cancer," Journal Of Cell Science, Vol. 119, No. 13, 2006, pp. 2635–41.

[68] Pajares, M. J., T. Ezponda, R. Catena, A. Calvo, R. Pio and L. M. Montuenga, "Alternative Splicing: An Emerging Topic In Molecular And Clinical Oncology," The Lancet Oncology, Vol. 8, No. 4, 2007, pp. 349–57.

[69] Jensen, C. J., B. J. Oldfield and J. P. Rubio, "Splicing, Cis Genetic Variation And Disease," Biochemical Society Transactions, Vol. 37, No. 6, 2009, pp. 1311–5.

[70] de la Mata, M., C. R. Alonso, S. Kadener, J. P. Fededa, M. Blaustein, F. Pelisch, P. Cramer, D. Bentley and A. R. Kornblihtt, "A slow RNA polymerase II affects alternative splicing in vivo," Molecular Cell, Vol. 12, No. 2, 2003, pp. 525–32.

[71] Sorek, R., R. Shamir and G. Ast, "How Prevalent Is Functional Alternative Splicing In The Human Genome?," Trends In Genetics, Vol. 20, No. 2, 2004, pp. 68–71.

[72] McGlincy, N. J. and C. W. Smith, "Alternative Splicing Resulting In Nonsense-mediated mRNA Decay: What Is The Meaning Of Nonsense?," Trends In Biochemical Sciences, Vol. 33, No. 8, 2008, pp. 385–93.

[73] Passos, D. O., M. K. Doma, C. J. Shoemaker, D. Muhlrad, R. Green, J. Weissman, J. Hollien, R. Parker, M. K. Doma and R. Parker, "Analysis

Of Dom34 And Its Function In No-go Decay," Molecular Biology Of The Cell, Vol. 20, No. 13, 2009, pp. 3025–32.
[74] Frischmeyer, P. A., A. van Hoof, K. O'Donnell, A. L. Guerrerio, R. Parker and H. C. Dietz, "An mRNA Surveillance Mechanism That Eliminates Transcripts Lacking Termination Codons," Science, Vol. 295, No. 5563, 2002, pp. 2258–61.
[75] Venables, J. P., "Unbalanced Alternative Splicing And Its Significance In Cancer," Bioessays : News And Reviews In Molecular, Cellular And Developmental Biology, Vol. 28, No. 4, 2006, pp. 378–86.
[76] Krawczak, M., N. S. Thomas, B. Hundrieser, M. Mort, M. Wittig, J. Hampe and D. N. Cooper, "Single Base-pair Substitutions In Exon-intron Junctions Of Human Genes: Nature, Distribution, And Consequences For mRNA Splicing," Human Mutation, Vol. 28, No. 2, 2007, pp. 150–8.
[77] Krawczak, M., J. Reiss and D. N. Cooper, "The mutational spectrum of single base-pair substitutions in mRNA splice junctions of human genes: causes and consequences," Human Genetics, Vol. 90, 1992, pp. 41–54.
[78] Ward, A. J. and T. A. Cooper, "The Pathobiology Of Splicing," The Journal Of Pathology, Vol. 220, No. 2, 2010, pp. 152–63.
[79] Holmila, R., C. Fouquet, J. Cadranel, G. Zalcman and T. Soussi, "Splice Mutations In The P53 Gene: Case Report And Review Of The Literature," Human Mutation, Vol. 21, No. 1, 2003, pp. 101–2.
[80] Sakurai, N., S. Iwamoto, Y. Miura, T. Nakamura, A. Matsumine, J. Nishioka, K. Nakatani and Y. Komada, "Novel P53 Splicing Site Mutation In Li-fraumeni-like Syndrome With Osteosarcoma," Pediatrics International : Official Journal Of The Japan Pediatric Society, Vol. 55, No. 1, 2013, pp. 107–11.
[81] Tanko, Q., B. Franklin, H. Lynch and J. Knezetic, "A Hmlh1 Genomic Mutation And Associated Novel mRNA Defects In A Hereditary Non-polyposis Colorectal Cancer Family," Mutation Research, Vol. 503, No. 1–2, 2002, pp. 37–42.
[82] Taylor, M. D., N. Gokgoz, I. L. Andrulis, T. G. Mainprize, J. M. Drake and J. T. Rutka, "Familial Posterior Fossa Brain Tumors Of Infancy Secondary To Germline Mutation Of The Hsnf5 Gene," American Journal Of Human Genetics, Vol. 66, No. 4, 2000, pp. 1403–6.
[83] Kurahashi, H., K. Takami, T. Oue, T. Kusafuka, A. Okada, A. Tawa, S. Okada and I. Nishisho, "Biallelic Inactivation Of The Apc Gene In Hepatoblastoma" Cancer Research, Vol. 55, No. 21, 1995, pp. 5007–11.

[84] Broeks, A., J. H. Urbanus, P. de Knijff, P. Devilee, M. Nicke, K. Klöpper, T. Dörk, A. N. Floore and L. J. van't Veer, "Ivs10–6t>g, An Ancient AtmGermline Mutation Linked With Breast Cancer," Human Mutation, Vol. 21, No. 5, 2003, pp. 521–8.

[85] Hoffman, J. D., S. E. Hallam, V. L. Venne, E. Lyon and K. Ward, "Implications Of A Novel Cryptic Splice Site In The Brca1 Gene," American Journal Of Medical Genetics, Vol. 80, No. 2, 1998, pp. 140–4.

[86] De Klein, A., P. H. Riegman, E. K. Bijlsma, A. Heldoorn, M. Muijtjens, M. A. den Bakker, C. J. Avezaat and E. C. Zwarthoff, "A G ? A transition Creates A Branch Point Sequence And Activation Of A Cryptic Exon, Resulting In The Hereditary Disorder Neurofibromatosis 2," Human Molecular Genetics, Vol. 7, No. 3, 1998, pp. 393–8.

[87] Liu, J., W. Lee, Z. Jiang, Z. Chen, S. Jhunjhunwala, P. M. Haverty, F. Gnad, Y. Guan, H. N. Gilbert, J. Stinson, C. Klijn, J. Guillory, D. Bhatt, S. Vartanian, K. Walter, J. Chan, T. Holcomb, P. Dijkgraaf, S. Johnson, J. Koeman, J. D. Minna, A. F. Gazdar, H. M. Stern, K. P. Hoeflich, T. D. Wu, J. Settleman, F. J. de Sauvage, R. C. Gentleman, R. M. Neve, D. Stokoe, Z. Modrusan, S. Seshagiri, D. S. Shames and Z. Zhang, "Genome and transcriptome sequencing of lung cancers reveal diverse mutational and splicing events," Genome Research, Vol. 22, No. 12, 2012, pp. 2315–27.

[88] Shibayama, M., S. Ohno, T. Osaka, R. Sakamoto, A. Tokunaga, Y. Nakatake, M. Sato and N. Yoshida, "Polypyrimidine Tract-binding Protein Is Essential For Early Mouse Development And Embryonic Stem Cell Proliferation," The Febs Journal, Vol. 276, No. 22, 2009, pp. 6658–68.

[89] Clarke, L. A., I. Veiga, G. Isidro, P. Jordan, J. S. Ramos, S. Castedo and M. G. Boavida, "Pathological Exon Skipping In An HnpccProband With Mlh1 Splice Acceptor Site Mutation," Genes, Chromosomes Cancer, Vol. 29, No. 4, 2000, pp. 367–70.

[90] Narla, G., A. Difeo, H. L. Reeves, D. J. Schaid, J. Hirshfeld, E. Hod, A. Katz, W. B. Isaacs, S. Hebbring, A. Komiya, S. K. McDonnell, K. E. Wiley, S. J. Jacobsen, S. D. Isaacs, P. C. Walsh, S. L. Zheng, B. L. Chang, D. M. Friedrichsen, J. L. Stanford, E. A. Ostrander, A. M. Chinnaiyan, M. A. Rubin, J. Xu, S. N. Thibodeau, S. L. Friedman and J. A. Martignetti, "A Germline DNA Polymorphism Enhances Alternative Splicing Of The Klf6 Tumor Suppressor Gene And Is Associated With Increased Prostate Cancer Risk," Cancer Research, Vol. 65, No. 4, 2005, pp. 1213–22.

[91] DiFeo, A., J. A. Martignetti and G. Narla, “The Role Of Klf6 And Its Splice Variants In Cancer Therapy,” Drug Resistance Updates : Reviews And Commentaries In Antimicrobial And Anticancer Chemotherapy, Vol. 12, No. 1–2,, pp. 1–7.

[92] Maas, S., S. Patt, M. Schrey and A. Rich, “Underediting Of Glutamate Receptor Glur-b mRNA In Malignant Gliomas,” Proceedings Of The National Academy Of Sciences Of The United States Of America, Vol. 98, No. 25, 2001, pp. 14687–92.

[93] Tacke, R. and J. L. Manley, “Functions Of SR And Tra2 Proteins In Pre-mRNA Splicing Regulation,” Proceedings Of The Society For Experimental Biology And Medicine. Society For Experimental Biology And Medicine, Vol. 220, No. 2, 1999, pp. 59–63.

[94] Ghigna, C., M. Moroni, C. Porta, S. Riva and G. Biamonti, “Altered Expression Of Heterogenous Nuclear Ribonucleoproteins And SR Factors In Human Colon Adenocarcinomas,” Cancer Research, Vol. 58, No. 24, 1998, pp. 5818–24.

[95] Mukherji, M., L. M. Brill, S. B. Ficarro, G. M. Hampton and P. G. Schultz, “A Phosphoproteomic Analysis Of The Erbb2 Receptor Tyrosine Kinase Signaling Pathways,” Biochemistry, Vol. 45, No. 51, 2006, pp. 15529–40.

[96] Pind, M. T. and P. H. Watson, “SR Protein Expression And CD44 Splicing Pattern In Human Breast Tumours,” Breast Cancer Research And Treatment, Vol. 79, No. 1, 2003, pp. 75–82.

[97] Stickeler, E., F. Kittrell, D. Medina and S. M. Berget, “Stage-specific Changes In SR Splicing Factors And Alternative Splicing In Mammary Tumorigenesis,” Oncogene, Vol. 18, No. 24, 1999, pp. 3574–82.

[98] Naor, D., S. Nedvetzki, I. Golan, L. Melnik and Y. Faitelson, “CD44 In Cancer,” Critical Reviews In Clinical Laboratory Sciences, Vol. 39, No. 6, 2002, pp. 527–79.

[99] Stickeler, E., F. Kittrell, D. Medina and S. M. Berget, “Stage-specific Changes In SR Splicing Factors And Alternative Splicing In Mammary Tumorigenesis,” Oncogene, Vol. 18, No. 24, 1999, pp. 3574–82.

[100] He, X., M. Pool, K. M. Darcy, S. B. Lim, N. Auersperg, J. S. Coon and W. T. Beck, “Knockdown Of Polypyrimidine Tract-binding Protein Suppresses Ovarian Tumor Cell Growth And Invasiveness In Vitro,” Oncogene, Vol. 26, No. 34, 2007, pp. 4961–8.

[101] Khan, D. H., S. Jahan and J. R. Davie, "Pre-mRNA Splicing: Role Of Epigenetics And Implications In Disease," Advances In Biological Regulation, Vol. 52, No. 3, 2012, pp. 377–88.

[102] Saito, Y., Y. Kanai, M. Sakamoto, H. Saito, H. Ishii and S. Hirohashi, "Overexpression Of A Splice Variant Of DNA Methyltransferase 3b, Dnmt3b4, Associated With DNA Hypomethylation On Pericentromeric Satellite Regions During Human Hepatocarcinogenesis," Proceedings Of The National Academy Of Sciences Of The United States Of America, Vol. 99, No. 15, 2002, pp. 10060–5.

[103] Lee, D. W., K. Zhang, Z. Q. Ning, E. H. Raabe, S. Tintner, R. Wieland, B. J. Wilkins, J. M. Kim, R. I. Blough and R. J. Arceci, "Proliferation-associated Snf2-like Gene (Pasg): A Snf2 Family Member Altered In Leukemia," Cancer Research, Vol. 60, No. 13, 2000, pp. 3612–22.

[104] Coulson, J. M., J. L. Edgson, P. J. Woll and J. P. Quinn, "A Splice Variant Of The Neuron-restrictive Silencer Factor Repressor Is Expressed In Small Cell Lung Cancer: A Potential Role In Derepression Of Neuroendocrine Genes And A Useful Clinical Marker," Cancer Research, Vol. 60, No. 7, 2000, pp. 1840–4.

[105] Reiter, R., A. Wellstein and A. T. Riegel, "An Isoform Of The Coactivator Aib1 That Increases Hormone And Growth Factor Sensitivity Is Overexpressed In Breast Cancer," The Journal Of Biological Chemistry, Vol. 276, No. 43, 2001, pp. 39736–41.

[106] Zhu, X., A. A. Daffada, C. M. Chan and M. Dowsett, "Identification Of An Exon 3 Deletion Splice Variant Androgen Receptor mRNA In Human Breast Cancer," International Journal Of Cancer. Journal International Du Cancer, Vol. 72, No. 4, 1997, pp. 574–80.

[107] Huang, R., Z. Xing, Z. Luan, T. Wu, X. Wu and G. Hu, "A Specific Splicing Variant Of Svh, A Novel Human Armadillo Repeat Protein, Is Up-regulated In Hepatocellular Carcinomas," Cancer Research, Vol. 63, No. 13, 2003, pp. 3775–82.

[108] Weiss, M., A. Baruch, I. Keydar and D. H. Wreschner, "Preoperative Diagnosis Of Thyroid Papillary Carcinoma By Reverse Transcriptase Polymerase Chain Reaction Of The Muc1 Gene," International Journal Of Cancer, Vol. 66, No. 1, 1996, pp. 55–9.

[109] Scheurlen, W. G. and L. Senf, "Analysis Of The Gap-related Domain Of The Neurofibromatosis Type 1 (Nf1) Gene In Childhood Brain Tumors," International Journal Of Cancer, Vol. 64, No. 4, 1995, pp. 234–8.

[110] Wang, L., L. Duke, P. S. Zhang, R. B. Arlinghaus, W. F. Symmans, A. Sahin, R. Mendez and J. L. Dai, "Alternative Splicing Disrupts A

Nuclear Localization Signal In Spleen Tyrosine Kinase That Is Required For Invasion Suppression In Breast Cancer," Cancer Research, Vol. 63, No. 15, 2003, pp. 4724–30.

[111] Luther, T., M. Kotzsch, A. Meye, T. Langerholc, S. Füssel, N. Olbricht, S. Albrecht, D. Ockert, B. Muehlenweg, K. Friedrich, M. Grosser, M. Schmitt, G. Baretton and V. Magdolen, "Identification Of A Novel Urokinase Receptor Splice Variant And Its Prognostic Relevance In Breast Cancer," Thrombosis And Haemostasis, Vol. 89, No. 4, 2003, pp. 705–17.

[112] Tanaka, S., K. Sugimachi, H. Saeki, J. Kinoshita, T. Ohga, M. Shimada, Y. Maehara and K. Sugimachi, "A Novel Variant Of Wisp1 Lacking A Von Willebrand Type C Module Overexpressed In Scirrhous Gastric Carcinoma," Oncogene, Vol. 20, No. 39, 2001, pp. 5525–32.

[113] Halin, C., L. Zardi, D. Neri, "Antibody-based targeting of angiogenesis," News in Physiological Sciences, Vol. 16, 2001, pp. 191–4.

[114] Walsh, G., G. Walsh, G. Walsh and G. Walsh, "Biopharmaceutical Benchmarks 2010," Nature Biotechnology, Vol. 28, No. 9, 2010, pp. 917–24.

[115] da Cunha, J. P., P. A. Galante, J. E. de Souza, R.F de Souza, P.M Carvalho, D. T. Ohara, R. P. Moura, S. M. Oba-Shinja, S. K. Marie, W. A. Silva Jr, R. O. Perez, B. Stransky, M. Pieprzyk, J. Moore, O. Caballero, J. Gama-Rodrigues, A. Habr-Gama, W. P. Kuo, A. J. Simpson, A. A. Camargo, L. J. Old and S. J. de Souza, "Bioinformatics construction of the human cell surfaceome," Proceedings Of The National Academy Of Sciences Of The United States Of America, Vol. 106, No. 39, 2009, pp. 16752–7.

[116] Kerbel, R. S., "Tumor Angiogenesis: Past, Present And The Near Future," Carcinogenesis, Vol. 21, No. 3, 2000, pp. 505–15.

[117] Aaltomaa, S., P. Lipponen, M. Ala-Opas, V. M. Kosma, S. Aaltomaa, P. Lipponen, J. Viitanen, J. P. Kankkunen, M. Ala-Opas and V. M. Kosma, "Expression And Prognostic Value Of CD44 Standard And Variant V3 And V6 Isoforms In Prostate Cancer," European Urology, Vol. 39, No. 2, 2001, pp. 138–44.

[118] Iczkowski, K. A., S. Bai and C. G. Pantazis, "Prostate Cancer Overexpresses CD44 Variants 7–9 At The Messenger RNA And Protein Level," Anticancer Research, Vol. 23, No. 4,, pp. 3129–40.

[119] Yeh, B. K., M. Igarashi, A. V. Eliseenkova, A. N. Plotnikov, I. Sher, D. Ron, S. A. Aaronson and M. Mohammadi, "Structural Basis By Which Alternative Splicing Confers Specificity In Fibroblast Growth

Factor Receptors," Proceedings Of The National Academy Of Sciences Of The United States Of America, Vol. 100, No. 5, 2003, pp. 2266–71.

[120] Jang, J. H., "Identification And Characterization Of Soluble Isoform Of Fibroblast Growth Factor Receptor 3 In Human Saos-2 Osteosarcoma Cells," Biochemical And Biophysical Research Communications, Vol. 292, No. 2, 2002, pp. 378–82.

[121] Vickers, S. M., Z. Q. Huang, L. MacMillan-Crow, J. S. Greendorfer and J. A. Thompson, "Ligand Activation Of Alternatively Spliced Fibroblast Growth Factor Receptor-1 Modulates Pancreatic Adenocarcinoma Cell Malignancy," Journal Of Gastrointestinal Surgery: Official Journal Of The Society For Surgery Of The Alimentary Tract, Vol. 6, No. 4, 2002, pp. 546–53.

[122] Jang, J. H., K. H. Shin and J. G. Park, "Mutations In Fibroblast Growth Factor Receptor 2 And Fibroblast Growth Factor Receptor 3 Genes Associated With Human Gastric And Colorectal Cancers," Cancer Research, Vol. 61, No. 9, 2001, pp. 3541–3.

[123] Kwabi-Addo, B., F. Ropiquet, D. Giri and M. Ittmann, "Alternative Splicing Of Fibroblast Growth Factor Receptors In Human Prostate Cancer," The Prostate, Vol. 46, No. 2, 2001, pp. 163–72.

[124] Markovic, D. and R. A. Challiss, "Alternative Splicing Of G Protein-coupled Receptors: Physiology And Pathophysiology," Cellular And Molecular Life Sciences, Vol. 66, No. 20, 2009, pp. 3337–52.

[125] Hellmich, M. R., X. L. Rui, H. L. Hellmich, R. Y. Fleming, B. M. Evers and C. M. Townsend, "Human Colorectal Cancers Express A Constitutively Active Cholecystokinin-b/gastrin Receptor That Stimulates Cell Growth," The Journal Of Biological Chemistry, Vol. 275, No. 41, 2000, pp. 32122–8.

[126] Lee, H. J., B. Wall and S. Chen, "G-protein-coupled Receptors And Melanoma," Pigment Cell Melanoma Research, Vol. 21, No. 4, 2008, pp. 415–28.

[127] Segal, E., N. Friedman, N. Kaminski, A. Regev and D. Koller, "From Signatures To Models: Understanding Cancer Using Microarrays," Nature Genetics, Vol. 37 Suppl, 2005, pp. S38–45.

[128] Gardina, P. J., T. A. Clark, B. Shimada, M. K. Staples, Q. Yang, J. Veitch, A. Schweitzer, T. Awad, C. Sugnet, S. Dee, C. Davies, A. Williams and Y. Turpaz, "Alternative Splicing And Differential Gene Expression In Colon Cancer Detected By A Whole Genome Exon Array," BMC Genomics, Vol. 7, 2006, pp. 325.

[129] Relógio, A., C. Ben-Dov, M. Baum, M. Ruggiu, C. Gemund, V. Benes, R. B. Darnell and J. Valcárcel, "Alternative Splicing Microarrays Reveal Functional Expression Of Neuron-specific Regulators In Hodgkin Lymphoma Cells," The Journal Of Biological Chemistry, Vol. 280, No. 6, 2005, pp. 4779–84.

2

Non-coding RNAs as Molecular Tools

**Renu Wadhwa, Yoshio Kato, Ran Gao
and Sunil C. Kaul**

Cell Proliferation Research Group and DBT-AIST International Laboratory for Advanced Biomedicine, Biomedical Research Institute, National Institute of Advanced Industrial Science & Technology (AIST), Tsukuba, Ibaraki, Japan

2.1 Introduction

Perception of RNA from a mere bridge between the genetic information and protein structure changed, with the discovery of "Ribozyme"– RNA molecules with enzymatic properties–nearly three decades ago. Haseloff and Gerlachdescovered RNA enzymes with specific endonuclease activity and named them hammerhead ribozymes (for resemblance of their two-dimensional structure to a hammerhead that requires their binding to metal ions)[1]. Subsequently, mechanism of action of these ribozymes including the catalytic domains, target specificity, intracellular stability, and accessibility to the target site and requirement of divalent metal ions have been largely demonstrated (Figure 2.1) [2–5].

In 2001, the world of small RNAs revolutioned to a new era with the discovery of regulation of gene expression by small RNAs [6–9]. These tiny RNA regulators were shown to originate from the large proportion of non-coding genome referred to as JUNK DNA and cause gene silencing by a highly coordinated and sequence-specific mechanism known as RNA interference (RNAi). Since then, the small RNAs, including short interfering RNAs (siRNAs) and microRNAs (miRNAs) have emerged as key components of an evolutionarily conserved system of RNA-based gene regulation involved in a variety of developmental, disease and defense mechanisms in eukaryotes. Briefly, the two kinds of small RNAs, i.e., siRNAs and miRNAs are produced by the cleavage of double-stranded RNA precursors by Dicer, a member of

Post-genomic Approaches in Cancer and Nano Medicine, 25–60.

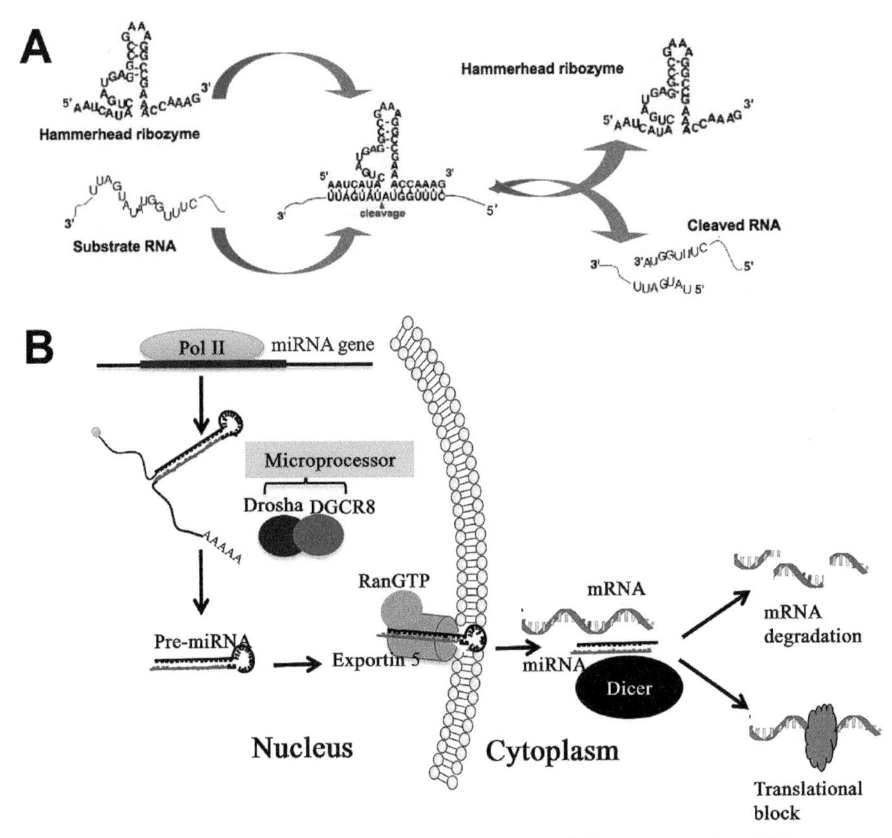

Figure 2.1 Schematic representation of action of ribozymes and miRNA.

the RNase III family of dsRNA-specific endonucleases [10–12]. Typically, siRNAs are 21 nt long double-stranded RNAs that contain 19 base pairs, with 2 nt, 3' overhanging ends. In contrast, miRNAs are single stranded products of endogenous, non-coding genes. These are produced by Dicer from an intermediate precursor RNA transcript (approximately 70 nt) that forms stem loop structure [6, 13]. In contrast to siRNA that can degrade mRNA mediated by RISC (RNA-induced silencing complex, functions as a siRNA directed endonuclease), miRNAs can also suppress the protein expression by inhibition of mRNA translation [12, 14]. Since both siRNA and miRNAs are found in similar, if not identical, complexes, it is suggested that RISC might be a bi-functional complex that mediates both cleavage and translational control [14]. Furthermore, some evidence suggested that the siRNAs and miRNAs are functionally interchangeable in which the target mRNA could be either cleaved or translationally repressed as determined by the degree of complementarity between the small RNA and its target [15, 16].

First miRNA (*lin-4*) was identified as a regulator of *Caenorhabditise-legans* development[17] following which several evolutionary conserved miRNAs that regulate development were identified as Let-7 miRNA family [6, 18]. Subsequent high-throughput screenings of miRNA have led to the identification of about 20,000 mature miRNAs from 168 species (miRbase database, http://www.mirbase.org/). Computational predictions of miRNA targets suggest that up to 60% of the human protein coding genes may be regulated by miRNAs. Some miRNAs show localized expression patterns with limited tissue, cell, and spatiotemporal specificities suggesting their important roles in biological processes. The most common and relatively well studied and established mechanism of action of miRNA involves base-pairing to the 3' untranslated region (UTR) of mRNA region (called miRNA recognition elements, MREs) that leads to the degradation of target mRNA transcripts. Briefly, miRNA genes are expressed in the cell nucleus by RNA polymerase II as a long double-stranded precursor called the primary miRNA or pri-miRNA (Figure 2.1). It contains a typical stem-loop structure that is processed to release a ~70 nucleotide (nt)-long precursor miRNA (pre-miRNA) by Drosha protein. The pre-miRNA is exported to the cytoplasm by the nuclear export protein Exportin 5. In the cytoplasm, the pre-miRNA is cleaved by enzyme Dicer into a ~20–25 nt double-stranded RNA, which is then separated into two single strands that form complex with the protein called RISC (RNA-Induced Silencing Complex). RISC targets mRNAs through direct base pairing. The 5' region of miRNA (nucleotides 1 through 8), known as as the "seed" region, determines its target specificity and function [19]. The miRNA target sites, located in the 3' UTR of messenger RNAs, are often imperfectly matched to the microRNA sequence. Since miRNAs do not require perfect complementarity for target recognition, a single miRNA is often able to regulate multiple messenger RNAs. On the other hand, most miRNAs exert subtle effects on each individual messenger RNA target and the combined effect of multiple miRNA is considered significant and produces measurable phenotypic results [20]. The other, relatively less understood, mechanism of action of miRNA involves transcriptional regulation by epigenetic gene silencing or translational regulation by binding to the heterogeneous ribonucleoproteins[15, 19, 21–23]. Although miRNAs have been implicated to almost all physiological processes including the immune response, cell cycle control, metabolism, viral replication, stem cell differentiation and development, identification of their target genes and their physiological relevance to a variety of diseases is a current challenge and has gained much attention due to their therapeutic potentials.

In this chapter, we review the potential of small RNAs as molecular tools, in identification and functional characterization of genes, using muscle differentiation and cancer as model biological processes in cell culture system. Whereas skeletal muscle differentiation is most commonly characterized by the terminal withdrawal of cells from the cell cycle and their fusion to form multinucleated cells that are enriched in muscle-specific proteins [24]. Mouse myoblast cell line, C_2C_{12}, is a well-established and convenient *in vitro* model to study myocyte differentiation. These cells can be induced to differentiate *in vitro* by serum deprivation, and offer a convenient way to study gene expression at various time points in a fixed or real time analysis [25]. The cancer cells (most simply defined by uncontrolled division of cells) is a highly complicated disorder in which multiplex of genetic and environmental factors contribute. For cancer studies, a variety of tumors derived or *in vitro* immortalized cell lines are used to elucidate the cancer signaling, key genes and drug targets.

2.2 Ribozyme as Molecular Tool

2.2.1 Muscle Differentiation as a Model

In order to explore the genes functionally involved in muscle differentiation, C_2C_{12} myoblasts were infected with randomized ribozyme library and induced to differentiate by low serum culture conditions (differentiation medium). Whereas control (empty vector infected) cells formed full myotubes in 72 to 96 h in differentiation medium (Figure 2.2), cells infected with the randomized ribozyme library showed abrogation of myotube formation [26] (Figure 2.2). Undifferentiating cells that continued to divide formed small colonies against the background of differentiating cells that fused to form myotubes. Ribozymes were recovered from such undifferentiating clones by RT-PCR using ribozyme vector-specific primers, cloned into a TA vector and subjected to sequence analysis. Identity of the ribozymes and their target genes were revealed by matching with the nucleotide database entries. By using this screening model, we identified 24 gene targets that could be categorized into four groups. The first group showed genes specifically expressed in muscles including MyoD1, Mylk, Myosin, Myogenin and Myf5. MyoD and Myf5 are well known as regulatory genes for myogenic determination, and myogenin and myosin (muscle structural proteins) are important for terminal differentiation and lineage maintenance [24, 25]. The second group contained genes involved in extracellular signaling, such as, fibronectin and FGFR4 implying their role in muscle differentiation as has also been reported in many other studies [27, 28]. The third group had tumor suppressors and cell cycle regulators including p27, p19^{ARF} and p21^{WAF1}. During muscle differentiation, transition of cells from the proliferative stage to form post-mitotic multinucleated myotubes and upregulation of muscle specific genes is regulated through highly ordered and temporally separable mechanisms [24]. In order to investigate the functional involvement of p19^{ARF} and p21^{WAF1} in muscle differentiation, we used specific ribozymes (Rz) and siRNA that knockdown the expression of these proteins by 30–50%. Interestingly, these cells showed delayed myotube formation, marked by decrease in myogenin; an established marker for myotube formation [26] (Figures 2.2B and 2.2C). The fourth group possessed gene encoding Fem (protein serine/threonine phosphatases of Type 2C (PP2C) that is enriched in adult heart and skeletal muscle, possesses unique tissue specific functions involved in male sexual development [29]. We found that the C_2C_{12} cells compromised for Fem, by transfection of specific siRNAs, resulted in an increase in proliferation of myoblasts and remarkable delay in the myotube formation. On the other hand, Fem overexpressing derivatives showed early fusion of cells to multinucleated myotubes[26] (Figure 2.2D). Altogether, the study demonstrated the use of ribozymes as molecular tools in the screening and characterization of the genes that are functionally involved in muscle differentiation.

2.2.2 Cancer Mechanism as a Model

Cancer is largely characterized by functional loss of tumor suppressor proteins that put the cells to replicative senescence and stress-induced senescence, apoptosis and cell death by activating the DNA damage responses and a variety of other signaling cascades. The INK4a locus on human chromosome 9p21 encodes two unrelated tumor suppressor proteins: p16^{INK4a}, an inhibitor of the cyclin D-dependent kinase that acts upstream of the retinoblastoma protein, pRb and p19ARF, an alternative reading frame protein that acts upstream of p53 [30–33]. These proteins play major role in the senescence of primary cells, activates pathways for cell cycle control and tumor suppression, and are often functionally inactivated in human tumors [31, 33, 34]. p16^{INK4a} inhibits cyclin dependent kinases and pRB phosphorylation [34] that is essential for cell cycle arrest function. The second protein from INK4a locus, p19ARF and

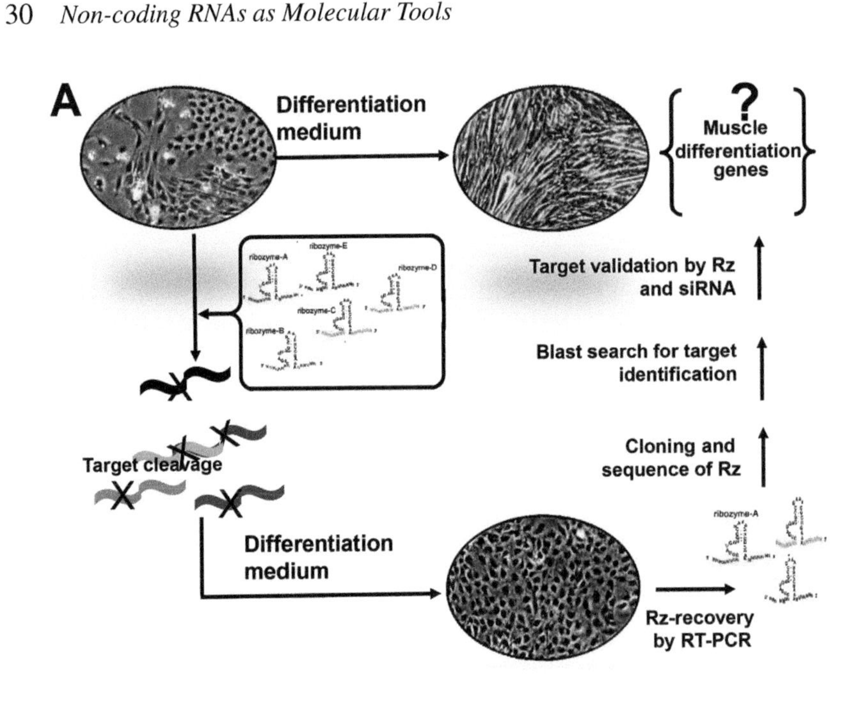

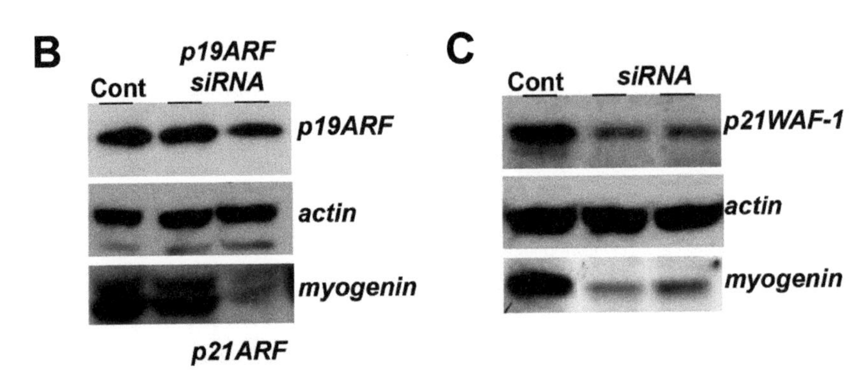

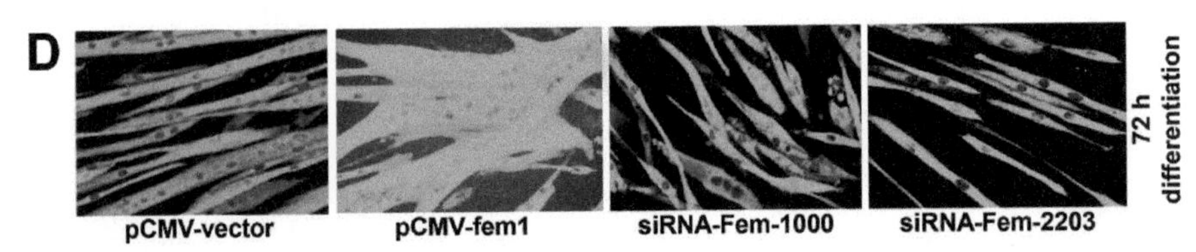

Figure 2.2 Schmatic representation of use of randomized ribozyme library for identification of genes involved in muscle differentiation: adopted from Wadhwa *et al.* [26].

its human homologue, p14ARF, activate p53 function by inactivation of p53 antagonist, HDM2 [35, 36]. Functional regulation of ARF is critical for cell cycle control in response to a variety of cellular and environmental signals. The mouse and human ARF proteins share only a limited homology at the cDNA and protein levels [30, 34] but the functional relevance of this genetic divergence has not been well studied.

By yeast two-hybrid interactive screen, we isolated Pex19p as an interacting partner of p19ARF (mouse ARF).Pex19p is a farnesylated cytosolic protein that acts as a soluble receptor or chaperone for targeting the peroxisomal membrane proteins [35, 36]. It plays an important role in peroxisomal biogenesis, membrane assembly and stabilization [37]. p19ARF–Pex19p interact in cell cytoplasm resulting in cytoplasmic retention of p19ARF and functional dampening of its p53 activation function [36]. Most intriguingly, we found that the human ARF which is shorter than the mouse ARF by 40 amino acids [38] did not interact with Pex19p in either the yeast or mammalian two-hybrid system [39] (Figure 2.3). Pex19p was shown to sequester p19ARF in the cytoplasm. p19ARF, but not p14ARF, colocalized with Pex19p in the cytoplasm. In agreement with this, p19ARF was detected first in the cytoplasm and subsequently moved to the nucleus and then to the nucleolus [39] (Figure 2.3). In contrast, p14ARF was visible in the nucleus even at the earliest time point (6 h) (Figure 2.3). Specific hammerhead ribozymes were used to knockdown Pex19p in NIH 3T3 cells that were stably expressing p53 dependent luciferase and metal inducible human/mouse ARF. Inactive versions of the ribozymes (change of nucleotide G^5 to A^5 within the catalytic domain of the ribozyme) were used as negative controls (Figure 2.3). Cells transfected with ribozymes were selected with puromycin, induced for ARF expression for 48 h and analyzed by luciferase assay. As expected, p19ARF or p14ARF resulted in upregulation of p53-dependent reporter activity. Coexpression of Pex19p ribozymes resulted in further enhancement of p19ARF-induced p53 activity, however, had no effect on p14ARF-dependent p53 transcriptional activation function [39](Figure 2.3). This study showed that the Pex19p interacts with mouse ARF protein and inactivates its function; human ARF by lacking a Pex19p binding region escapes from such inactivation.

It has long been known that there are substantial differences in regulation of cellular senescence in mouse and human cells, and that mouse cells become immortalized much more readily than human cells [40–42]. The underlying reasons and mechanisms for these differences are not well understood. Our study showed that the ARF protein, a major player in cellular senescence and tumor suppression, is controlled differently in mouse and human cells.

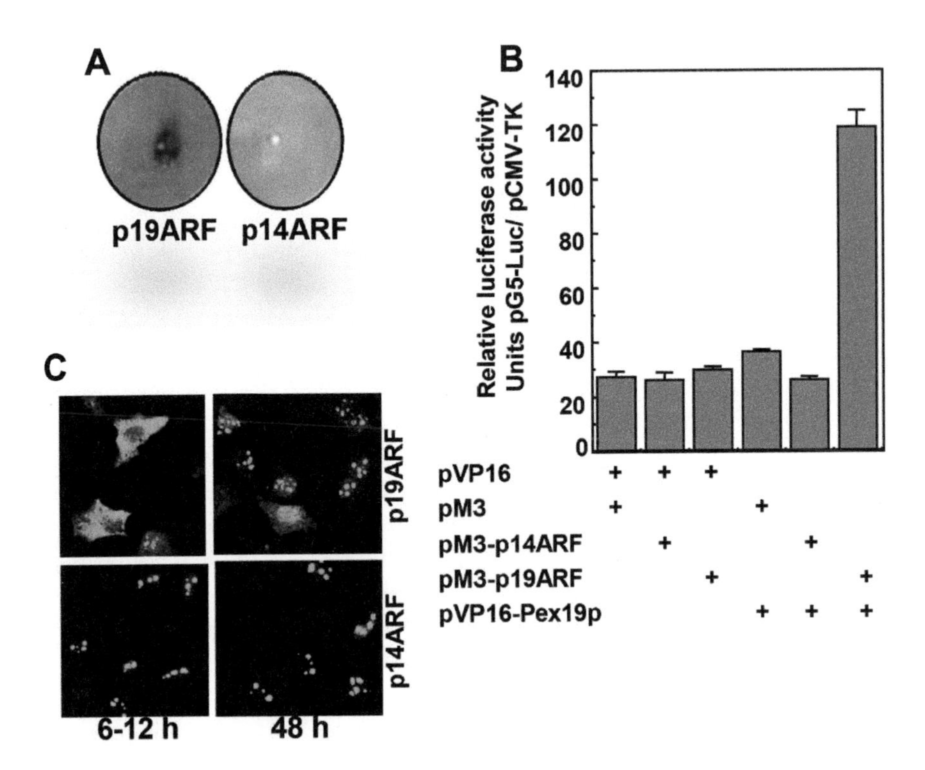
A
p19ARF
p14ARF
B
Relative luciferase activity
Units pG5-Luc/ pCMV-TK
140
120
100
80
60
40
20
0
pVP16
pM3
pM3-p14ARF
pM3-p19ARF
pVP16-Pex19p
C
p19ARF
p14ARF
6-12 h
48 h

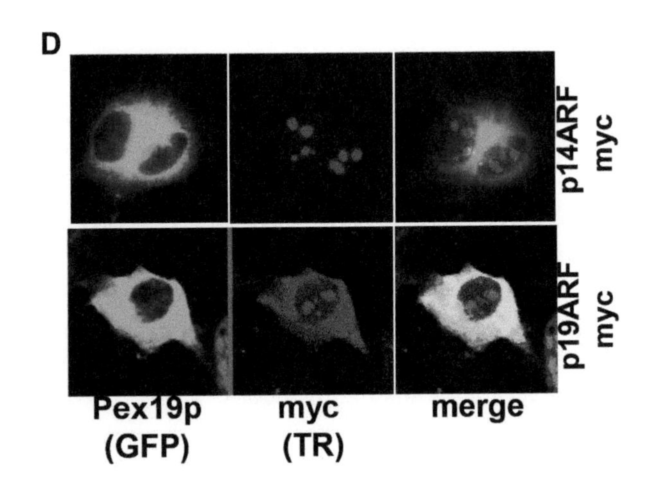
D
p14ARF
myc
p19ARF
myc
Pex19p
(GFP)
myc
(TR)
merge

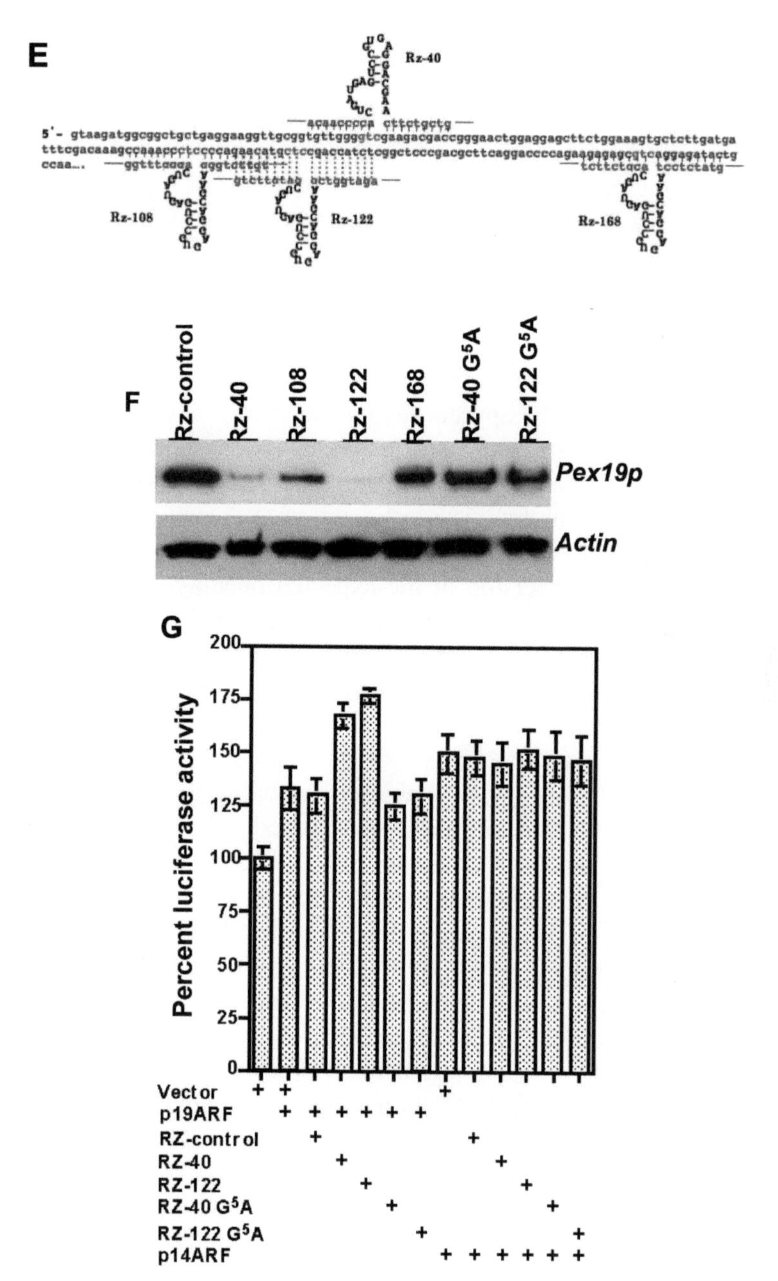

Figure 2.3 Demonstration of ARF-Per19p interaction in mouse(p19ARF), but not in human (p14ARF) cells. Pex19p specific ribozymes increased the activity of p19ARF only as shown in G. Adopted from Wadhwa *et al.* [39].

Its activity is decreased because of its interactions with Pex19p in the cytoplasm of mouse, but not of human cells. This difference may provide a mechanistic explanation of the more stringent imposition of senescence in human cells and their resistance to immortalization. Other studies describing the use of ribozymes as molecular tools in understanding of protein functions in carcinogenesis and metastasis include Wadhwa *et al.* [43], Suyama *et al.*[44][45].

2.2.3 Cancer Targets as a Model

Identification and validation of cancer targets is an extremely important aspect in cancer therapy. One of the approaches for identification of cancer targets is the "loss-of-function" assay in which the genes are knocked down and the abrogation of a cell death by drugs is monitored in cell-based assays. This can be achieved through one-by-one knockdown assays in which a single gene knockdown response can be monitored or it can be achieved in a pool in which the population showing reversion of the expected phenotype (cell survival or a reporter based readout) is analyzed and characterized for compromised gene expression. The strategy is described here using a model study in which cancer cells were killed by alcoholic extract of Ashwagandha (Indian traditional plant of solanaceae family) leaves (i-Extract) [46]. The viewpoint that the cancer cell killing effect of i-Extract was likely to be mediated by more than one genes/pathways, an unbiased screening to identify cellular targets using hammerhead ribozymes was adopted. Ribozyme population rescued from randomized ribozyme library-infected cancer cells that survived the i-Extract treatment was characterized by cloning and sequence analysis (Figures 2.4A and 2.4B). Gene targets for the isolated ribozyme sequences were determined by database search (http://blast.ncbi.nlm.nih.gov/Blast.cgi). In order to validate the involvement of these genes in i-Extract-induced cancer cell killing, two-way analyses were performed; first–validation of the targets by gene specific shRNA-mediated silencing and second–bioinformatics and systems biology-directed pathway identification.

In the first approach, specific shRNAs were prepared against seven genes (IGF2R, SREBF2, AKAP11, TFAP2A, LHX3, TPX2 and ING1), identified as candidate cellular targets in ribozyme screening. Cells were first transfected with shRNAs followed by treatment with either i-Extract or its purified components, withanone and withaferin A. These assays revealed that the knockdown of four (Group 1 - TPX2, ING1, TFAP2A and LHX3), out of the seven, genes resulted in significant ($\sim 40\%$) reversion of i-Extract-induced cell

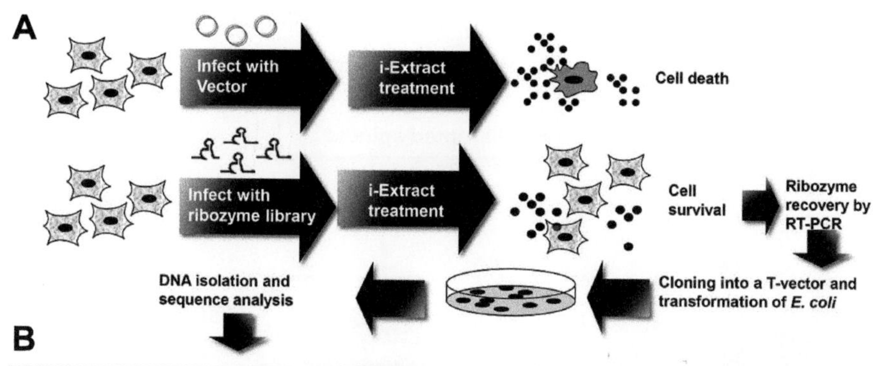

Sr. No	Gene Name	Gene Symbol
1	Inhibitor of growth family, member 1	ING1
2	**Cyclin-dependent kinase inhibitor 1A-p21, Cip1**	**CDKN1A**
3	Transcription factor AP-2 alpha	TFAP2A
4	Damage-specific DNA binding protein 2	DDB2
5	Cyclin-dependent kinase inhibitor 2B	CDKN2B
6	Heat shock 70kDa protein 9	HSPA9
7	CDKN2A interacting protein	CDKN2AIP
8	Ankyrin repeat and SOCS box-containing 15	ASB15
9	Lymphocyte antigen 86	LY86
10	Zinc finger protein 543	ZNF543
11	Insulin-like growth factor 2 receptor	IGF2R
12	**Microtubule-associated protein**	**TPX2**
13	Sterol regulatory element binding transcription factor 2	SREBF2
14	A kinase (PRKA) anchor protein 11	AKAP11
15	LIM homeobox 3	LHX3

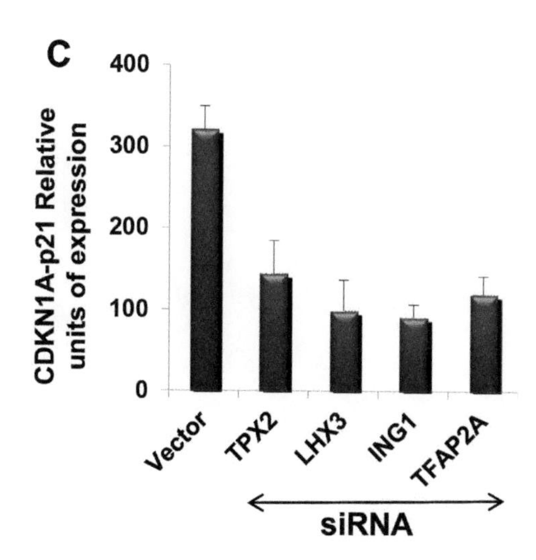

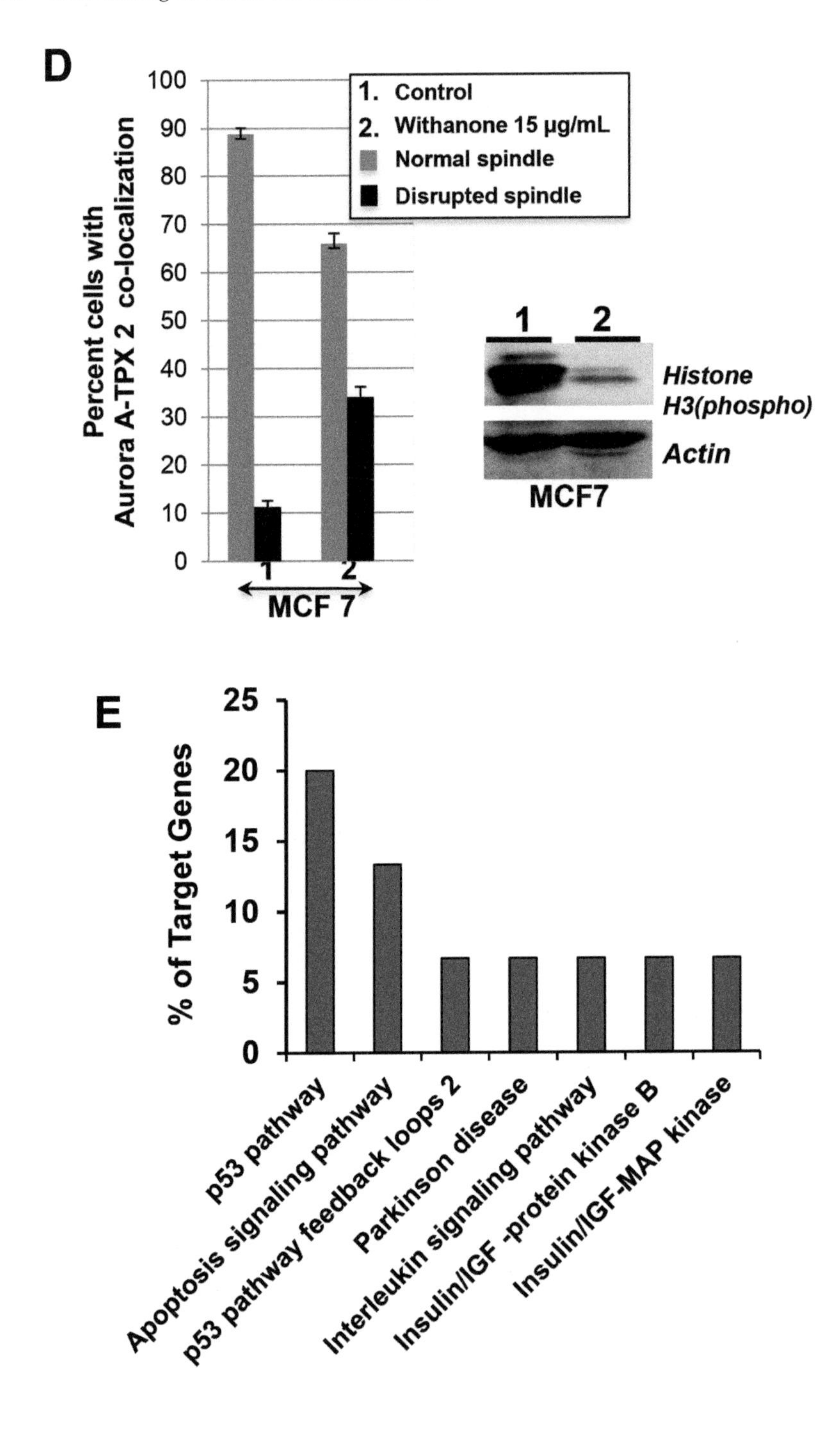
D
100
90
80
70
60
50
40
30
20
10
0
Percent cells with
Aurora A-TPX 2 co-localization
1. Control
2. Withanone 15 µg/mL
Normal spindle
Disrupted spindle
1
2
MCF 7
1
2
Histone
H3(phospho)
Actin
MCF7
E
25
20
15
10
5
0
% of Target Genes
p53 pathway
Apoptosis signaling pathway
p53 pathway feedback loops 2
Parkinson disease
Interleukin signaling pathway
Insulin/IGF -protein kinase B
Insulin/IGF-MAP kinase

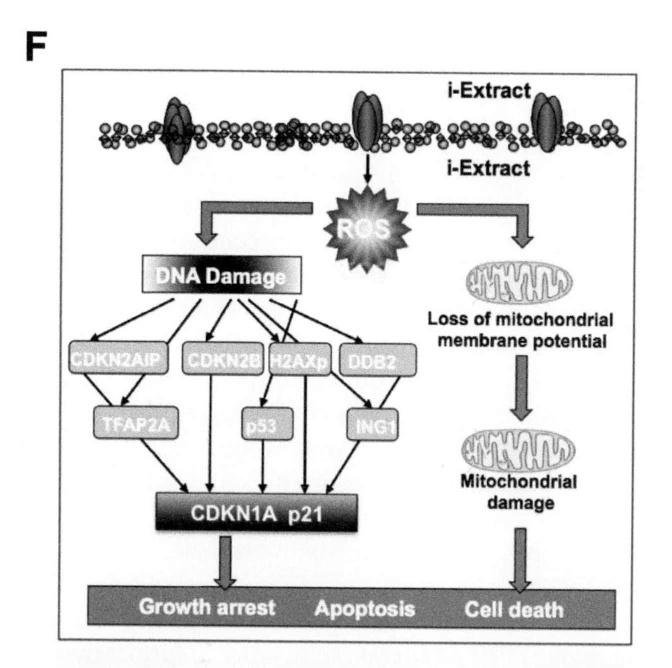

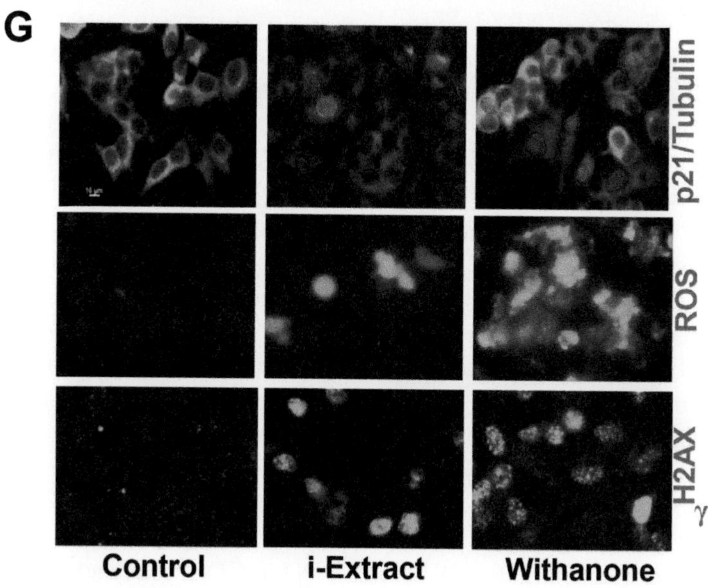

Figure 2.4 Use of randomized for identification of genes involved in killing od cancer cells by Ashwagandha leaf extract (i-Extract)and its pure phytochemical, Withanone. Adopted from Widodo *et al.* [46].

death. Interestingly, it was also accompanied by decrease in p21 expression, shown to mediate i-Extract-induced cancer cell death [47] (Figure 2.4C). The other three genes (Group 2 - IGF2R, SREBF2 and AKAP11) showed only minor (6%) reversion suggesting the crucial involvement of Group 1 proteins in i-Extract, withanone or withaferin A induced cancer cell killing. TPX2 is a microtubule-associated protein. It functions as an allosteric regulator of Aurora-A, an oncogene with essential role in centrosome maturation and chromosome segregation during mitosis [48, 49]. It is overexpressed in multiple human cancers and has been proposed as an attractive anticancer target [49]. Involvement of TPX2-Aurora A in i-Extract-induced cell death was further investigated by bioinformatics and wet experimental approaches. Bioinformatics and molecular dynamics analyses suggested that TPX2-Aurora A is a potential target of withanone[50]. Human cancer cells treated with withanone were then examined for TPX2-Aurora A interactions by biochemical and visual assays. Immunofluorescence revealed that the two proteins co-localized in cancer cells. When cells were treated with withanone, the number of cells with co-localized TPX2 and Aurora A proteins decreased by 50% (Figure 2.4D) endorsing that withanone abrogates TPX2-Aurora A complex formation and hence inhibits cancer cell proliferation. In order to further investigate the functional significance of TPX2-Aurora A complex disruption by withanone, kinase activity of the complex was examined in control and withanone treated cells. Phosphorylation of Aurora A and its substrate histone H3 decreased in the latter (Figure 2.4D). For its essential role in the spindle formation and mitosis, Aurora A is regarded as cancer target, and its kinase inhibitors have been considered as potential anticancer drugs [49]. Specificity of the drugs that block ATP binding site of Aurora A kinase has been a major issue due to cross-reactivity to other kinases. An alternate approach to interfere the ability of Aurora A to interact with TPX2 may overcome these concerns and hold promise to identify novel anticancer drugs.

In the second approach, the gene targets for i-Extract-mediated cancer cell killing identified from the ribozyme screening were studied by bioinformatics approach. The pathway analysis of the fifteen genes, identified in library screening (Figure 2.4), revealed that these genes are involved in several kinds of biological processes including oncogenesis, cell cycle, DNA repair and nucleic acid metabolism. The top two identified pathways were p53 tumor suppressor (gene targets – DDB2, CDKN1A, CDKN2B) and apoptosis (gene targets – IGF2R and HSPA9). Ras, insulin/IGF, angiogenesis and cytoskeleton regulation pathways that are tightly linked with apoptosis and tumor development were also identified (Figure 2.4E). These results revealed that the i-Extract-mediated cell killing may involve DNA damage, cell cycle arrest and apoptosis pathways that in turn take place by ROS-mediated damage either at the mitochondria or nucleus (Figure 2.4F). As predicted by pathway analysis, the involvement of DNA damage and repair signaling pathway in i-Extract-induced cancer cell killing was investigated. Induction of reactive oxygen species (ROS) (an oxidative stress marker) and γH2AX (DNA damage marker) was detected in response to the treatment with i-Extract and withanone in breast cancer cells (Figure 2.4G). Cancer cells frequently exhibit high oxidative stress and increased generation of ROS. Several reagents, such as avocado extract [51], beta-phenylethylisothiocyanate (PEITC) [52, 53], thalidomide analogs [54] and MKT077 [55–57] have been shown to involve ROS pathway in their selective cancer cell toxicity. Taken together with our ribozyme-based screening, higher level of ROS has been proposed as a selective therapeutic target for cancer.

2.3 siRNA as a Molecular Tool

2.3.1 Anticancer siRNAs and Cancer Targets

It has been well established that normal human cells possess only a limited proliferative capacity *in vitro* and *in vivo*. When cultured, they undergo a fixed number of divisions and enter a stage of permanent growth arrest. During this stage (called Hayflick limit or replicative senescence) [58], cells remain viable and metabolically active, however, are refractory to proliferation stimuli providing a barrier to immortalization, a crucial step in human carcinogenesis [30–34, 42]. Senescent cells are characterized by altered cell morphology such as flat, irregular and giant size, increased cytoplasmic granularity and large nucleus, G1 arrest, expression of β-galactosidase, increased expression of tumor suppressor proteins including p53, $p21^{WAF1}$, $p16^{INK4A}$, pRB protein [30–34, 42]. Genetic alterations that override senescence have been shown to be tightly associated with the development of tumors providing evidence that senescence is an innate anti-tumor mechanism in normal cells and offers new perspectives for cancer therapy [42]. According to this viewpoint, in order to form tumors, incipient cancer cells must evade senescence and attain additional characteristics such as (i) continued proliferation, independent of growth factors, anchorage, (ii) resistant to contact inhibition, (iii) increased motility, invasion and angiogenic capacity, (iv) deregulation of apoptosis and (v) drug resistance [31–34, 40–42]. While all these features may be regulated by miRNAs and serve as good targets for cancer therapy, induction of senescence/permanent growth arrest has been predicted to yield beneficial outcomes for the reasons that most of these features are reversed in senescing cell populations [41, 42, 56]. In addition to the wide recognition of senescence as an innate tumor suppressor mechanism, induction of cellular senescence has been demonstrated as a good model to identify components and understand the functioning of tumor suppression mechanisms [40, 41] leading to the development of new anticancer drugs [56].

Using cancer cells as models, we used immunostaining of mortalin as an endogenous reporter for identification of senescence-inducing siRNAs. Mortalin is a essential protein and belongs to the heat shock 70 (hsp70) family. Mortalin performs numerous functions including mitochondrial import, enery generation, intracellular trafficking and chaperoning [59]. It is enriched in cancer cells, interacts with p53 in cell cytoplasm and nucleus causing inhibition of its transcriptional activation and control of centrosome duplication functions, respectively [59–61]. Mortalin knockdown in cancer cells caused their growth arrest or apoptosis through activation of p53 function. Besides upregulation, mortalin has also been found to be differentially distributed in cancer cells. Whereas it is widely distributed in the cytoplasm of normal cells, it shows perinuclear staining pattern in immortalized and tumor derived cells [62]. Induction of senescence in human transformed cells by single chromosomes and chromosome fragments, drugs (MKT-077, BrdU, Withaferin A, H_2O_2, 5-Aza-dC, Epoximicin) led to shift in staining pattern from perinuclear to pancytoplamic type [47, 57, 61, 63]. Furthermore, the concentration of drugs that exhibit shift in mortalin staining pattern was 10 to 100 fold lower than their IC50 suggesting that the low dose of drugs could induce senescence in cancer cells and hence may offer a safe anticancer strategy compared to the high doses that are cytotoxic to normal cells [64]. Since shift in staining pattern of mortalin was also marked by activation of senescence speicificβ-gal staining in cancer cells induced to senescence, it could be used as a reliable assay for induced senescence (i-Senescence) in cancer cells. Hence, mortalin staining could be useful for cell-based screening of anticancer compounds including chemical drugs, peptides, small molecules and bio reagents [64]. In one study, we used library of shRNA expression plasmids in pcPUR hU6 vector, designed to express a short hairpin RNA driven by human U6 promoter. Randomly selected 768 genes, two target sites for each gene, using an algorithm (www.igene-therapeutics.co.jp) were used. The cells were plated in 96-well plates and transfected with shRNA expression plasmids [65] (Figure 2.5). The transfected cells were selected in a medium supplemented with puromycin and were then fixed in methanol:acetone (1:1) after 24–48 h. The fixed cells were stained for mortalin and scanned under the automated scanning system [65]. As shown in Figure 2.5, the shRNAs that caused shift in mortalin staining from the perinuclear to pancytoplasmic were selected for the second round of screening. The cycle was repeated four times and resulted in the identification of 22 shRNAs [65] with potential to induce shift in mortalin staining pattern in cancer cells. The induction of senescence (i-Senescence) effect of the 22 selected shRNAs was examiend for six other cancer (four breast carcinomas and two gall bladder carcinomas) cells [65]. It was found that 9 of the 22 shRNAs induced shift in mortalin staining pattern and i-Senescence in a variety of cancer cells. Interestingly, many of these genes (BCL2A1, MAP3K7, MAPK1,TERT-mRNA, Gp96, TNKS1BP1, IL1A, KPNA2, CALR and USP10) are enriched in cancers [65]. Bionformatics approach using MetaCore Pathway analysis and Data Mining (http://www.genego.com/genego_lp.php) to analyze the pathways mediated by the selected gene targets revealed that the selected genes mediate the pathways involved in apoptosis, telomere maintenance, DNA damage regulation, immune response, protein degradation and protein transport suggesting their potential as cancer targets. Taken together, the study demonstrated the use of siRNA library as a molecular tool for identification of anticancer siRNAs, cancer targets and pathways for therapeutic development of either the biological reagents including armed oncolytic viruses, antibodies, chemicals including inibitors and antagonists.

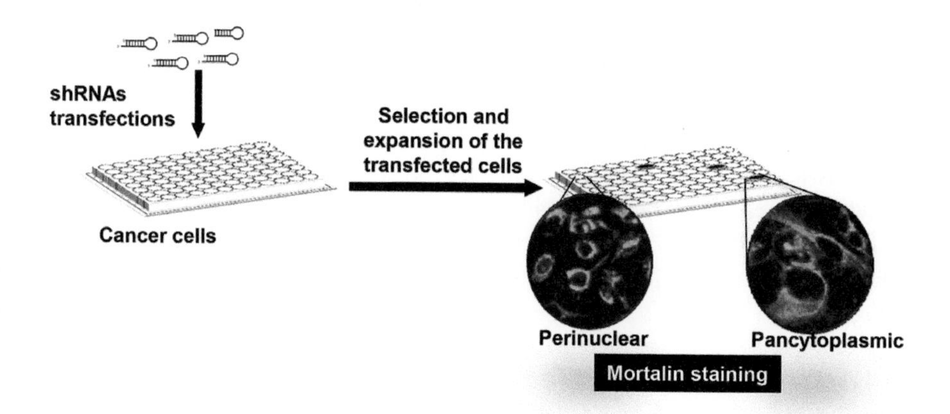

Figure 2.5 Use of mortalin staining as a reporter for induction of senescence in cancer cells and hence the identification of anticancer siRNAs. Adopted from Gao *et al.* [65].

2.3.2 Cancer Drug Target Screening

In an independent approach, siRNA library was used to identify cellular targets of anticancer alcoholic extract of Ashwagandha leaves (i-Extract) [47]. shRNA expression plasmids were transfected into the cells cultured in 96-well plates. The transfected cells were then treated with i-Extract. The cells that escaped the effect of i-Extract were the ones transfected with shRNAs including (i) cyclin-dependent kinase inhibitor 2B, (ii) NIMA (never in mitosis arrest)-related kinase 2, (iii) cyclin-dependent kinase 8, (iv) tumor suppressor protein p53, (v) inhibitor of apoptosis protein 1/baculoviral IAP repeat-containing 3 protein and (vi) cyclin-dependent kinase 5 [47]. Biochemical analysis of i-Extract treated cancer cells revealed an upregulation of p53 and change in mortalin staining pattern from perinuclear to pancytoplasmic type endorsing an induction of senescence in cancer cells.

2.4 miRNA as a Molecular Tool

2.4.1 Detection of miRNAs

Characterization of miRNAs in different biological samples can provide clues to their normal or abnormal physiological functions. However, knowledge regarding the characterization and expression patterns of miRNAs is currently restricted to a subset of known miRNAs. This is due, in part, to the limitations on current miRNA detection assays. In this section, we first provide a summary of various detection methods for miRNAs. Starting from the specimen which is often tissues or cells (Figure 2.6), they are subjected to the fixation, extraction of RNAs, or fractionation of small RNAs. Detection methods for miRNAs are classified into three categories based on; RT-PCR, hybridization and imaging. In order to compare the advantages or disadvantages of various methods, researcher can experimentally use a "pseudo-miRNA", which is a synthetic sequence and non-homologous to any endogenous miRNAs. The simplified description of the principles for each of the three technologies are described here along with a study result on the role of miR-296 in human carcinogenesis.

2.4.1.1 RT-PCR and Related Technologies

The main technical problem to detect miRNAs in cellular RNAs is often due to their small size. Different from relatively longer messenger RNAs (mRNAs), the detection of shorter miRNA needs additional procedures to bridge the conventional RNA detection methods. In the case of RT-PCR, a

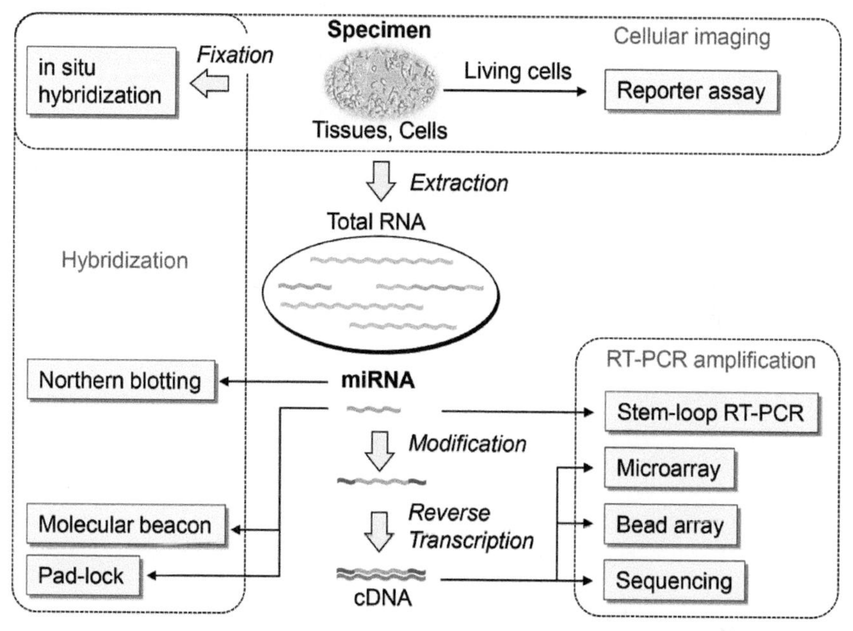

Figure 2.6 A representative flowchart of miRNA detection. Three major approaches to detect miRNAs; hybridization, RT-PCR and cellular imaging are shown.

pair of primers (usually >15 nt each in length) needs at least 30 nt to bind its template target. miRNA, due to its short size (approximately 20–25 nt), can not serve as template and hence pose hurdles in detection. This is technically resolved by several methods, such as (i) ligation of adaptor DNA/RNA, (ii) addition of adenosine tracts on RNA, or (iii) addition of cytosine tracts on cDNA.

For the ligation of adaptors (or linker) onto miRNAs, one needs to consider the terminal structure of adaptors and miRNAs, where miRNAs generally have 5'-monophospate (5'-P) and 3'-hydroxyl group (3'-OH) (Figure 2.7). Commercially available T4 RNA ligase that reacts a 5'-monophosphate RNA and ATP to yield 5'-adenylated RNA (5'-App-RNA) as an intermediate is used. The intermediate is then attacked with 3'-OH of miRNAs to yield the single-stranded ligated product with native phosphor-diester linkage. In order to avoid side reactions such as self-ligation or multimerization, the 3' terminus of the downward adaptor is often blocked with modifications, such as 3'-H (dideoxy), 3'-NH2 or 3'-biotin. The 5' terminus of the downward adaptor is designed to have 5'-P, or 5'-App (commercially available) especially when

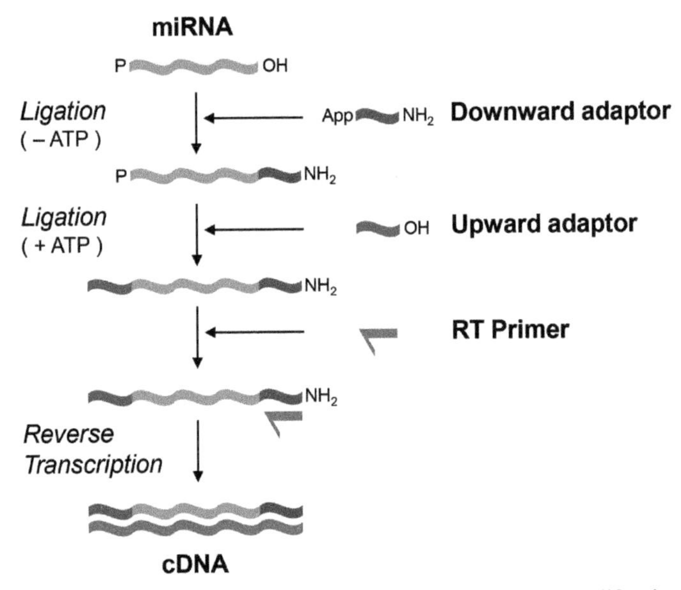

Figure 2.7 Flowchart describing the modification of miRNAs for amplification is shown. In order to amplify miRNA pools from the total RNA, miRNAs can be tagged with adaptors based on their chemical characteristics as shown.

the reaction is performed in the absence of ATP. On the other hand, upward adaptor is modified to possess 3'-OH instead of 5'-P. Unreacted excess amount of adaptors are removed by polyacrylamide gel electrophoresis (PAGE).

Addition of adenosine tracts on RNA is a much simpler approach. *E. Coli* poyl(A) polymerase is used to catalyze the reaction for adding oligo adenosines to the 3' terminus of RNA. Poly(A)-tailed miRNAs are then reverse-transcribed with an oligo-dT primer to yield cDNAs for miRNAs. During the course of cDNA synthesis, reverse transcriptases have the characteristics to attach several cytosines on the terminus of cDNA after the template-dependent reverse transcription to yield 3' overhang with cytosine tracts. This process is called switching mechanism at the 5' end of the RNA transcript (SMART), and provides dC-tailed cDNAs, which can be amplified with oligo-dG primers at the final step of RT-PCR. The procedure is applied for a miRNA amplification profiling (mRAP) [66].

Once the miRNAs are amplified, they are subjected to various detection assays. TaqMan-based qRT-PCR can quantify the amount of miRNAs. In this assay, a DNA probe having two different fluorophores at two ends, called as a TaqMan probe binds to the middle of the miRNA sequence of interest.

TaqMan probe is designed to possess FRET (fluorescent resonance energy transfer) activity. PCR polymerases digest the probe with its 5' exonuclease activity resulting in displacing of the probe from the template DNA and loss of FRET. Since such loss in FRET corresponds to the annealing of the primers to the miRNA target sequences, it enables the identification and quantification, at least in part, of miRNA targets. The combination of TaqMan probe with a stem-loop primer commercialized by Applied Biosystems produces systematic approach to detect various miRNAs, around 96 miRNAs, only with a qRT-PCR machine at the same time [67]. The stem-loop primer approach allows a faster and simpler analyses for the specific miRNAs, since one needs only a pair of specific primers against a target miRNA. The additional procedures such as adaptor ligation or tailing reaction can be avoided.

For the profiling of the miRNA expression pattern in different samples, miRNA microarrays on glass slide or beads have been invented. In both the cases, probe DNAs complementary to the sequence of the individual miRNAs are immobilized to the surface of solid phase. Probe DNAs are either dot spotted (printed) or *in situ* synthesized onto the glass slides, the latter is higher density of the probe and provides higher sensitivity [68]. A microarray slide typically has >10,000 spots and covers large majority of human miRNAs. miRNA pool extracted from two different samples are separately amplified by RT-PCR as noted above and labeled with two different fluorescent dyes such as fluorescein, rhodamine or Cy-dyes. The differentially labeled miRNAs are then subjected onto the same microarray. After the hybridization, the fluorescent intensity on the individual dots are quantified and analyzed by a fluorescence scanner. The differentially expressing miRNAs between two samples can be screened with such microarray analysis. The bead array is almost the same as the microarray, except that probes are immobilized on beads with 100 nm diameter instead of glass slides, and the address of the probes are identified with the beads emitting >100 different intensity of the fluorescence, instead of the position on the spot on the glass slide for microarrays [69].

Recently, high-throughput sequencing technologies have been developed that also enable miRNA expression analysis [70]. The high-throughput sequencing can read the sequence more than 1,000,000,000 kinds of sequence query with read length over 50 nt. With specific sequence tagging, more than 10 different samples could be analyzed at the same time in a sequencing run. The result of data analysis displays massive sequences of miRNAs and the frequencies that correspond to the expression levels of individual miRNAs. In comparison to other profiling strategies, the high-throughput sequencing has much more information especially on the sequence of miRNAs with a nucleotide level accuracy and sometimes on the discovery of unidentified miRNAs. Therefore, the high-throughput sequencing now can be applied to reveal the role of miRNAs in different stages or tissues in diverse biological processes.

2.4.1.2 Hybridization-Based Technologies

In addition to PCR-based technologies for miRNA detection, the hybridization-based technologies have been developed as seen in other detection methods for nucleic acids. However, short RNAs, such as miRNAs, are difficult to be bound to the probe with complementary sequence due to the thermodynamic instability. Since RNA–DNA hybridization is weaker and have lower thermostability (Tm) than that of RNA–RNA or RNA–LNA (locked nucleic acids), LNA probes are preferred, whereever possible, for the hybridization against miRNAs.

Northern blotting analysis is the standard way for the detection and the quantitation of miRNA levels. Total RNAs are extracted and purified from tissues or cells followed by denaturing PAGE and transferred to a nylon or nitrocellulose membrane. After fixation of the RNA onto the membrane, labeled probes are subjected to hybridization with the miRNA of interest and mark as the band. Probe DNAs or RNAs are labeled with fluorescence or enzymes, however, radio isotopes, ^{32}P, are often used to detect RNA with higher sensitivity. Northern analysis can also be used to clarify the size of RNAs, and discriminates the mature RNAs from their precursors.

miRNAs are initially synthesized as precursor, and then processed to their physiologically active species. Detection of only the active mature miRNAs, without interference from their precursor forms, is extremely important for evaluating their functional and physiological significance, since the expression levels of mature miRNAs and their precursors do not exhibit the same tendencies among several kinds of miRNAs, similar to the alternative splicing observed in the regulation of mRNAs. However, it is unavoidable that a probe against a mature RNA also hybridizes to and detects the precursor RNA, since the precursor always includes the sequence of the mature RNA. Therefore, such probes cannot be used to distinguish mature RNAs from their precursors without electrophoretic separation or to determine the intracellular localizations of mature RNAs by hybridization. In order to overcome this problem, we have developed a novel type of probe, designated the molecular spotter probe, for detecting mature miRNAs (Figure 2.8). The principle of this type of probe is based on the molecular beacon probe[71] and involves the use of a single-stranded nucleic acid molecule that acquires a stem-loop structure.

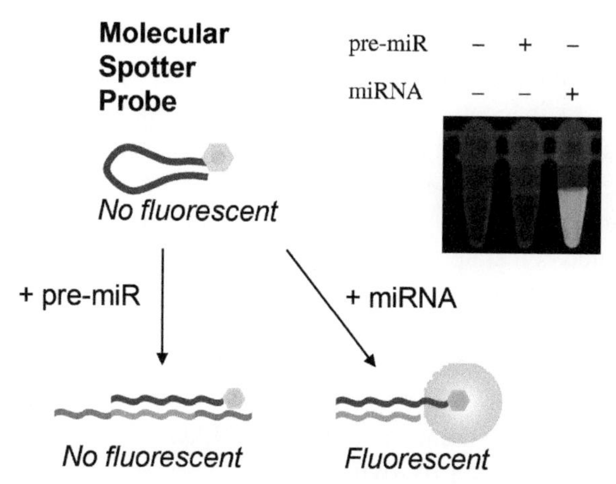

Figure 2.8 Schematic representation of molecular spotter probe.The probe is designed to be quenched in the presence of precursor RNA. It emits fluorescence only upon hybridization to the mature RNA.

The loop portion of the molecule is a probe sequence that is complementary to a predetermined sequence in the target miRNA. In the molecular spotter probe, the sequences of either part of the stem are selectively designed to be complementary to the sequences of the precursor but not those of the mature miRNA. Since the probe discriminates the mature miRNA from the precursor without electrophoretic separation, enzymatic reaction or washing out of excess amount of probes, it could be useful for real-time detection of the mature miRNAs[72].

Hybridization-based detection of miRNAs such as molecular beacon probes often lack the signal amplification due to its 1:1 binding of the target and the probe. Since some miRNAs are not expressed abundantly, signal amplification of the binding should be amplified. In the case of pad-lock method, an miRNA of interest is circulized with the splint DNA, and then amplified with rolling circle by reverse transcriptase [73]. The amplified product is a single-stranded long cDNA, which could be detected by fluorescence of intercalating dyes such as SyBr green.

2.4.1.3 Imaging of miRNAs in Cells

To determine the expression level of miRNAs, Northern blotting and RT-PCR with microarray analysis are often carried out for the direct detection of miRNA. For miRNA visualization, *in situ* hybridization analysis is conventionally performed using specific probes and fixed tissues. Indirect detection of miRNAs entails the use of reporter genes whose UTRs are connected with the target sequence of the miRNAs [74]. In this system, if the target sequence of miRNAs is located downstream of the reporter genes, including β-galactosidase or luciferase, miRNAs induce a decrease in reporter signaling by reducing protein translation. However, there are few reports on monitoring the dynamic function of miRNAs in intact cells or organs among the mixture of closely associated cell state. Recently, Brown *et al.* reported that a lentiviral vector encoding green fluorescent protein (GFP) connected to a target sequence allowed them to visualize the activity of miR-142-3p followed by immunostaining of an internal control gene using fixed tissues [74]. Sample processing such as including homogenization or fixation of cells or tissues prevents the detection of dynamic action of miRNAs.

We have developed novel retroviral or lentiviral vectors to monitor the specific miRNA activity in living cells for the first time [75, 76]. By using red and green fluorescent proteins under the control of bidirectional two distinct promoters, miRNA expression was successfully visualized and quantified by the reduction in green fluorescence when 3' UTRs were connected to the target sequences of various miRNAs in living cells (Figure 2.9).We demonstrated the use of such vector to identify an increase in the expression

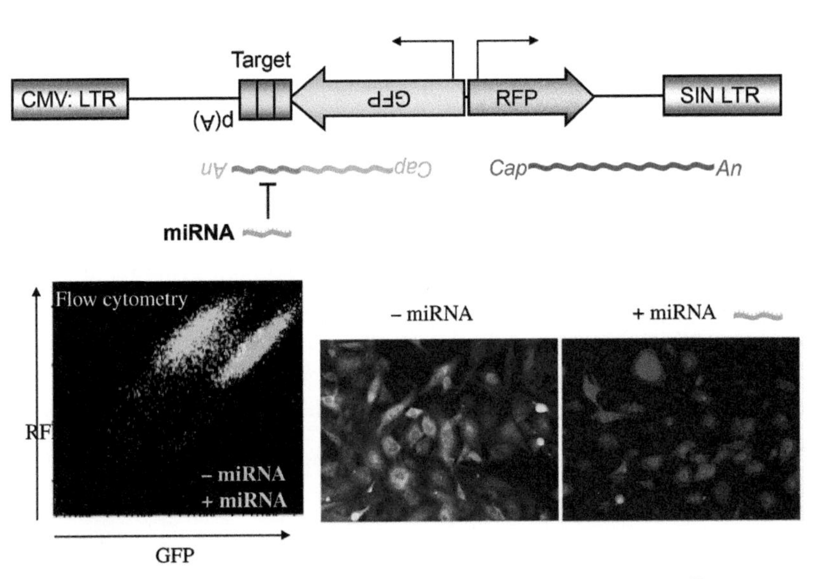

Figure 2.9 Schematic representation of the dual-color sensor vector system. Reporter system enables the real-time, quantitative detection of miRNA in single cells using dual color fluorescent proteins.

of miR-133 during myogenesis. We found that the conversion of myoblasts to myotubes that was accompied by upregulation of miR-133 resulted in the reduction of GFP signal [75]. Furthermore, adenoviral infection and consequent miRNA expression was successfully visualized and quantified by the reduction in green fluorescence when 3' UTRs were connected to the target sequences of the adenovirus-derived miRNAs [76]. These data endorsed that the fluoresence-based imaging provides a rapid and reliable method to track miRNAsasscociated with any physiological function in living cells.

2.4.2 miRNA in Cancers

Human81 cancers are often associated with a widespread abnormal expression of miRNAs suggesting their link between tumor suppressor or oncogenenic pathways. Several miRNA genes have been located at fragile sites in the genome regions that are either amplified or deleted in human cancers, and have been shown to regulate cancer-related processes including cell growth, tissue differentiation and apoptosis, tumor suppression and oncogenesis [77]. Table 2.1 provides a list of miRNA that are recently shown to be involved in carcinogenesis. For example, reduction in let-7 family of miRNAs in lung cancers has been linked to RAS overexpression resulting in increased proliferation and tumorignesis. At the same time, another miRNA of let-7 family, miR-84, was shown to suppress the multivulva phenotype of activating RAS mutations [78]. Based on these findings, *let-7* has been suggested as a tumor suppressor miRNA. The expression of miR-143 and -145 was reported to be severely downregulated in the majority of human cancer cells derived from colorectal, lung pancreatic, cervical, gastric, bladder and prostate cancers [79, 80]. These were shown to be induced by DNA damage and oncogenic stress in a p53-dependent manner and to modulate p53-mediated apoptosis, cell cycle arrest and senescence by suppression of BCL2, MYCN, SIRT1 and E2F functions [81–83]. p53 was also shown to regulate miRNA expression and processing several of them by interacting with Drosha, suggests that there is a complex network of interactions in the regulation of tumor suppressors, their effectors and regulators [84, 85].On the other hand, MiR-21 is one of the first microRNA to be described as an oncomir. It has been shown to target tumor suppressor gene including PTEN, Pdcd4 and RECK [86, 87]. The miR-34 family of miRNAs is downregulated in several types of cancers [84].

While there can be several different approaches to identify cancer related miRNAs, we chose to compare those, in human normal cells and their immortalized derivatives, that upregulate telomerase, an established hallmark

Table 2.1 microRNAs associated with human cancers

miRNA	Gene loci	Cancer association	Function	References
let-7 family members	Multiple loci	Negatively regulate the Ras oncogenes; direct cell proliferation and differentiantion; decreased abundance in lung cancer	TS	[78]
miR-15a, miR-16-1	Chromosome 13q14	Frequently deleted or downregulated in B-cell chronic lymphocytic leukemia; negatively regulates the anti-apoptotic gene BCL2	TS	[82, 83]
miR-17-19b cluster	Chromosome 13q31-32	Upregulated by MYC; negatively modulates the E2F1 oncogene; loss of heterozygosity of this cluster is found in hepatocellular carcinoma; overexpressed in B-cell lymphomas	TS/OG	[89]
miR-21	Chromosome 17q23.2	Anti-apoptotic factor; upregulated in glioblastomas, breast cancer and ovarian cancer	OG	[90]
miR-31	Chromosome 9p21.3	Activates the RAS pathway and functions as an oncogenic microRNA	OG	[91]
miR-93	Chromosome 7q22.1	miR-93 expression was significantly decreased in tumoral compared with nontumoral colon tissues ($P<0.001$)	TS	[92]
miR-101	Chromosome 1p31.3	miR-101 were found to be lost both in clinically localize PCa (37.5%) and in metastatic cancer (66.7%)	TS	[93]
miR-143, miR-145	Chromosome 5q32–33	Decreased abundance in colorectal cancer; downregulated in breast, prostate, cervical and lymphoid cancer cell lines; miR-145 is decreased in breast cancer	TS	[94]

(Continued)

Table 2.1 Continued

miRNA	Gene loci	Cancer association	Function	References
miR-155	Chromosome 21q21	Upregulated in paediatric Burkitt, Hodgkin, primary mediastinal and diffuse large-B-cell lymphomas; upregulated in human breast cancer	OG	[90]
miR-183	Chromosome 7q32.2	Upregulated in colorectal cancer, prostante cancer, bladder cancer, lung cancer and hepatocellular carcinoma	OG	[95, 96]
miR-200a/b	Chromosome Xp11.3	MiR-200a promotes anoikis resistance and metastasis by targeting YAP1 in human breast cancer; miR-200b is overexpressed in endometrial adenocarcinomas and enhances MMP2 activity by downregulating TIMP2 in human endometrial cancer cell line HEC-1A	OG	[97, 98]
miR-221	Chromosome 17q23.2	Serum miR-221 was upregulated in patients with EOC (epithelial ovarian cancer, n = 96) compared with healthy controls (n = 35)	OG	[99]
miR-224	Chromosome Xq28	Upregulated in colorectal cancer and cervical cancer	OG	[100, 101]
miR-335	Chromosome 7q32.2	MiR-335 was downregulated in the ovarian cancer cell lines relative to normal ovarian epithelium tissues; decreased expression in breast cancer specimens	TS	[102, 103]
miR-449a	Chromosome 5q11.2	Acts as a tumor suppressor in human bladder cancer through the regulation of pocket proteins	TS	[104]
miR-497	Chromosome 17p13.1	Targets insulin-like growth factor 1 receptor and has a tumour suppressive role in human colorectal cancer	TS	[105]

TS: tumor suppressor; OG: oncogene

of carcinogenesis. Although the process of immortalization and telomerase activation are anticipated to involve miRNA regulation, no miRNAs with such function have yet been identified. A comparative miRNA array analysis on normal human fibroblasts and their SV40 T antigen-immortalized telomerase positive derivatives revealed an upregulation of miR-296 in telomerase positive cells [88]. By an independent experiment on genomic analysis of cancer cells we found that the chromosome region (20q13.32) of miR-296 was amplified in 28/36 cell lines, and most of these showed enriched miR-296 expression suggesting that it may have a role in telomerase–mediated carcinogenesis [88]. Overexpression of miR-296 in human cancer cells, with and without telomerase activity, was then undertaken. We noticed that the telomerase activity of the cells remained unaffected in miR-296 overexpressing cells. In order to understand its role in carcinogenesis, we investigated the effect of miRNA on tumor suppressor protein, p53 and its downstream effector, $p21^{WAF1}$, involved in growth arrest function. We found that the miRNA 296 targets p53 and $p21^{WAF1}$ [88] (Figure 2.10) and plays a key role in

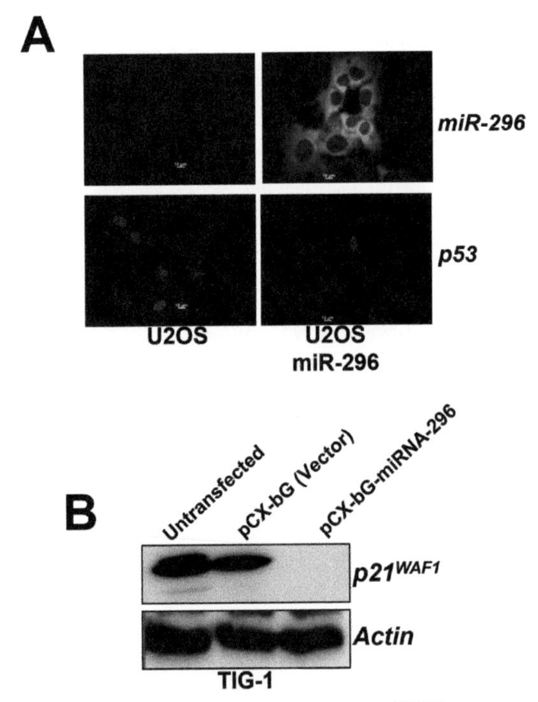

Figure 2.10 Demonstration of targeting of p53 and $p21^{WAF1}$ by miR-296.Adopted from Yoon et al [88].

the control of cell cycle, apoptosis and DNA repair. Since these functions are largely deregulated in human cancers, miR-296 was proposed as a candidate therapeutic target for cancer therapy.

References

[1] Haseloff, J. and W.L. Gerlach, Simple RNA enzymes with new and highly specific endoribonuclease activities. Nature, 1988. **334** (6183): p. 585–91.

[2] Rossi, J.J., Ribozymes, genomics and therapeutics.ChemBiol, 1999. **6**(2): p. R33–7.

[3] Serganov, A. and D.J. Patel, Ribozymes, riboswitches and beyond: regulation of gene expression without proteins. Nat Rev Genet, 2007. **8**(10): p. 776–90.

[4] Mulhbacher, J., P. St-Pierre, and D.A. Lafontaine, Therapeutic applications of ribozymes and riboswitches.CurrOpinPharmacol. 2010. **10**(5): p. 551–6.

[5] Hammann, C., et al., The ubiquitous hammerhead ribozyme. RNA. 2012. **18**(5): p. 871–85.

[6] Ambros, V., microRNAs: tiny regulators with great potential. Cell, 2001. **107**(7): p. 823–6.

[7] Couzin, J., Breakthrough of the year. Small RNAs make big splash. Science, 2002. **298**(5602): p. 2296–7.

[8] Whalley, K., Breakthrough for systemic RNAi. Nat. Rev. Genet., 2006. **7**: p. 1.

[9] Haussecker, D., The Business of RNAi Therapeutics in 2012. MolTher Nucleic Acids. **1**: p. e8.

[10] Bernstein, E., et al., Role for a bidentateribonuclease in the initiation step of RNA interference. Nature, 2001. **409**(6818): p. 363–6.

[11] Bartel, D.P., MicroRNAs: genomics, biogenesis, mechanism, and function. Cell, 2004. **116**(2): p. 281–97.

[12] Carthew, R.W. and E.J. Sontheimer, Origins and Mechanisms of miRNAs and siRNAs. Cell, 2009. **136**(4): p. 642–55.

[13] Lim, L.P., et al., Vertebrate microRNA genes. Science, 2003. **299**(5612): p. 1540.

[14] Mourelatos, Z., et al., miRNPs: a novel class of ribonucleoproteins containing numerous microRNAs. Genes Dev, 2002. **16**(6): p. 720–8.

[15] Doench, J.G., C.P. Petersen, and P.A. Sharp, siRNAs can function as miRNAs. Genes Dev, 2003. **17**(4): p. 438–42.
[16] Hutvagner, G. and P.D. Zamore, A microRNA in a multiple-turnover RNAi enzyme complex. Science, 2002. **297**(5589): p. 2056–60.
[17] Lee, R.C., R.L. Feinbaum, and V. Ambros, The C. elegansheterochronic gene lin-4 encodes small RNAs with antisense complementarity to lin-14. Cell, 1993. **75**(5): p. 843–54.
[18] Rougvie, A.E., Control of developmental timing in animals. Nat Rev Genet, 2001. **2**(9): p. 690–701.
[19] Lewis, B.P., C.B. Burge, and D.P. Bartel, Conserved seed pairing, often flanked by adenosines, indicates that thousands of human genes are microRNA targets. Cell, 2005. **120**(1): p. 15–20.
[20] Krek, A., et al., Combinatorial microRNA target predictions. Nat Genet, 2005. **37**(5): p. 495–500.
[21] Yan, H., et al., Identification and functional analysis of epigenetically silenced microRNAs in colorectal cancer cells.PLoS One. 2011. **6**(6): p. e20628.
[22] Pasquinelli, A.E., MicroRNAs: deviants no longer. Trends Genet, 2002. **18**(4): p. 171–3.
[23] Provost, P., et al., Ribonuclease activity and RNA binding of recombinant human Dicer. EMBO J, 2002. **21**(21): p. 5864–74.
[24] Taylor, M.V., Muscle differentiation: signalling cell fusion.CurrBiol, 2003. **13**(24): p. R964–6.
[25] McKinsey, T.A., C.L. Zhang, and E.N. Olson, Signaling chromatin to make muscle.CurrOpin Cell Biol, 2002. **14**(6): p. 763–72.
[26] Wadhwa, R., et al., Use of a randomized hybrid ribozyme library for identification of genes involved in muscle differentiation. J BiolChem, 2004. **279**(49): p. 51622–9.
[27] Pownall, M.E., M.K. Gustafsson, and C.P. Emerson, Jr., Myogenic regulatory factors and the specification of muscle progenitors in vertebrate embryos.Annu Rev Cell DevBiol, 2002. **18**: p. 747–83.
[28] Cameron, T.L., et al., Global comparative transcriptome analysis of cartilage formation in vivo. BMC DevBiol, 2009. **9**: p. 20.
[29] Bouzeghrane, F., et al., Alpha8beta1 integrin is upregulated in myofibroblasts of fibrotic and scarring myocardium. J Mol Cell Cardiol, 2004. **36**(3): p. 343–53.
[30] Quelle, D.E., et al., Alternative reading frames of the INK4a tumor suppressor gene encode two unrelated proteins capable of inducing cell cycle arrest. Cell, 1995. **83**(6): p. 993–1000.

[31] Kamijo, T., et al., Tumor suppression at the mouse INK4a locus mediated by the alternative reading frame product p19ARF. Cell, 1997. **91**(5): p. 649–59.

[32] Sherr, C.J., The INK4a/ARF network in tumour suppression. Nat Rev Mol Cell Biol, 2001. **2**(10): p. 731–7.

[33] Bracken, A.P., et al., The Polycomb group proteins bind throughout the INK4A-ARF locus and are disassociated in senescent cells. Genes Dev, 2007. **21**(5): p. 525–30.

[34] Serrano, M., et al., Role of the INK4a locus in tumor suppression and cell mortality. Cell, 1996. **85**(1): p. 27–37.

[35] Snyder, W.B., et al., Pex19p interacts with Pex3p and Pex10p and is essential for peroxisome biogenesis in Pichiapastoris.MolBiol Cell, 1999. **10**(6): p. 1745–61.

[36] Sugihara, T., et al., Pex19p dampens the p19ARF-p53-p21WAF1 tumor suppressor pathway. J BiolChem, 2001. **276**(22): p. 18649–52.

[37] Gotte, K., et al., Pex19p, a farnesylated protein essential for peroxisome biogenesis.Mol Cell Biol, 1998. **18**(1): p. 616–28.

[38] Bates, S., et al., p14ARF links the tumour suppressors RB and p53. Nature, 1998. **395**(6698): p. 124–5.

[39] Wadhwa, R., et al., A major functional difference between the mouse and human ARF tumor suppressor proteins. J BiolChem, 2002. **277**(39): p. 36665–70.

[40] DeGregori, J., Evolved tumor suppression: why are we so good at not getting cancer? Cancer Res. 2011. **71**(11): p. 3739–44.

[41] Rodier, F., J. Campisi, and D. Bhaumik, Two faces of p53: aging and tumor suppression. Nucleic Acids Res, 2007. **35**(22): p. 7475–84.

[42] Kaul, S.C., Wadhwa, R. (Eds), Mortalin Biology: Life, Stress and Death. Springer, 2012.11

[43] Wadhwa, R., et al., Targeting mortalin using conventional and RNA-helicase-coupled hammerhead ribozymes. EMBO Rep, 2003. **4**(6): p. 595–601.

[44] Suyama, E., et al., Identification of genes involved in cell invasion by using a library of randomized hybrid ribozymes.ProcNatlAcadSci U S A, 2003. **100**(10): p. 5616–21.

[45] Suyama, E., et al., Identification of metastasis-related genes in a mouse model using a library of randomized ribozymes. J BiolChem, 2004. **279**(37): p. 38083–6.

[46] Widodo, N., et al., Selective killing of cancer cells by Ashwagandha leaf extract and its component Withanone involves ROS signaling.PLoS One, 2010. **5**(10): p. e13536.
[47] Widodo, N., et al., Selective killing of cancer cells by leaf extract of Ashwagandha: identification of a tumor-inhibitory factor and the first molecular insights to its effect.Clin Cancer Res, 2007. **13**(7): p. 2298–306.
[48] Furukawa, T., et al., AURKA is one of the downstream targets of MAPK1/ERK2 in pancreatic cancer. Oncogene, 2006. **25**(35): p. 4831–9.
[49] Warner, S.L., et al., Validation of TPX2 as a potential therapeutic target in pancreatic cancer cells.Clin Cancer Res, 2009. **15**(21): p. 6519–28.
[50] Grover, A., et al., Ashwagandha derived withanone targets TPX2-Aurora A complex: computational and experimental evidence to its anticancer activity.PLoS One, 2012. **7**(1): p. e30890.
[51] Ding, H., et al., Selective induction of apoptosis of human oral cancer cell lines by avocado extracts via a ROS-mediated mechanism.Nutr Cancer, 2009. **61**(3): p. 348–56.
[52] Wu, X.J. and X. Hua, Targeting ROS: selective killing of cancer cells by a cruciferous vegetable derived pro-oxidant compound. Cancer BiolTher, 2007. **6**(5): p. 646–7.
[53] Trachootham, D., et al., Selective killing of oncogenically transformed cells through a ROS-mediated mechanism by beta-phenylethylisothiocyanate. Cancer Cell, 2006. **10**(3): p. 241–52.
[54] Ge, Y., et al., Selective leukemic-cell killing by a novel functional class of thalidomide analogs. Blood, 2006. **108**(13): p. 4126–35.
[55] Modica-Napolitano, J.S., et al., Selective damage to carcinoma mitochondria by the rhodacyanine MKT-077. Cancer Res, 1996. **56**(3): p. 544–50.
[56] Modica-Napolitano, J.S. and J.R. Aprille, Delocalized lipophilic cations selectively target the mitochondria of carcinoma cells.Adv Drug Deliv Rev, 2001. **49**(1–2): p. 63–70.
[57] Wadhwa, R., et al., Selective toxicity of MKT-077 to cancer cells is mediated by its binding to the hsp70 family protein mot-2 and reactivation of p53 function. Cancer Res, 2000. **60**(24): p. 6818–21.
[58] Hayflick, L. and P.S. Moorhead, The serial cultivation of human diploid cell strains.Exp Cell Res, 1961. **25**: p. 585–621.

[59] Kaul, S.C., et al., Activation of wild type p53 function by its mortalin-binding, cytoplasmically localizing carboxyl terminus peptides. J BiolChem, 2005. **280**(47): p. 39373–9.

[60] Wadhwa, R., K. Taira, and S.C. Kaul, Mortalin: a potential candidate for biotechnology and biomedicine.HistolHistopathol, 2002. **17**(4): p. 1173–7.

[61] Kaul, S.C., et al., Mortalin: present and prospective.ExpGerontol, 2002. **37**(10-11): p. 1157–64.

[62] Wadhwa, R., et al., Differential subcellular distribution of mortalin in mortal and immortal mouse and human fibroblasts.Exp Cell Res, 1993. **207**(2): p. 442–8.

[63] Michishita, E., et al., 5-Bromodeoxyuridine induces senescence-like phenomena in mammalian cells regardless of cell type or species. J Biochem, 1999. **126**(6): p. 1052–9.

[64] Kaul, Z., et al., Quantum dot-based mortalin staining as a visual assay for detection of induced senescence in cancer cells. Ann N Y AcadSci, 2007. **1100**: p. 368–72.

[65] Gao, R., Yaguchi, T., Horimoto, K., Ishii, T., Kaul, S. C., Wadhwa, R., Identification of Anti-cancer shRNAs Based on the Staining Pattern mortalin.Tiss. Cult. Res. Commu, 2010. **29**: p. 147–53.

[66] Takada, S. and H. Mano, Profiling of microRNA expression by mRAP. Nat Protoc, 2007. **2**(12): p. 3136–45.
Chen, C., et al., Real-time quantification of microRNAs by stem-loop RT-PCR. Nucleic Acids Res, 2005. **33**(20): p. e179.

[67] Callari, M., et al., Comparison of microarray platforms for measuring differential microRNA expression in paired normal/cancer colon tissues.PLoS One, 2012. **7**(9): p. e45105.

[68] Lu, J., et al., MicroRNA expression profiles classify human cancers. Nature, 2005. **435**(7043): p. 834–8.

[69] Kircher, M. and J. Kelso, High-throughput DNA sequencing–concepts and limitations.Bioessays, 2010. **32**(6): p. 524–36.

[70] Tyagi, S. and F.R. Kramer, Molecular beacons: probes that fluoresce upon hybridization. Nat Biotechnol, 1996. **14**(3): p. 303–8.

[71] Kato, Y., An efficient fluorescent method for selective detection of mature miRNA species. Nucleic Acids SympSer, 2008. **52**: p. 71–2.

[72] Jonstrup, S.P., J. Koch, and J. Kjems, A microRNA detection system based on padlock probes and rolling circle amplification. RNA, 2006. **12**(9): p. 1747–52.

[73] Brown, B.D., et al., Endogenous microRNA can be broadly exploited to regulate transgene expression according to tissue, lineage and differentiation state. Nat Biotechnol, 2007. **25**(12): p. 1457–67.

[74] Kato, Y., et al., Real-time functional imaging for monitoring miR-133 during myogenic differentiation.Int J Biochem Cell Biol, 2009. **41**(11): p. 2225–31.

[75] Kato, Y., S.Y. Sawata, and A. Inoue, A lentiviral vector encoding two fluorescent proteins enables imaging of adenoviral infection via adenovirus-encoded miRNAs in single living cells. J Biochem, 2010. **147**(1): p. 63–71.

[76] Calin, G.A., et al., Human microRNA genes are frequently located at fragile sites and genomic regions involved in cancers.ProcNatlAcadSci U S A, 2004. **101**(9): p. 2999–3004.

[77] Johnson, S.M., et al., RAS is regulated by the let-7 microRNA family. Cell, 2005. **120**(5): p. 635–47.

[78] Noguchi, S., et al., MicroRNA-143 functions as a tumor suppressor in human bladder cancer T24 cells. Cancer Lett, 2011. **307**(2): p. 211–20.

[79] Li, X., et al., Identification of new aberrantly expressed miRNAs in intestinal-type gastric cancer and its clinical significance.Oncol Rep, 2011. **26**(6): p. 1431–9.

[80] Zheng, J., et al., miR-21 downregulates the tumor suppressor P12 CDK2AP1 and stimulates cell proliferation and invasion. J Cell Biochem, 2011. **112**(3): p. 872–80.

[81] Calin, G.A., et al., Frequent deletions and down-regulation of micro-RNA genes miR15 and miR16 at 13q14 in chronic lymphocytic leukemia.ProcNatlAcadSci U S A, 2002. **99**(24): p. 15524–9.

[82] Cimmino, A., et al., miR-15 and miR-16 induce apoptosis by targeting BCL2. ProcNatlAcadSci U S A, 2005. **102**(39): p. 13944–9.

[83] Corney, D.C., et al., MicroRNA-34b and MicroRNA-34c are targets of p53 and cooperate in control of cell proliferation and adhesion-independent growth. Cancer Res, 2007. **67**(18): p. 8433–8.

[84] Vousden, K.H. and G.F. Woude, The ins and outs of p53. Nat Cell Biol, 2000. **2**(10): p. E178–80.

[85] Folini, M., et al., miR-21: an oncomir on strike in prostate cancer.Mol Cancer, 2010. **9**: p. 9–12.

[86] Reis, S.T., et al., miR-21 may acts as an oncomir by targeting RECK, a matrix metalloproteinase regulator, in prostate cancer. BMC Urol, 2012. **12**: p. 14.

[87] Yoon, A.R., et al., MicroRNA-296 is enriched in cancer cells and downregulates p21WAF1 mRNA expression via interaction with its 3' untranslated region. Nucleic Acids Res, 2011. **39**(18): p. 8078–91.

[88] He, L., et al., A microRNA polycistron as a potential human oncogene. Nature, 2005. **435**(7043): p. 828–33.

[89] Iorio, M.V., et al., MicroRNA gene expression deregulation in human breast cancer. Cancer Res, 2005. **65**(16): p. 7065–70.

[90] Sun, D., et al., MicroRNA-31 activates the RAS pathway and functions as an oncogenic MicroRNA in human colorectal cancer by repressing RAS p21 GTPase activating protein 1 (RASA1). J BiolChem, 2013. **288**(13): p. 9508–18.

[91] Xiao, Z.G., et al., Clinical significance of microRNA-93 downregulation in human colon cancer.Eur J GastroenterolHepatol, 2013. **25**(3): p. 296–301.

[92] Pang, Y., C.Y. Young, and H. Yuan, MicroRNAs and prostate cancer.ActaBiochimBiophys Sin (Shanghai), 2010. **42**(6): p. 363–9.

[93] Michael, M.Z., et al., Reduced accumulation of specific microRNAs in colorectal neoplasia.Mol Cancer Res, 2003. **1**(12): p. 882–91.

[94] Zhang, Q.H., et al., Meta-analysis of microRNA-183 family expression in human cancer studies comparing cancer tissues with noncancerous tissues. Gene, 2013. **527**(1): p26–32.

[95] Ahmed, F.E., et al., Diagnostic MicroRNA markers to screen for sporadic human colon cancer in stool: I. Proof of Principle. Cancer Genomics Proteomics, 2013. **10**(3): p. 93–113.

[96] Yu, S.J., et al., MicroRNA-200a promotes anoikis resistance and metastasis by targeting YAP1 in human breast cancer.Clin Cancer Res, 2013. **19**(6): p. 1389–99.

[97] Dai, Y., et al., MicroRNA-200b is overexpressed in endometrial adenocarcinomas and enhances MMP2 activity by downregulating TIMP2 in human endometrial cancer cell line HEC-1A cells. Nucleic Acid Ther, 2013. **23**(1): p. 29–34.

[98] Hong, F., et al., Prognostic significance of serum microRNA-221 expression in human epithelial ovarian cancer. J Int Med Res, 2013. **41**(1): p. 64–71.

[99] Liao, W.T., et al., MicroRNA-224 promotes cell proliferation and tumor growth in human colorectal cancer by repressing PHLPP1 and PHLPP2.Clin Cancer Res, 2013.

[100] Shen, S.N., et al., Upregulation of microRNA-224 is associated with aggressive progression and poor prognosis in human cervical cancer.DiagnPathol, 2013. **8**: p. 69.

[101] Cao, J., et al., miR-335 represents an invasion suppressor gene in ovarian cancer by targeting Bcl-w.Oncol Rep, 2013. **30**(2): p. 701–6.

[102] Heyn, H., et al., MicroRNA miR-335 is crucial for the BRCA1 regulatory cascade in breast cancer development.Int J Cancer, 2011. **129**(12): p. 2797–806.

[103] Chen, H., et al., MicroRNA-449a acts as a tumor suppressor in human bladder cancer through the regulation of pocket proteins. Cancer Lett, 2012. **320**(1): p. 40–7.

[104] Guo, S.T., et al., MicroRNA-497 targets insulin-like growth factor 1 receptor and has a tumour suppressive role in human colorectal cancer. Oncogene, 2013. **32**(15): p. 1910–20.

3

PPAR Responsive Regulatory Modules in Breast Cancer

Meena K Sakharkar[1], Babita Shashni[2], Karun Sharma[2], Ramesh Chandra[3,4] and Kishore R Sakharkar[5]

[1]College of Pharmacy and Nutrition, University of Saskatchewan, SK, Canada
[2]Graduate School of Life and Environmental Sciences, University of Tsukuba, Japan
[3]Department of Chemistry, Delhi University, India
[4]B. R. Ambedkar Center for Biomedical Research, University of Delhi, India
[5]OmicsVista, Singapore

3.1 Introduction

The peroxisome proliferator-activated receptors (PPARs) are ligand activated transcription factors, belonging to the nuclear receptor superfamily, that control the expression of genes involved in organogenesis, inflammation, cell differentiation, proliferation, lipid, and carbohydrate metabolism [1–2]. PPARs activated by their selected ligands, heterodimerize with retinoid X receptor (RXR), then bind to peroxisome proliferator response elements (PPREs) and/or PPAR-associated conserved motif (PACM) motifs, specific sequences in their target genes (Figure 3.1). The consensus PPRE site consists of a direct repeat of the sequence AGGTCA separated by a single/double nucleotide, which is designated as DR-1 site/DR-2 site [3]. Recently, a PACM motif of width 15 bp with the consensus TTCATTTGGACATTG was discovered. It is reported that PACM motifs are more common than PPREs [4].

Three isoforms (α, β and γ) for PPAR have been identified so far in Xenopus, mouse, human, rats and hamsters. Each major isoform of PPAR

Post-genomic Approaches in Cancer and Nano Medicine, 61–84.

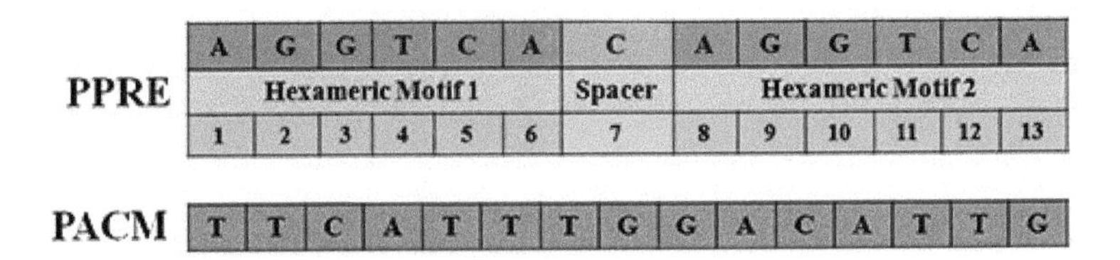

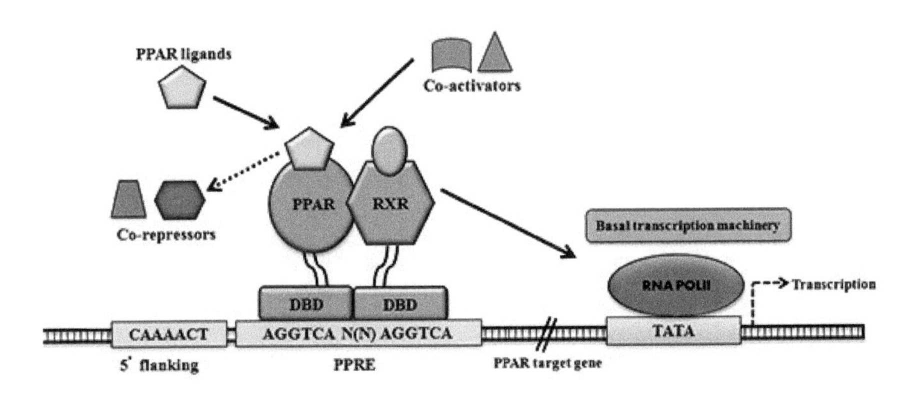

Figure 3.1 PPARγ activation mechanism. Upon ligand activation PPARγ heterodimerizes with Retinoid X Receptor (RXR) in nucleus and binds to PPRE and/or PACM motifs in the promoter region and modulates the expression of genes downstream. The consensus PPRE site consists of a direct repeat of the sequence AGGTCA separated by a single/double nucleotide, which is designated as DR-1 site/DR-2 site and PACM site consist of 15 bp consensus sequence, TTCATTTGGACATTG. The PACM motifs are reported to be more common than PPREs.

(PPARα, PPARβ/δ, and PPARγ) is encoded by a different gene, performs different functions and exhibits different tissue localizations in many parts of the human body [5]. Like other nuclear receptors, PPARγ exhibits a modular structure consisting of distinct functional domains and the receptor exerts many of its effects by regulating target gene transcription in a ligand-dependent manner.

The peroxisome proliferator-activated receptor γ (PPARγ) is the most extensively studied subtype of the PPARs [6]. PPARγ is expressed in adipose tissue, colon, immune system, hematopoietic cells, and retina involved in lipid anabolism, adipocyte differentiation, control of inflammation, macrophage maturation, embryo implantation, and molecular targets of antidiabetic thiazolidinediones [7] (Figure 3.2). Differential promoter usage and alternate splicing of the gene generates three mRNA isoforms: PPARγ1, PPARγ2 and PPARγ3. PPARγ1, and PPARγ3 encode the same protein product; the PPARγ2 isoform contains an additional 28 amino acids at its N-terminus (Figure 3.3). PPARγ1 exhibits widespread expression, although at low levels, while PPARγ2 and PPARγ3 are highly expressed in adipose tissue. The 28 additional N-terminal amino acids in PPARγ2 confer a 5- to 6-fold increase in transcription-stimulating activity of the ligand-independent activation function-1 domain [8–10]. Following 'docking'/'binding' of the PPARγ-RXR heterodimer to specific DNA sequences (response elements - PPREs) located in the target gene promoter, binding of cognate or exogenous ligands mediates cofactor recruitment, which in turn leads to transcriptional regulation (Figure 3.4). The identity(ies) of the endogenous ligand(s) for PPARγ remains a matter of debate (hence its classification as an 'orphan receptor'), although several naturally occurring compounds have been shown to be capable of activating the receptor at concentrations comparable to physiological levels, including a variety of polyunsaturated fatty acids (e.g. linoleic acid, linolenic acid and arachidonic acid) and eicosanoids (e.g. prostaglandin J2 derivatives) [11–12]. Currently however, synthetic ligands (e.g. TZDs and tyrosine agonists) remain the most potent known activators of PPARγ. Ligand binding is also reported to lead to preferential recruitment of

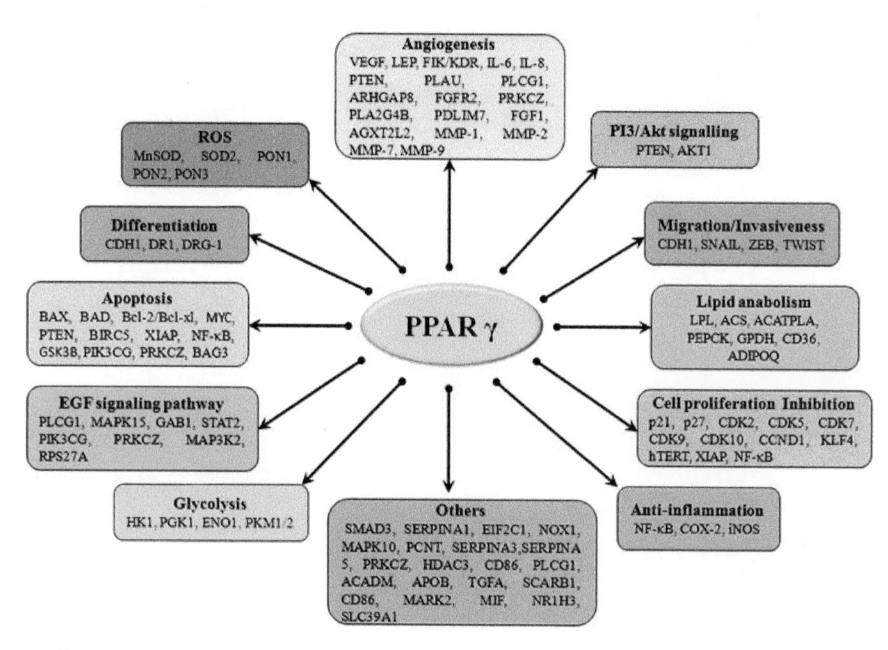

Figure 3.2 Molecular targets of PPARγ and pathways associated [Adapted from 27].

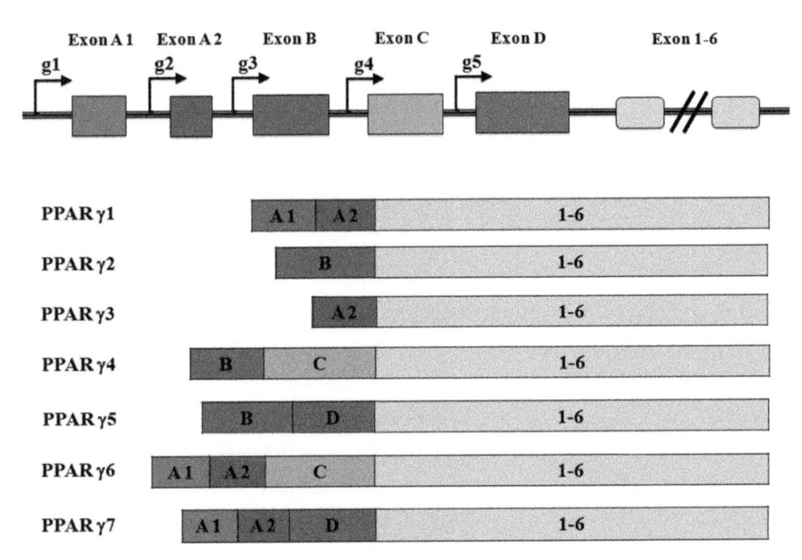

Figure 3.3 Genomic structure of the human PPAR gamma gene (5' end) and PPARγ mRNA splicing forms and protein variants. There are seven isoforms of PPARγ with common exons 1–6.

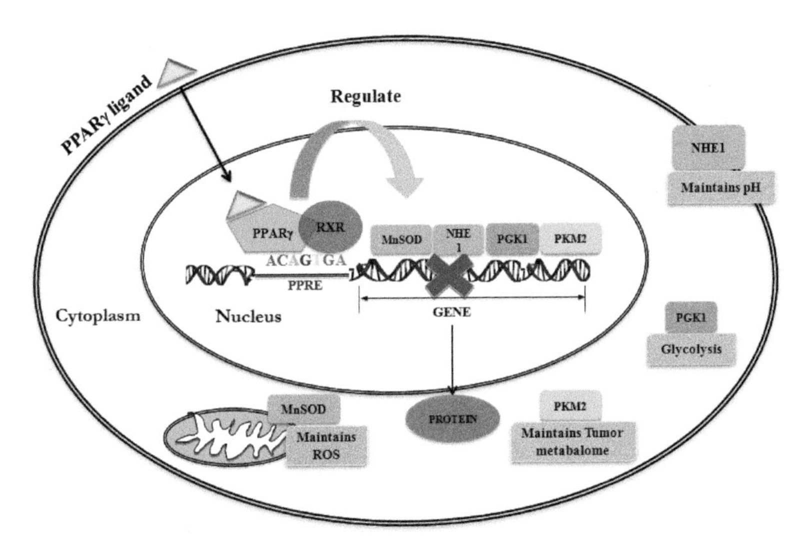

Figure 3.4 Transcriptional regulation of PPARγ gene targets. A PPAR protein binds PPRE/PACM motifs in combination with retinoid X receptors (RXRs) upon ligand activation. The two paired up proteins then regulate transcription of various genes e.g. PPARγ upon activation is known to down regulate glycolytic genes - PGK1 and PKM2, pH regulator -NHE1, anti-oxidant enzyme - MnSOD in breast cancer cells.

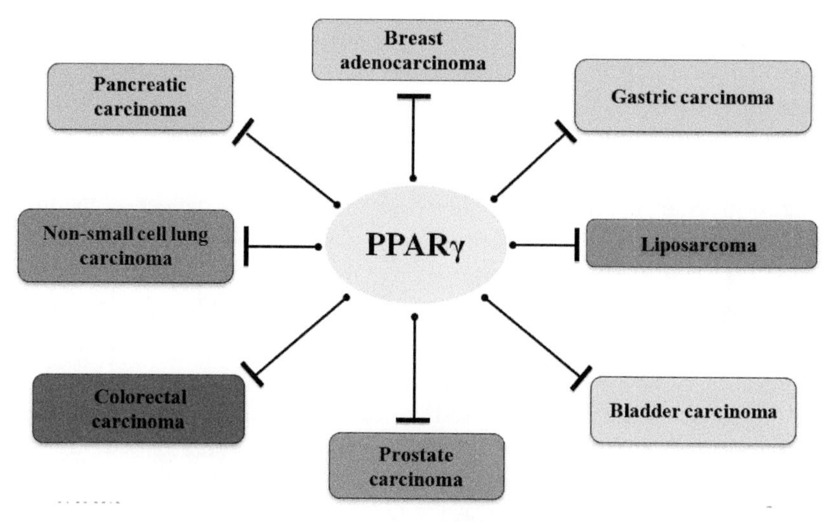

Figure 3.5 PPARγ activation inhibits many malignancies.

chromatin-decondensing cofactors complexes and favors dismissal of the corepressor complex. In addition, PPARs may influence gene expression indirectly, and usually negatively, through competition with other transcription factors [13]. Its role in cancer development and potential as a target for cancer prevention and treatment strategies has been noted in recent years. Activation of PPARγ could possibly be an approach to induce differentiation in cells thereby inhibiting proliferation of a variety of cancers. This antiproliferative effect has been reported in many different cancer cell lines including breast [14], colon [15], prostate [16], and non-small-cell lung cancer [17] (Figure 3.5). The receptor is also reported to play a critical role in fat cell differentiation, inducing the expression of adipocyte-specific genes, and promoting the formation of mature lipid-laden adipocytes [18–19].

3.2 PPAR Gamma, Breast Cancer and Energy Metabolism

Breast cancer is the fifth most common cancer globally and accounts for the highest morbidity and mortality. It is the second highest occurring cancer in women and one of the leading causes of death [20]. Although antiestrogens have provided an effective endocrine therapy, a significant proportion of patients have acquired resistance to these drugs, others are intrinsically resistant [21]. Hence, there is a requirement for alternative therapeutics to treat breast cancer. Development of selective anticancer agents based on the biological differences between normal and cancer cells is essential to improve therapeutic selectivity, sensitivity, and specificity.

The metabolic phenotype of cancer is distinct from that of normal cells. Differences in energy metabolism between normal and cancer cells are reported and alterations in cellular bioenergetics are one of the hallmarks of cancer [22]. The general principles of metabolic control analysis can be effective for cancer management as abnormal energy metabolism and biological disorder are characteristics of tumors [23]. In line with this, increased aerobic glycolysis and elevated oxidative stress are two prominent biochemical features frequently observed in cancer cells, as shown by the Warburg hypothesis. In comparison to the normal aerobic glucose metabolism pathway which uses mitochondrial oxidation, cancer cells develop Warburg effect, in which aerobic glycolysis is increased and for which drug-driven disruption might lead to minimal side effects [24]. Altenberg *et al.*, found overexpression of glycolytic genes in 24 cancer cell lines [25]. Hypoxic tumours have generally been known to evolve into invasive and metastatic tumors as compared to those with normal oxygen levels due to increased glycolysis requirements in hypoxia. These results demonstrate the clinical importance of glucose metabolism in the treatment of cancer. Hence, disruption of Warburg effect may help resolve the malignant process independently of tumor origin, and is likely to have broad therapeutic implications [26].

Apart from the effects of PPARγ on cell death, none of the studies on the effect of PPARγ ligands on breast cancer demonstrates the molecular mechanisms, pathways or genes that are modulated by PPARγ to cause cell death. Towards this end, the identification of genes with PPRE signal could help identify genes in the energy pathway that could be possibly regulated by PPARγ. Our analyses showed that several genes in the glycolytic pathway have PPRE sites (Figure 3.6). We further focused on energy generation pathways and selected two glycolytic genes PGK1 and PKM2 for experimental validation on their regulation by PPARγ [27]. These genes were selected because they were reported to be involved in breast cancer in literature, they had PPRE motif but were not earlier reported to be regulated by PPARγ. Also, they were at important steps in biological pathway and were hubs. PGK1 and PKM2 catalyze sixth and ninth step, respectively in glycolysis generating 2 ATP each. PGK1 and PKM2 are reported to be overexpressed

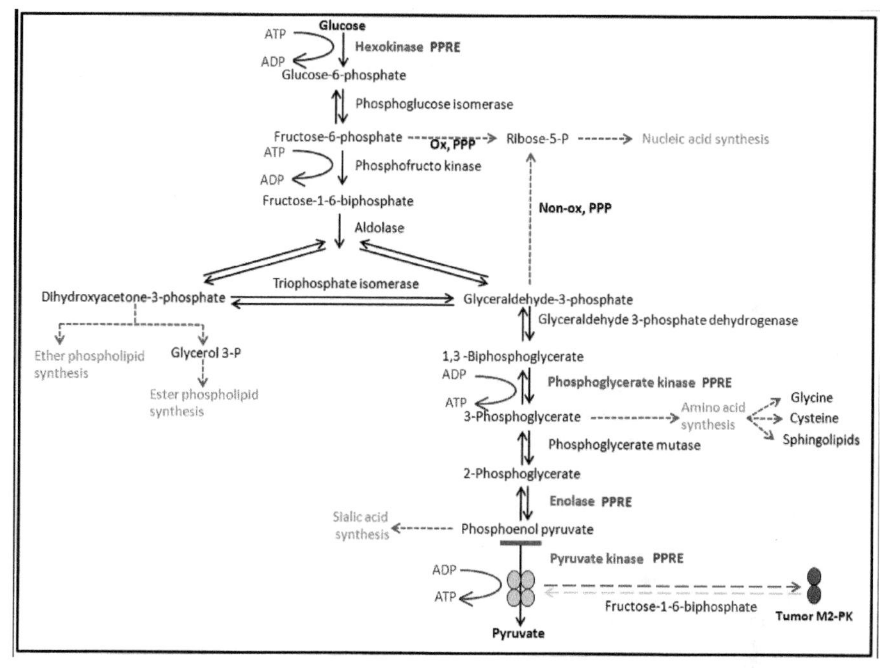

Figure 3.6 Metabolic targets of PPARγ. Many glycolytic enzymes are over expressed in cancers. Glycolytic enzyme pyruvate kinase-muscle 2 (PKM2) is a key regulator of tumor metabolism which promotes tumor growth and Warburg effect by switching between its dimeric form the active one, which has higher affinity for substrate Phosphoenol pyruvate (PEP) to tetrameric form the inactive form, with lower affinity for substrate PEP and vice-versa. This switching behavior of PKM2 keeps a balance of activation of many pathways including, glycerol, serine/glycine, ether/ester phospholipid pyrimidine biosynthesis (in green) and oxidative metabolism for energy production, thereby promoting tumor growth and tumor cell proliferation.

in many cancers [25, 28] accounting for their pivotal role in most fundamental metabolic alterations [24]. A reduction in expression of PGK1 and PKM2 (Figure 3.7) was observed with increase in 15d-PGJ2 concentration (a natural ligand of PPARγ). Since both the enzymes are involved in ATP synthesis, with repression in PGK1 and PKM2, the ATP level was found to be altered/lowered in cytosol and mitochondria in both the breast cancer cell lines. ATP depletion is a strong apoptotic signal. Caspase 8 activation is one of the early events leading to apoptosis. It has earlier been elucidated that PPARγ synthetic ligands activate intrinsic and extrinsic apoptotic cascade [29–30]. ATP depletion due to activation of PPARγ causes caspase 8 activation (decrease in caspase 8 proenzyme expression). This demonstrates that the

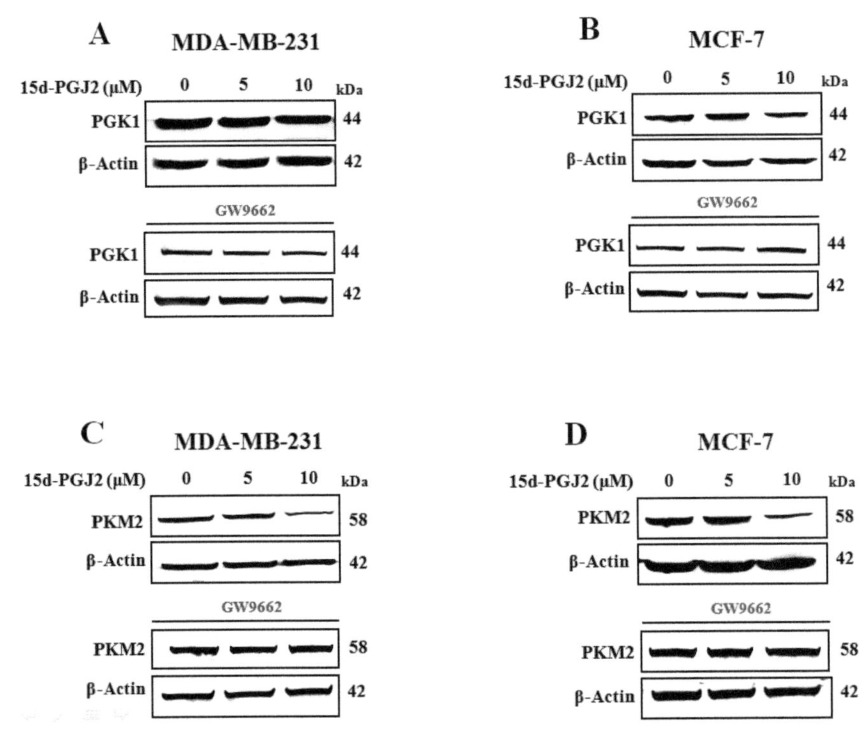

Figure 3.7 Regulation of glycolytic genes - PGK1 and PKM2 by PPARγ in human breast cancer cell lines – MDA-MB-231 and MCF-7. The human breast cancer cells were exposed to 10 μM of PPARγ inhibitor GW9662 for 4 h followed by 5 μM and 10 μM of 15d-PGJ2 for 48 h at 37 °C. PPARγ activation by 15d-PGJ2 down regulated the expression of glycolytic enzyme, PGK1 and PKM2 in breast cancer cell lines - MDA-MB-231 (7A and 7C) and MCF-7 (7B and 7D). Inhibiting the activation of PPARγ by the PPARγ inhibitor GW9662, did not affect the expression of PGK1 and PKM2, suggesting the transcriptional regulation of these glycolytic genes by PPARγ [Adapted from 27].

caspase 8 proenzyme is being cleaved to active caspase 8 fragments in a dose dependent fashion of 15d-PGJ2 (Figure 3.8A and 15.8B). Apoptosis leads to activation of series of biochemical events that alter cell characteristics. One of these important changes is chromatin condensation and nuclear fragmentation. To further examine these changes, Hoechst stain was used and increase in condensed nuclei (apoptotic cells) in dose dependent fashion was observed (Figure 3.8C and 15.8D). Examination on mitochondrial damage by analyzing mitochondrial membrane potential with mitochondria fluorescent dye JC-1 confirmed loss of mitochondrial membrane potential, coinciding with caspase 8 activation (Figure 3.9) [27].

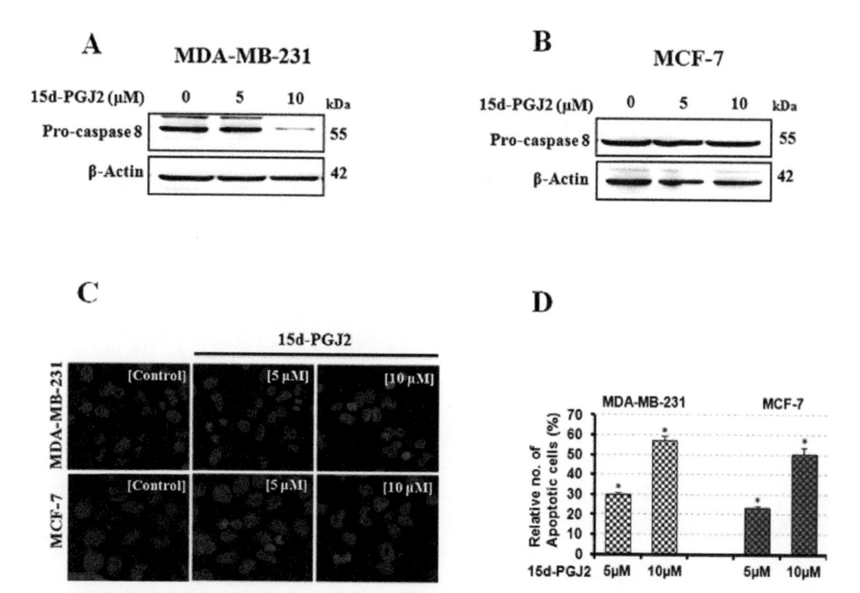

Figure 3.8 Apotosis and PPARγ. Apoptosis was initiated in human breast cancer cells lines MDA-MB-231 and MCF-7 upon 15d-PGJ2 activation of PPARγ. The breast cancer cells were exposed to 5 μM and 10 μM of 15d-PGJ2 for 48 h at 37 °C. Caspase dependent apoptosis was confirmed in MDA-MB-231 and MCF-7 by expression studies for active caspase 8 (8A and 8B) and chromatin condensation as assessed by nuclear specific dye Hoechst (8C and 8D) [Adapted from 27].

3.3 Glycolysis, Cell pH and NHE1

There is plenty of research showing link between acidic pH and cancer. Cancer thrives in an acidic environment. One major difference between many solid tumors and surrounding normal tissue is the nutritional and metabolic environment [26]. The functional vasculature of tumors is often inadequate to supply the nutritional needs of the expanding population of tumor cells, leading to deficiency of oxygen and many other nutrients. The production of lactic acid under anaerobic conditions and the hydrolysis of ATP in an energy-deficient environment contribute to the acidic microenvironment which is observed in many types of tumors. Deficiency of nutrients and acid conditions has been reported to contribute to cell death and necrosis within solid tumors. Hence, hypoxia and acidity represent two factors that might be exploited therapeutically to destroy cancer cells.

Normal functioning of cell metabolism occurs within a restricted intracellular pH (pHi) range. Cells have also developed several membrane-based

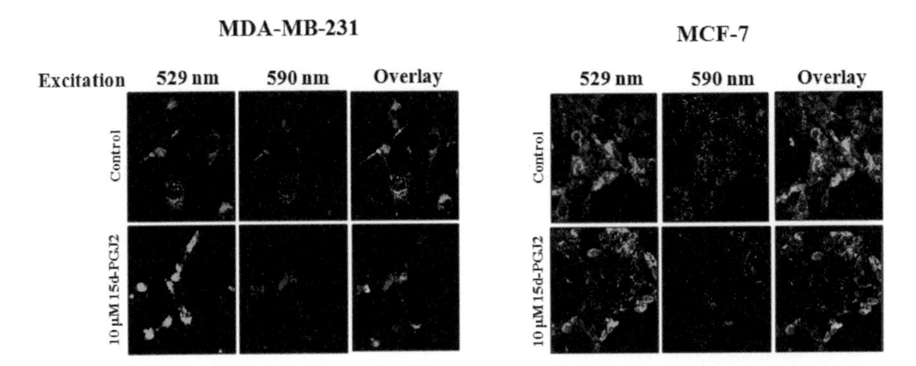

Figure 3.9 Disruption of mitochondrial potential in human breast cancer cells by PPARγ ligand. The human breast cancer cells were exposed to 5 μM and 10 μM of 15d-PGJ2 for 48 h at 37 °C. PPARγ ligand, 15d-PGJ2 induced loss of mitochondrial potential in human breast cancer cell lines -MDA-MB-231 and MCF-7 as assessed by potential dependent dye, JC-1. In healthy/non-apoptotic cells, JC-1 exists as a monomer in the cytosol (green) and accumulates as J-aggregates in the active mitochondria, which appear red. In apoptotic cells, due to loss mitochondrial potential relatively lower dye J-aggregates accumulates in the mitochondria than cytoplasm where it is remains as a monomer. Control breast cancer cell lines had higher Red/Green flouresence ratio than 15d-PGJ2 treated test cells, suggesting the loss of mitochondrial potential by PPARγ ligand, 15d-PGJ2 [Adapted from 27].

ion transport mechanisms for regulating pH. Major transport mechanisms which are known to contribute to regulation of pH in many types of cell include the Na^+/H^+antiport, the Na^+ - dependent HCO^-/Cl^- exchanger, and the cation-independent HCO^-/Cl^- exchanger [31–34]. The first two of these are involved primarily with regulation of pH in acid-loaded cells, whereas the latter probably participates in lowering pH of alkaline-loaded cells. Other membrane transporters may contribute to the regulation of pH, by extruding protons from cells; examples include H^+ (ATPase) pumps often found in specialized epithelia and the lactate:protonsymport [35]. The metastatic dissemination of tumour cells, which is the predominant cause of death from cancer, is facilitated by dysregulated pH. An increased pH is necessary for the directed migration of multiple mammalian cell types, and a decreased pH promotes degradation of the ECM for cell invasion [36].

NHE1 was observed to have 3 PPRE motifs in its UTR. Here, we will discuss on the regulation of NHE1 by PPARγ and its involvement in cell proliferation. Regulation of pHi is accomplished via active extrusion of H^+ by the Na^+/H^+ Exchangers (NHEs), a membrane antiporter expressed in a variety of cell types. The NHE family consists of 10 isoforms, NHE1 to NHE10 [37]. Interestingly, apart from its role as a principal regulator of pHi and cell volume, the ubiquitously expressed NHE1 has been implicated in cell proliferation and transformation. Conversely, tumor cells deficient in NHE1 either fail to grow or show severely retarded growth when implanted in immuno-deficient mice [38]. More recently, it was reported that down regulation of NHE1 expression by direct silencing of the gene expression or H_2O_2 treatment leads to cells' growth arrest and sensitization to etoposide or staurosporine [38–39].

Interestingly, a recent report implicated the pH regulator NHE1 in tumor cell growth is arrested by activated PPARγ, an interesting connotation considering that the activation of NHE1 is an oncogenic signal necessary for the development and maintenance of the transformed phenotype [40–42]. Also, it was recently reported that decrease in NHE1 expression led to tumor cell growth arrest, intracellular acidification, and sensitization to death stimuli [43]. These data support down-regulation of NHE1 as a possibility for inducing growth arrest in cancer cells [44]. In light of the increased expression of PPARγ in breast cancer cell lines and its association with acidic intracellular pH, we hypothesized that, in addition to inhibiting NHE1 activity, ligand-induced activation of PPARγ could regulate NHE1 gene expression. Interestingly, the results corroborate with the data which reported that exposure of breast cancer cell lines expressing high levels of PPARγ to natural ligands of PPARγ significantly inhibited NHE1 gene expression compared with noncancerous cells or cancer cell lines expressing low levels of PPARγ [11, 44] (Figure 3.10).

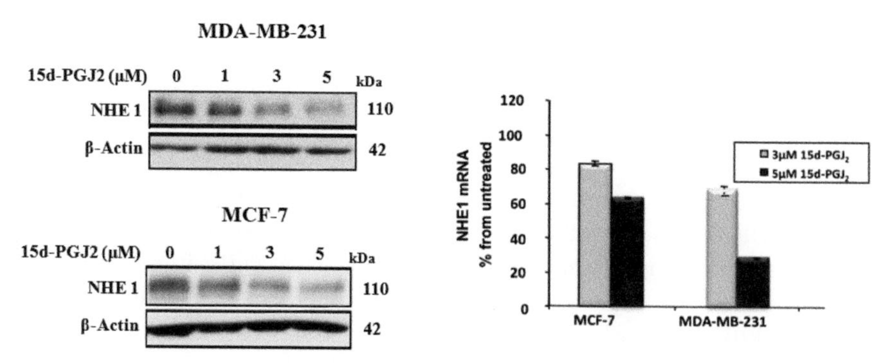

Figure 3.10 PPARγ activation by 15d-PGJ2 represses NHE1 mRNA and protein levels in human breast cancer cell lines - MDA-MB-231 and MCF-7. The breast cancer cells were exposed to 1 μM, 3 μM and 5 μM of 15d-PGJ2 for 24 h at 37 °C. A significant down regulation of NHE1 expression was observed. A similar pattern was followed at transcriptional level, suggesting PPARγ regulation of NHE1 in breast cancer cell lines [Adapted from 44].

In contrast to the above evidence for the involvement of the Na^+/ H^+antiport (or cellular alkalinization), several studies have demonstrated that activation of the antiport and elevation of pH are probably not sufficient and may not even be necessary for proliferation of many types of cells [45].

3.4 ROS and Breast Cancer

As shown above, experimental evidences link upregulation of the pH regulator Na^+/H^+ exchanger 1 (NHE1) to the development and progression of carcinogenesis [46–50] and its down regulation to inhibition of cells' growth and enhanced apoptotic sensitivity [38–39]. It is also suggested that cancer cells are generally under reactive oxygen species (ROS) stress [51]. Reactive oxygen species (ROS) are constantly generated and eliminated in the biological system, and play important roles in a variety of normal biochemical functions and abnormal pathological processes. Growing evidence suggests that cancer cells exhibit increased intrinsic ROS stress, due in part to oncogenic stimulation, increased metabolic activity, and mitochondrial malfunction. The increased amounts of ROS in cancer cells may have significant consequences, such as stimulation of cellular proliferation, promotion of mutations and genetic instability, and alterations in cellular sensitivity to anticancer agents. It is logical to speculate that the biochemical and molecular changes caused by ROS may contribute to the development of a heterogeneous cancer cell population and the emergence of drug-resistant cells during disease progression. Since the mitochondrial respiratory chain (electron transport complexes) is a major source of ROS generation in the cells, the vulnerability of the mitochondrial DNA to ROS-mediated damage appears to be a mechanism to amplify ROS stress in cancer cells. The escalated ROS generation in cancer cells serves as an endogenous source of DNA-damaging agents that promote genetic instability and development of drug resistance. Malfunction of mitochondria also alters cellular apoptotic response to anticancer agents. Despite the negative impacts of increased ROS in cancer cells, it is possible to exploit this biochemical feature and develop novel therapeutic strategies to preferentially kill cancer cells through ROS-mediated mechanisms. Recently, a novel PPARγ ligand, hydroxy hydroquinone (HHQ) induced significant intracellular ROS formation in human breast cancer cell lines, MDA-MB-231 and MCF-7 in dose dependent manner was reported. The effective enhancement of ROS production by HHQ correlate to its cytotoxic nature (Figure 3.11) [52].

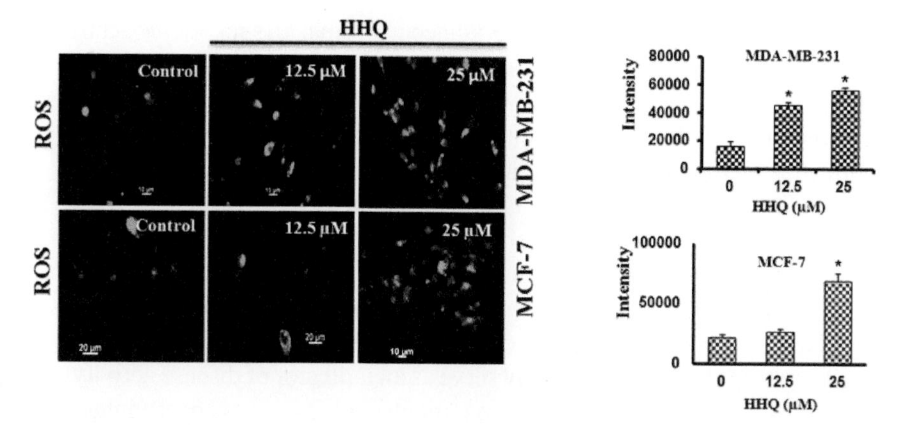

Figure 3.11 PPARγ novel ligand, Hydroxy hydroquinone (HHQ) induces intracellular Reactive Oxygen Species (ROS) formation in human breast cancer cell lines - MDA-MB-231 and MCF-7. The breast cancer cells were treated with 12.5 μM and 25 μM of HHQ for 48 h at 37 °C. Intracellular ROS formation was found to be significantly increased in HHQ treated cells as compared to control cells in a dose dependent manner was reported. The effective enhancement of ROS production by HHQ correlates to its cytotoxicity nature [Adapted from 52].

Cells have evolved several antioxidant defenses, including repair and detoxifying enzymes, and small scavenger molecules, such as glutathione. The intracellular ROS-scavenging system includes superoxide dismutases (SOD), glutathione peroxidase (GPx), peroxiredoxins (PRDXs), glutaredoxins, thioredoxins (TRXs), and catalases. In mitochondria, superoxide anions can be dismutated to hydrogen peroxide (H_2O_2) by two enzymes, namely, copper-zinc superoxide dismutase (CuZnSOD) and manganese superoxide dismutase (MnSOD), that are present in the mitochondrial matrix and in the intermembrane space, respectively [53]. Once generated, H_2O_2 can be quenched by GPx in mitochondria, or by catalase in the cytosol. The expression of antioxidant enzymes is regulated by complex mechanisms, oxidative stress being a major factor that induces the adaptive expression of these enzymes [53]. Thus, increased ROS stress in cancer cells is likely to cause increased expression of SOD and other antioxidant enzymes. In fact, analysis of SOD protein expression in primary tissues from adenocarcinomas of the stomach and squamous cell carcinomas of the oesophagus showed a significantly higher MnSOD expression in the cancer cells compared to normal mucosa cells [54]. The activities of SOD, glutathione peroxidase (GPx), and glutathione-S-transferase (GST) were increased significantly in the mitochondria of colorectal cancer tissues compared to adjacent normal tissues of the same subjects [55]. Increased SOD levels were also observed in breast cancer tissue from 23 patients [55]. Detection of MnSOD using specific antibody showed positive reactions with ovarian carcinomas and malignant brain tumors, but not with the respective normal control tissues [57]. Some studies further demonstrated a significantly increased expression of CuZnSOD (SOD1), MnSOD (SOD2), and catalase in chronic lymphocytic leukemia cells and ovarian cancer cells [58]. Increases in SOD1 and SOD2 have been observed in blood samples from patients with various types of leukemia [59]. Interestingly, leukemia regression was accompanied by a decrease in the serum MnSOD, suggesting that MnSOD in serum may serve as an indicator of disease activity. Analysis of SOD in blood samples from patients with ovarian cancer yielded conflicting results [60].

As mitochondrial respiration is the main source of O2• generation in the cells, MnSOD (manganese superoxide dismutase) is of prime importance in maintaining cellular ROS balance and mitochondrial integrity in cells. This is specifically true for tumor cells which are constantly under ROS stress due to its increased metabolic processes [61]. Development of breast cancer has also been correlated to oxidative stress, brought about by alterations to the delicate balance between reactive oxygen species (ROS) and oxidative defences [62–63]. ROS stress seems to render cancer cells more dependent on MnSOD to protect them by maintaining cellular ROS balance. The importance of MnSOD became clearer when genetic knockout studies in mice indicate that MnSOD, but not other SODs, is essential for cell survival. Several recent

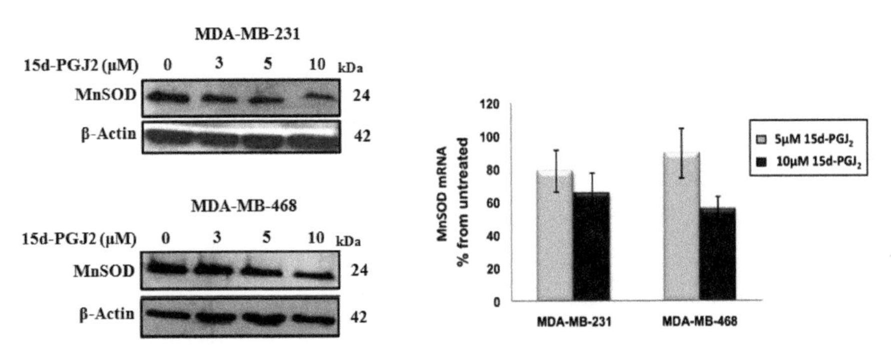

Figure 3.12 PPARγ activation by 15d-PGJ2 represses MnSOD mRNA and protein levels in human breast cancer cell lines - MDA-MB-231 and MDA-MB-468. The breast cancer cells were exposed to 3 μM, 5 μM and 10 μM of 15d-PGJ2 for 24 h at 37 °C. A significant repression of MnSOD level was observed in dose-dependent manner [Adapted from 44].

studies reported that forced suppression of MnSOD expression by siRNA leads to decrease in breast cancer cells invasive property [64] and to sensitization of ovarian cancer cells to anti-cancer drugs [65]. Although the mouse MnSOD gene has been shown to be a PPARγ target gene [66], a direct association between PPARγ and human MnSOD from tumor cells' perspective has not been shown. We confirmed human MnSOD as a PPARγ target gene and the downregulation of MnSOD gene expression *in vitro* by PPARγ agonists [44] (Figure 3.12).

3.5 Natural and Synthetic Ligands of PPARγ

PPARγ receptor can be activated by endogenous ligands, for example, prostaglandin D2 (PGD2), 15-deoxy prostaglandin J2 (15dPGJ2), or 15-hydroxyeicosatetraenoic acid (15-HETE) [67–68]. Recently, an oxidized phosphatidylcholine has also been identified as a potent alternative (patho) physiological natural ligand of PPARgamma. Synthetic ligands for PPARγ include insulin sensitizing antidiabetic thiazolidinediones (TZD); troglitazone (TGZ), rosiglitazone (RGZ), ciglitazone (CGZ), or pioglitazone (PGZ) [69–71], and non-steroidal anti-inflammatory compounds indomethacin, ibuprofen, flufenamic acid, or fenoprofen [72] are commonly known as PPARγ ligands (Table 3.1). Due to high levels of toxicity associated with the first generation TZDs, troglitazone (Rezulin), rosiglitazone (Avandia), and pioglitazone (Actos), there is a renewed search for newer PPAR drugs that exhibit better efficacy but lesser toxicity. Towards this end, developing known dietary components (nutraceuticals) that bind and activate PPARγ with more efficacy and safety, while promoting health benefits has become an absolute necessity [73]. The majority of nutraceuticals are of plant origin and hence these compounds are considered safe and are popular among consumers. Dietary components that act as ligands of PPARγ include dietary lipids such as n-3 and n-6 fatty acids and their derivatives, isoflavones and flavonoids. Though a plethora of ligands are known to activate pparγ, what is lacking, however, is the delineation of the metabolic effects that are specific to this PPAR gamma activation (Table 3.1). These investigations "have provided" a cause and effect relationship between ligand activation of PPAR gamma and its physiological function and will help in effective therapeutic options in several pathophysiological conditions.

Table 3.1 List of natural and synthetic ligands of PPARγ

Natural Ligands	Synthetic Ligands
Abietic acid [PMID: 12935909]	Azelaoylphosphatidylcholine (in oxidized LDL) [PMID: 11279149]
Biochanin A [PMID: 16549448]	Arachidonic acid (20:4, n-3) [PMID: 11422732]
Bixin [PMID: 21307572]	Balsalazide [PMID: 18077625]
Capsaicin [PMID: 15383218]	Bezafibrate [PMID: 12676649]
Citronellol [PMID: 21597168]	CAY10599 [PMID: 19301897]
Daidzein [PMID: 19775880]	Ciglitazone [PMID: 8576907]
Dehydroabietic acid (DAA) [PMID: 18267111]	Conjugated linoleic acid isomers (CLA) [PMID: 15986437]
Equol [PMID: 19775880]	Docosahexaenoic acid (DHA) [PMID: 18193404]
Farnesol [PMID: 21307572]	DRF 2519 [PMID: 15140637]
Genistein [PMID: 12421816]	Gliquidone [PMID: 17082235]
Geraniol [PMID: 21597168]	Glipizide [PMID: 16168052]
Geranylgeraniol [PMID: 18608213]	GW 1929 [PMID: 23100239]
Glycyrrhizic acid [PMID: 20011054]	Icosapent [PMID: 11552681]
Hesperidin [PMID: 18086153]	Indomethacin [PMID: 20665425]
Hydroxy Hydroquinone [Ref:]	Linoleic acid [PMID: 15701701]
Lipoxygenase (LOX) [PMID: 15698583]	Methyl-8-hydroxy-8-(2-pentyl-oxyphenyl)-oct-5-ynoate [PMID: 20518620]Mitiglinide [PMID: 17082235]
Phytol [PMID: 21629877]	Muraglitazar [PMID: 23594962]
Psi-baptigenin [PMID: 18086153]	Nateglinide [PMID: 17082235]
Quercetin [PMID: 22226987]	Nitroalkenes (NO2-FA) [PMID: 20097754]
Resveratrol [PMID: 22792089]	nTZDpa [PMID: 12554792]
Rosmarinic acid [PMID: 22226987]	Omega-3 (or n-3) polyunsaturated fatty acids (PUFAs) [PMID: 18769551]
2'-Hydroxy chalcone [PMID: 22226987]	PAz-PC [PMID: 11279149]
	PGD2 [PMID: 11422732]
	PGJ2 [PMID: 11422732]
	Pioglitazone [PMID: 11422732]
	Ripaglinide [PMID: 17082235]
	Rosiglitazone (BRL49653) [PMID: 11422732]
	Telmisartan [PMID: 23829789]
	Tesaglitazar [PMID: 17166340]
	Troglitazone [PMID: 11422732]
	WY-14643 [PMID: 11422732]
	4-Hydroxy docosahexaenoic acid (4-HDHA) [PMID: 18193404]
	4-Oxodocosahexaenoic acid (4-oxo-DHA) [PMID: 18193404]
	9-HODE [PMID: 11422732]

(Continued)

Table 3.1 Continued

Natural Ligands	Synthetic Ligands
	9/10-NO2-linoleic acid [PMID: 19105608]
	12-NO2-linoleic acid [PMID: 19105608]
	13-NO2-linoleic acid [PMID: 19105608]
	13-HODE [PMID: 11422732]
	15-Deoxy-12,14-prostaglandin J2 [PMID: 11422732]

3.6 Conclusions

PPAR gamma agonists have been reported as new and potentially efficacious treatment of inflammation, diabetes, obesity, cancer, AD, and schizophrenia [74]. The use of synthetic PPARγ ligands as drugs and their recent withdrawal/restricted usage highlight the lack of understanding of the molecular basis of these drugs, their off-target effects, and their network. These data further underscore the complexity of nuclear receptor signalling mechanisms. Thus, there is a need to continue enhancing our understanding of the complexities of nuclear receptor pharmacology and a need to view the functions of this family of transcription factors in detail supported by clinical trials and adverse side effects data as tenable well beyond traditional discreet categories of agonism and antagonism which will open new doors to using PPAR gamma as a target.

References

[1] B. P. Kota, T. H. Huang, and B. D. Roufogalis, "An overview on biological mechanisms of PPARs," Pharmacological Research, Vol. 51, No. 2, pp. 85–94, 2005.

[2] M. Tous, N. Ferré, A. Rull et al., "Dietary cholesterol and differential monocyte chemoattractant protein-1 gene expression in aorta and liver of apo E-deficient mice," Biochemical and Biophysical Research Communications, Vol. 340, No. 4, pp. 1078–1084, 2006.

[3] H. Castelein, T. Gulick, P. E. Declercq, G. P. Mannaerts, D. D. Moore, and M. I. Baes, "The peroxisome proliferator activated receptor regulates malic enzyme gene expression," The Journal of Biological Chemistry, Vol. 269, No. 43, pp. 26754–26758, 1994.

[4] D.G. Lemay DG and D.H. Hwang, "Genome-wide identification of peroxisome proliferator response elements using integrated computational genomics,"J Lipid Res, Vol. 47, pp. 1583–1587, 2006.

[5] M. Van Bilsen, G. J. Van Der Vusse, A. J. Gilde, M. Lindhout, and K. A. J. M. Van Der Lee, "Peroxisome proliferator-activated receptors: lipid binding proteins controling gene expression,"Molecular and Cellular Biochemistry, Vol. 239, No. 1–2, pp. 131–138, 2002.

[6] J. J. Mansure, R. Nassim, and W. Kassouf, "Peroxisome proliferator-activated receptor gamma in bladder cancer: a promising therapeutic target in cancer," Cellular and Genetic Practices for Translational Medicine, Vol. 8, No. 7, pp. 169–195, Research Signpost, 2011.

[7] J. I. Park, "The role of 15d-PGJ2, a natural ligand for peroxisome proliferator-activated receptor γ (PPARγ), in cancer," Pharmacological Research, Vol. 51, No. 2, pp. 85–94, 2005.

[8] P. Tontonoz, E. Hu, R.A. Graves, A.I. Budavari and B.M. Spiegelman, "mPPAR gamma 2: tissue-specific regulator of an adipocyte enhancer," Genes Dev. Vol. 8, pp. 1224–1234, 1994.

[9] A.J. Vidal-Puig, et al., "Peroxisome proliferator-activated receptor gene expression in human tissues. Effects of obesity, weight loss, and regulation by insulin and glucocorticoids," J. Clin. Invest, Vol. 99, pp. 2416–2422, 1997.

[10] L. Fajas, et al., "The organization, promoter analysis, and expression of the human PPAR gamma gene," J. Biol. Chem, Vol. 272, pp. 18779–18789, 1997.

[11] B.M. Forman, P. Tontonoz, J. Chen, R.P. Brun, B.M. Spiegelman and R.M. Evans, "15-Deoxy-delta 12,14 prostaglandin J2 is a ligand for the adipocyte determination factor PPAR???" Cell, Vol. 83, pp. 803–812, 1997.

[12] S.A. Kliewer, J.M. Lenhard, T.M. Wilson, I. Patel, D.C. Morris and J.M. Lehmann, "A prostaglandin J2 metabolite binds peroxisomeproliferator-activated receptor gamma and promotes adipocyte differentiation," Cell, Vol. 83, pp. 813–819, 1995.

[13] R. Cunard, et al., "Repression of IFN-gamma expression by peroxisome proliferator-activated receptor gamma," J. Immunol, Vol. 172, pp. 7530–7536, 2004.

[14] C. C. Woo, S. Y. Loo, V. Gee et al., "Anticancer activity of thymoquinone in breast cancer cells: possible involvement of PPAR-γ pathway," Biochemical Pharmacology, Vol. 82, No. 5, pp. 464–475, 2011.

[15] Y. L. Lu, G. L. Li, H. L. Huang, J. Zhong, and L. C. Dai, "Peroxisome proliferator-activated receptor-γ 34C $>$ G polymorphism and colorectal cancer risk: a meta-analysis," World Journal of Gastroenterology, Vol. 16, No. 17, pp. 2170–2175, 2010.

[16] G. Venkatachalam, A. P. Kumar, L. S. Yue, S. Pervaiz, M. V. Clement, and M. K. Sakharkar, "Computational identification and experimental validation of PPRE motifs in NHE1 and MnSOD genes of human," BMC Genomics, Vol. 10, Supplement 3, Article S5, 2009.

[17] Y. Jeong, Y. Xie, W. Lee et al., "Research resource: diagnostic and therapeutic potential of nuclear receptor expression in lung cancer," Molecular Endocrinology, Vol. 26, No. 8, pp. 1443–1454, 2012.

[18] P. Tontonoz, R.A. Graves, A.I. H. BromageErdjumentBudavari, M. Lui, E. Hu, P. Tempst and B.M. Spiegelman, "Adipocyte-specific transcription factor ARF6 is a heterodimeric complex of two nuclear hormone receptors, PPARγ and RXRα," Nucleic Acids Res. Vol. 22, pp. 5628–5634, 1994a.

[19] P. Tontonoz, E. Hu, R.A. Graves, A.I. Budavari and B.M. Spiegelman, "PPARγ2: Tissue-specific regulator of an adipocyte enhancer," Genes &Dev, Vol. 8, pp. 1224–1234, (1994b).

[20] S. Mukhopadhyay, S. K. Das and S. Mukherjee, "Expression of Mn-superoxide dismutase gene in non-tumorigenic and tumorigenic human mammary epithelial cells," Journal of Biomedicine and Biotechnology, Vol. 2004, No. 4, pp. 195–202, 2004.

[21] P. Fedele, N. Calvani, A. Marino et al., "Targeted agents to reverse resistance to endocrine therapy in metastatic breast cancer: where are we now and where are we going?" Critical Reviews in Oncology/Haematology, Vol. 51, No. 2, pp. 85–94, 2012.

[22] T. N. Seyfried and L. M. Shelton, "Cancer as a metabolic disease," Nutrition & Metabolism, Vol. 7, Article 7, 2010.

[23] R. J. Klement and U. Kämmerer, "Is there a role for carbohydrate restriction in the treatment and prevention of cancer? " Nutrition & Metabolism, Vol. 8, Article 75, 2011.

[24] O. Warburg, "On the origin of cancer cells," Science, Vol. 123, pp. 309–314, 1956.

[25] B. Altenberg and K.O. Greulich, "Genes of glycolysis are ubiquitously-overexpressed in 24 cancer classes," Genomics, Vol. 84, pp. 1014–1020, 2004.

[26] C.V. Dang, "Links between metabolism and cancer," Genes Dev, Vol. 26(9), pp. 877–890, May 2012.

[27] B. Shashni, K.R. Sakharkar, Y. Nagasaki and M.K. Sakharkar, "Glycolytic enzymes PGK1 and PKM2 as novel transcriptional targets of PPARγ in breast cancer pathophysiology," J Drug Target, Vol. 21(2), pp. 161–74, February 2013.

[28] K. Jiang, B. He, L. Lai, Q. Chen, Y. Liu, Q. Guo, Q. Wang, "CyclosporineA inhibits breast cancer cell growth by downregulating theexpression of pyruvate kinase subtype M2," Int J Mol Med, Vol. 30, pp. 302–308, 2012.

[29] D.L. Crowe and R.A. Chandraratna, "A retinoid X receptor (RXR)-selective retinoid reveals that RXR-alpha is potentially atherapeutic target in breast cancer cell lines, and that it potentiatesantiproliferative and apoptotic responses to peroxisomeproliferator-activated receptor ligands," Breast Cancer Res, Vol. 6, pp. R546–R555, 2004.

[30] D. Bonofiglio, S. Aquila, S. Catalano, S. Gabriele, M. Belmonte, E. Middea, H. Qi, C. Morelli, M. Gentile, M. Maggiolini and S. Ando, "Peroxisome proliferator-activated receptor-gamma activates p53gene promoter binding to the nuclear factor-kappaB sequence inhuman MCF7 breast cancer cells," MolEndocrinol, Vol. 20, pp. 3083–3092, 2006.

[31] A. Roos and W.F. Boron, "Intracellular pH," Physiol. Rev, Vol. 61 pp. 296–434, 1981.

[32] W.H. Moolenaar, L.G.J. Tertoolen and De Laat, "The regulation of cytoplasmic pH in human fibroblasts," J. Biol. Chem., Vol. 259, pp. 7563–7569, 1984.

[33] Frelin, C. Vigne, P. Ladoux and M. Lazdunski, "The regulation of the intracellular pH in cells from vertebrates," Eur. J. Biochem, Vol. 174, pp. 3–14, 1988.

[34] I.H. Madshus, "Regulation of intracellular pH in eukaryotic cells," Biochem. J, Vol. 250, pp. 1–8, 1988.

[35] M.S. Anwer and K. Nolan, "Characterization of proton efflux pathways in 84 rat hepatocytes," Hepatology (Baltimore), Vol. 8, pp. 728–734, 1988.

[36] Thistlethwaite, A.J. Alexander, G.A. Moylan, D.J. HI and D.B. Leeper, "Modification of human tumor pH by elevation of blood glucose," Int. J. RadiÃ¢t.Oncol. Biol. Phys., Vol. 13, pp. 603–610, 1987.

[37] D. Rotin, D. Steele-Norwood, S. Grinstein and I. Tannock, "Requirement of the Na^+/H^+ exchanger for tumor growth," Cancer research, Vol. 49(1), pp. 205–211, 1989.

[38] S. Akram, H.F. Teong, L. Fliegel, S. Pervaiz and M.V. Clement, "Reactive oxygen species-mediated regulation of the Na^+/H^+ exchanger 1 gene expression connects intracellular redox status with cells' sensitivity to death triggers," Cell death and differentiation, Vol. 13(4), pp. 628–641, 2006.

[39] A.P. Kumar, M.K. Chang, L. Fliegel, S. Pervaiz and M.V. Clement, "Oxidative repression of NHE1 gene expression involves iron-mediated caspase activity," Cell death and differentiation, Vol. 14(10), pp.1733–46, 2007.

[40] S.J. Reshkin, A. Bellizzi, S. Caldeira, et al., "Na^+/H^+ exchanger-dependent intracellular alkalinization is an early event in malignant transformation and plays an essential role in the development of subsequent transformation-associated phenotypes," The FASEB Journal, Vol. 14(14), pp. 2185–2197, 2007.

[41] J. Pouyssegur J, A. Franchi A and G. Pages, "pHi, aerobic glycolysis and vascular endothelial growth factor in tumour growth," Novartis Foundation Symposium, Vol. 240, pp. 186–196, 2001.

[42] F. Turturro, E. Friday, R. Fowler, D. Surie and T. Welbourne, "Troglitazone acts on cellular pH and DNA synthesis through a peroxisome proliferator-activated receptor γ-independent mechanism in breast cancer-derived cell lines," Clinical Cancer Research. Vol. 10(20), pp. 7022–7030, 2004.

[43] S. Akram, H.F.C. Teong, L. Fliegel, S. Pervaiz and M.V. Clément, "Reactive oxygen species-mediated regulation of the Na^+–H^+ exchanger 1 gene expression connects intracellular redox status with cells' sensitivity to death triggers," Cell Death and Differentiation, Vol. 13(4), pp. 628–641, 2006.

[44] G. Venkatachalam, A. P. Kumar, L. S. Yue, S. Pervaiz, M. V. Clement, and M. K. Sakharkar, "Computational identification and experimental validation of PPRE motifs in NHE1 and MnSOD genes of human," BMC Genomics, Vol. 10, Supplement 3, Article S5, 2009.

[45] I.F. Tannock and D. Rotin, "Acid pH in tumours and its potential for therapeutic exploitation," Cancer Res. Vol. 49(16), pp. 4373–84, August 1989.

[46] J. Noel and J. Pouyssegur, "Hormonal regulation, pharmacology, and membrane sorting of vertebrate Na^+/H^+ exchanger isoforms," The American journal of physiology, Vol. 268(2 Pt 1), pp. C283–296, 1995.

[47] S. Wakabayashi, M. Shigekawa and J. Pouyssegur, "Molecular physiology of vertebrate Na^+/H^+ exchangers," Physiol Rev, Vol. 77(1), pp. 51–74, 1997.

[48] S.J. Reshkin, A. Bellizzi, S. Caldeira, V. Albarani, I. Malanchi, M. Poignee, M. Alunni-Fabbroni, V. Casavola and M. Tommasino, "Na^+/H^+ exchanger-dependent intracellular alkalinization is an early event in

malignant transformation and plays an essential role in the development of subsequent transformation-associated phenotypes," Faseb J, Vol. 14(14), pp. 2185–2197, 2000.

[49] J. Pouyssegur, A. Franchi and G. Pages, "pHi, aerobic glycolysis and vascular endothelial growth factor in tumour growth," Novartis Found Symp, Vol. 240, pp.186–196, 2001.

[50] S.M. Bell, C.M. Schreiner, P.J. Schultheis, M.L. Miller, R.L. Evans, C.V. Vorhees, G.E. Shull and W.J. Scott, "Targeted disruption of the murine Nhe1 locus induces ataxia, growth retardation, and seizures," The American journal of physiology, Vol. 276(4 Pt 1), pp. C788–795, 1999.

[51] E.O. Hileman, J. Liu, M. Albitar, M.J. Keating and P. Huang, "Intrinsic oxidative stress in cancer cells: a biochemical basis for therapeutic selectivity," Cancer ChemotherPharmacol, Vol. 53(3), pp. 209–219, 2004.

[52] B. Shashni, K. Sharma, R. Singh, K. R. Sakharkar, S. K. Dhillon, Y. Nagasaki and M. K. Sakharkar "Coffee component hydroxyl hydroquinone (HHQ) as a putative ligand for PPAR gamma and implications in breast cancer," BMC Genomics, Vol. 14 (Suppl 5): S6, 2013.

[53] K. B. Storey, "Oxidative stress: animal adaptations in nature," Brazilian Journal of Medical and Biological Research, Vol. 29, No. 12, pp. 1715–1733, 1996.

[54] A. M. L. Janssen, C. B. Bosman, W. Van Duijn et al., "Superoxide dismutases in gastric and esophageal cancer and the prognostic impact in gastric cancer," Clinical Cancer Research, Vol. 6, No. 8, pp. 3183–3192, 2000.

[55] O. Kanbagli, G. Ozdemirler, T. Bulut, S. Yamaner, G. Aykaç-Toker and M. Uysal, "Mitochondrial lipid peroxides and antioxidant enzymes in colorectal adenocarcinoma tissues," Japanese Journal of Cancer Research, Vol. 91, No. 12, pp. 1258–1263, 2000.

[56] K. Punnonen, M. Ahotupa, K. Asaishi, M. Hyöty, R. Kudo and R. Punnonen, "Antioxidant enzyme activities and oxidative stress in human breast cancer," Journal of Cancer Research and Clinical Oncology, Vol. 120, No. 6, pp. 374–377, 1994.

[57] C. S. Cobbs, D. S. Levi, K. Aldape, and M. A. Israel, "Manganese superoxide dismutase expression in human central nervous system tumors," Cancer Research, Vol. 56, No. 14, pp. 3192–3195, 1996.

[58] E. A. Hileman, G. Achanta and P. Huang, "Superoxide dismutase: an emerging target for cancer therapeutics," Expert Opinion on Therapeutic Targets, Vol. 5, No. 6, pp. 697–710, 2001.

[59] T. Nishiura, K. Suzuki, T. Kawaguchi et al., "Elevated serum manganese superoxide dismutase in acute leukemias," Cancer Letters, Vol. 62, No. 3, pp. 211–215, 1992.

[60] S. Senthil, R. M. Veerappan, M. Ramakrishna Rao and K. V. Pugalendi, "Oxidative stress and antioxidants in patients with cardiogenic shock complicating acute myocardial infarction," ClinicaChimicaActa, Vol. 348, No. 1–2, pp. 131–137, 2004.

[61] T.D. Oberley, "Mitochondria, manganese superoxide dismutase, and cancer," Antioxid Redox Signal, Vol. 6(3), pp. 483–487, 2004.

[62] D. St Clair, X. Wan, M. Kuroda, S. Vichitbandha, E. Tsuchida and M. Urano, "Suppression of tumour metastasis by manganese superoxide dismutase is associated with reduced tumorigenicity and elevated fibronectin," Oncol Rep, Vol. 4, pp. 753–757, 1997.

[63] C. Ambrosone, "Oxidants and antioxidants in breast cancer," Antioxid Redox Signal, Vol. 2, pp. 903–917, 2000.

[64] Z. Kattan, V. Minig, P. Leroy, M. Dauca and P. Becuwe, "Role of manganese superoxide dismutase on growth and invasive properties of human estrogen-independent breast cancer cells," Breast Cancer Res Treat, Vol. 108(2), pp. 203–215, 2008.

[65] B.H.Y. Yeung, K.Y. Wong, M.C. Lin, C.K.C. Wong, T. Mashima, T. Tsuruo, A.S.T. Wong, "Chemosensitisation by manganese superoxide dismutase inhibition is caspase-9 dependent and involves extracellular signal-regulated kinase 1//2," British journal of cancer, Vol. 99(2), pp. 283–293, 2008.

[66] G. Ding, M. Fu, Q. Qin, W. Lewis, H.W. Kim, T. Fukai, M. Bacanamwo, Y.E. Chen, M.D. Schneider, D.J. Mangelsdorf, et al., "Cardiac peroxisome proliferator-activated receptor gamma is essential in protecting cardiomyocytes from oxidative damage," Cardiovascular research, Vol. 76(2), pp. 269–279, 2007.

[67] J. T. Huang, J. S. Welch, M. Ricoteet al., "Interleukin-4-dependent production of PPAR-γ ligands in macrophages by 12/15-lipoxygenase," Nature, Vol. 400, No. 6742, pp. 378–382, 1999.

[68] L. Nagy, P. Tontonoz, J. G. A. Alvarez, H. Chen and R. M. Evans, "Oxidized LDL regulates macrophage gene expression through ligand activation of PPARγ," Cell, Vol. 93, No. 2, pp. 229–240, 1998.

[69] J. Berger, P. Bailey, C. Biswaset al., "Thiazolidinediones produce a conformational change in peroxisomal proliferator-activated receptor-γ: binding and activation correlate with antidiabetic actions in db/db mice," Endocrinology, Vol. 137, No. 10, pp. 4189–4195, 1996.

[70] J. M. Lehmann, L. B. Moore, T. A. Smith-Oliver, W. O. Wilkison, T. M. Willson and S. A. Kliewer, "An antidiabetic thiazolidinedione is a high affinity ligand for peroxisome proliferator-activated receptor γ (PPARγ)," The Journal of Biological Chemistry, Vol. 270, No. 22, pp. 12953–12956, 1995.

[71] K. G. Lambe and J. D. Tugwood, "A human peroxisome-proliferator-activated receptor-γ is activated by inducers of adipogenesis, including thiazalidinedione drugs," European Journal of Biochemistry, Vol. 239, No. 1, pp. 1–7, 1996.

[72] R. D. Unwin, R. A. Craven, P. Harndenet al., "Proteomic changes in renal cancer and co-ordinate demonstration of both the glycolytic and mitochondrial aspects of the Warburg effect," Proteomics, Vol. 3, No. 8, pp. 1620–1632, 2003.

[73] S. Mukhopadhyay, S. K. Das and S. Mukherjee, "Expression of Mn-superoxide dismutase gene in nontumorigenic and tumorigenic human mammary epithelial cells," Journal of Biomedicine and Biotechnology, Vol. 2004, No. 4, pp. 195–202, 2004.

[74] S. Tyagi, P. Gupta, A. S. Saini, C. Kaushal, and S. Sharma, "The peroxisome proliferator-activated receptor: a family of nuclear receptors role in various diseases," Journal of Advanced Pharmaceutical Technology & Research, Vol. 2, No. 4, pp. 236–240, 2011.

4

Animal Model of Cancer and Infection

Karun Sharma[1], Babita Shashni[1], Meena K Sakharkar[2], Kishore R Sakharkar[3] and Ramesh Chandra[4,5]

[1]Graduate School of Life and Environmental Sciences, University of Tsukuba, Japan
[2]College of Pharmacy and Nutrition, University of Saskatchewan, SK, Canada
[3]OmicsVista, Singapore
[4]Department of Chemistry, Delhi University, India
[5]B. R. Ambedkar Center for Biomedical Research, University of Delhi, India

4.1 Introduction

Although they appear very different, species as diverse as yeast, flies, worms, zebra fish, dogs and mice share a lot of genes and molecular pathways with humans. The extensive similarities between the genomes of human and these so-called 'model organisms', are the foundation of much of modern biology, with model organism experimentation permitting valuable insight into the biological function and aetiology of human disease. Mice show remarkable biological similarity to people and share majority of the same genes as humans, making them an important model for understanding how genes work in health and disease. Indeed, 99% of mouse genes have an equivalent in humans, making mice ideal for studying the function of human genes in health as well as in diseases such as cancer, cardiovascular diseases and diabetes. Even though the mouse genome is 19% smaller than the human genome, the vast majority of human genes have counterparts in the mouse, especially genes related to human disease. Concurrently, analysis of the finished genome sequence of mouse has proved essential in understanding the full range of biology of both the mouse and human. The development and use of an increasing number and variety of natural, transgenic, induced mutant, and

Post-genomic Approaches in Cancer and Nano Medicine, 85–100.

genetically engineered mouse, knockout and mouse/human hybrids models have been imperative in understanding the gene-driven cellular and molecular mechanisms and pathophysiological manifestations (phenotypes) of several disorders that are similar to mice and humans. In this chapter, we describe the use of mouse models in understanding cancer and infectious diseases.

4.2 Mouse Models for Cancer

Cancer is one of the most exploited areas of research, however, due to a lack of understanding of cancer pathophysiology, no promising drug is available so far which is permanently able to cure cancer. Cancer is characterized by uncontrolled proliferation of cells that have transformed from normal cells of the body. Cancer cells can invade distant organs and distant tissues through normal circulation or via blood flow. Metastatic spread of cancer cells is the main cause of death in most of the cases in cancer patients. In advanced stages cancer, patients may die as a result of either improper medical treatment or improper diagnosis of cancer. Anticancer drugs have been developed from variety of sources which range from natural, that is, plants and microbes, to synthetic molecules. However, the widely used drugs result in severe adverse side effects which can lead to death in some patients. Common adverse effects of anticancer agents are due to high toxicity, which could lead to bone marrow suppression, alopecia, nausea and vomiting. Therefore, there is a need to identify novel and effective therapeutic strategies which could help in treatment of human cancers without causing as many or as severe side effects. Studies of cancer rely on the use of primary tumours [1, 3], paraffin-embedded samples [1], cancer cell lines [1, 3, 4], xenografts [2, 5, 6], tumour primary cell cultures [3, 4] and/or genetically engineered mice [2]. As certain types of manipulations for the genetic and DNA methylation analysis and drug testing are ethically and practically difficult to perform, in animals each of the above methods are used for different studies. Although cancer cell lines have been widely used for research purposes and are a useful tool in the genetic approaches, and their characterization shows that they are an excellent model for the study of the biological mechanisms involved in cancer [1], there are some issues that make animal models a necessity. Some manuscripts have raised issues on whether they are representative of the original tumour [5, 14]. On the other hand, some authors agree with the idea that there is a high (but not perfect) genomic similarity between the original tumour and the cancer cell line derived from it [8, 13, 15–17]. None-the-less, most of the novel drug entities generally undergo animal trials prior to entering clinical trials and so it is essential to understand screening methods for such entities in *in-vivo* animal models (specifically mouse), with focus on their strengths and drawbacks.

4.3 *In-vivo Methods*

Accurate and reliable diagnostic tests for the evaluation of adverse drug reactions remain elusive. A truly conclusive method for determining the contribution of drugs to adverse effects is essential. However, there are significant disadvantages in exposing patients to drugs that may cause serious adverse effects. *In vivo* models are advantageous over *in vitro* methods as they detect the host-mediated activity, are relatively predictable and estimate therapeutic ratio. The *in vitro* methods have low sensitivity, are less cost effective, and time consuming. A large number of *in vitro* samples cannot be handled and are very difficult to manage. Animal models are used both for toxicological studies and for detecting preclinical anticancer efficacy. They are able to detect the agents irrespective of their mechanism of action. Drugs with high degree of efficacy and a broad spectrum of action in animal model are usually expected to be effective in studies on clinical cancer, however, there are exceptions, which could be due to metabolic differences and heterogeneity of cancer cell between human and rodent. Despite these differences, animal models are widely used to support the results obtained from *in vitro* studies. Generally, the most promising drug is tested in more than one animal model. Dose response relationship, combined effect of drugs, mode of their anticancer mechanism and organ specificity can be obtained by *in vivo* methods. The selective animal model should be representative of high incidence of human cancer [2]. *In vivo* anticancer drug screening requires preparation of representative cancer models which can be done either by the use of chemical agents or cell line/tumor grafts (xenograft).

4.3.1 Tumor Model by use of Chemical Agents

Chemical agents such as factory pollutants or chemical carcinogens are known to be the cause for more than 80% of all cancers and thus they can be used to induce cancer in animal models. Carcinogens require metabolic activation before they can induce carcinogenesis. The epithelial studies indicate that human carcinogenesis occur through multiple steps in the same way as in mouse skin. Experimental carcinogenesis involves three steps.

- Initiation: Exposure of normal cells to carcinogens or chemical agents which transforms them to cancer cells. This may be due to damage to the genome or due to the disruption of cellular metabolic processes. Several radioactive substances are considered carcinogens, but their carcinogenic activity is attributed to radiation which they emit, for example, gamma rays and alpha particles.
- Promotion: Involves proliferation of the uncontrolled growth of targeted cells.
- Malignant transformation: It is metastasis, and involves the movement of cancer cells from the original site to other organs of the body by circulation, thus forming secondary tumors in the body.
- Although the exact cellular sequence of cellular, biochemical and molecular genetic events may differ between tissues and species, the overall concept seems to be directly applicable to clinical cancer, and thus the metastatic mouse skin carcinogenesis model is of immense use in understanding epithelial cancer in human [3]. Below, we show some drugs that have been used in mice for induction of tumors.

4.3.1.1 DMBA- induced mouse skin papillomas

Mouse skin is generally very sensitive to chemical carcinogens. SENCAR and BALB/c nude mice are highly sensitive to DMBA(7,12- dimethylbenz[a]anthracene) induced skin tumors. Rats, hamsters and rabbits are less sensitive to this drug, whereas guinea pigs are very resistant to it. DMBA acts as an initiator of skin papillomas, and 12-O-tetradecanoyl-phorbol-13-acetate (TPA) a promoter. Mice are topically applied with a single dose of 2.5–3μg of DMBA in acetone on their shaved back, followed by 10μg of TPA in 0.2ml acetone twice a week. Papillomas begin to appear after 6 or 7 weeks of application of TPA. Control groups are set with test and standard drug. In this, control means saline treated or carboxymethyl cellulose treated, whereas test means the new drug and standard means previously established drugs that are in the market, or generic drugs. Weekly observations are made to monitor tumor development. The experiment is carried out for about 18 weeks. Percentage tumor growth and multiplicity of treatment group is compared to control group. The tumor success rate in this model is usually 100 percent. Repeated use of DMBA can induce tumors [5, 6].

4.3.1.2 N-methyl, N-nitrosurea (MNU) induced Rat mammary gland carcinogenesis

MNU induces hormone dependant tumors [7]. In MNU induced rat mammary gland cancer model, single intravenous injection of 50mg/kg body weight of MNU (pH 5.0) is given to Sprague dawley rats, at 50 days of age. The incidence of tumor occurence in this model is 75–90% within 180 days post challenge. MNU induced tumors are invasive. Because MNU does not require metabolic activation, this model cannot detect inhibition of carcinogen activation. This model is referred to as a better model of human breast cancer because it gives a clear picture of breast cancer mechanism. MNU initiated cancers are comparable to human cancers in many ways such as high frequency of hormone dependence and pathological progression from ductal hyperplasia and ductal carcinoma in-situ. Test drug efficacy can be measured as percentage reduction of adenocarcinoma as compared to carcinogen control.

4.3.1.3 DMBA induced rat mammary gland carcinogenesis

In this model female Sprague dawley rats are given a single intra-gastric injection of 12mg/kg DMBA at the age of 50 days. This dose is sufficient to produce 80–100 % incidence of total mammary gland tumor within 120 days post challenge. This model can be used to detect the agents or drugs that inhibit carcinogen activation, for example, drugs that are used to inhibit cytochrome P450. Tumors are activated by *ras* gene. Test drug efficacy can be measured by percentage reduction in adenoma incidence as compared to carcinogen control.

4.3.1.4 MNU-induced tracheal squamous cell carcinoma in hamster

In this model, 5% solution of MNU in normal saline is administered once a week for 15 weeks, usually by a specially designed catheter which exposes a defined area of the trachea to the carcinogen. 15 weeks after MNU administration, this produces a tumor in 40–50% of hamsters. Test drug efficacy can be measured as percentage reduction of tumor incidence as compared to control.

4.3.1.5 Azoxymethane(AOM) induced aberrant crypt Foci in Rat

Aberrant crypt foci are single or multiple colonic crypt. Aberrant crypt foci are a cluster of dysplastic tube like glands in the lining of the colon and rectum in the advanced colon cancer. These are precancerous lesions and are known to be biomarkers for colon cancer in rodents. These crypt foci can be produced by a single injection of AOM 30mg/kg body weight in rats. At the end of the experiment, animals are sacrificed and the frequency of aberrant crypt foci is determined by histopathologic examinations.

4.3.1.6 1,2 Dimethylhydralazine(DMH) induced colorectal adenocarcinomain rat and mouse

Intraperitoneal injection of DMH produces colorectal adenocarcinoma in both rats and mice. DMH is first activated to azoxymethane and then to methylazoxymethanol. In rats, a single subcutaneous injection dose of 30 mg/kg body weight given to 7 week old mice produces colon adenomas and adenocarcinomas within 40 weeks.

4.3.2 Models Involving Cell Lines or Tumor Pieces (Xenograft)

4.3.2.1 Cell lines

Cell lines, if inoculated in specific number into sensitive or immunocompromised mouse strains, can lead to rapid and fast development of tumors as compared to chemical carcinogen induced tumors. MDA-MB231, L-1210 and P-388 cell lines are frequently used. MDA-MB231 cell lines are obtained from human mammary gland derived from metastatic site. It is aneuploid female with chromosome counts in the near-triploid range. Normal chromosomes N8 and N15 are absent. L-1210 and P-388 cell lines are derived from mouse lymphocytic leukemia and have 100% growth fraction. Tumor implanted animal dies on reaching the tumor burden. Hence, the effective drug is aimed at retarding the growth of the tumor and prolonging the life span of the animal. A drug which prolongs the life span of animal by 20% is taken for subsequent studies involving testing on other transplantable tumors. Some other cell lines which are used to induce tumor are lewis lung carcinoma and sarcoma-180. The host mouse strain of above cell lines is BDF1, except Swiss albino mice for sarcoma -180. While P-388 and L-1210 cell lines are inoculated intraperitoneally, lewis lung cancer carcinomas is given intramuscularly and sarcoma -180 subcutaneously. The experiment takes about 10 days for completion. [9]

Parameters to monitor:

Mean survival time (MST): Mean survival time (MST): It is the chance of surviving beyond the time (50%).

T/C% =

$$\frac{\text{MST of treated animal (T) x 100}}{\text{MST of control animals (C)}}$$

Tumor inhibiting activity: It is the capacity of the test drug or standard drug to help prevent growth in the size of tumor or prevent its metastasis.

$$\frac{\text{Average tumor weight of treated animal (T) x 100}}{\text{Average tumor weight of control animals (C)}}$$

4.3.2.2 Xenograft

Small hollow fibres tubes made of the plastic polyvinylidene fluoride containing cells of human tumors are inserted underneath the skin or in body cavity of the mouse, thus initiating tumors. Each test drug is administered at required dosage and is tested against the target tumor cell. More than 20 compounds a week can be screened by this method.

4.3.3 Transgenic Mice

Transgenic mouse models have been developed with increasing sophistication over the years. Mouse models that use widespread oncogene-expression or tumor suppression inactivation rarely survive long enough to form mammary tumor, because these mice often die due to due to metastasis in other tissues. In order to overcome this problem, researchers target gene expression in many cases by placing the regulation of transgene expression under the control of mouse mammary tumor virus (MMTV) promoter [10]. Meta mouse is a genetically engineered animal (Meta mice are the green fluorescent and red fluorescent protein bearing mice). In this mouse, tumor pieces of patients are transplanted into the organ of primary growth. In this particular mouse, metastasis and weight loss occurs in the same way as in humans. Certain human cancers which can be developed well in meta mouse include liver, pancreas, head, neck, bladder, stomach, ovarian, colon and lymphomas. However, the metastasis of breast and prostate is very slow in this mouse. The important indication of meta mouse is to test new routes, doses and indications of old drugs. This model is very useful to clinicians for making better therapeutic choices by first testing the drugs on patient tumor growth in meta mouse. The main drawback of these investigations is that they require long time (40–50 days). [11]

For transgenesis, DNA can be introduced in mice by several methods:

1. Using retroviral vectors that infect the cells at an early stage embryo prior to implantation into receptive female.
2. Microinjection into the enlarged sperm nucleus of fertilised egg.

4.3.3.1 Retroviral vector method

This model has the advantage of being an effective means of injecting the transgene into the genome of recipient cell [12]. The limitation is that this can only transfer small pieces of DNA, which due to the size constraint, may lack essential adjacent sequences for regulating the expression of the transgene. As more precise alternative methods have now become available, retroviral vectors are rarely used for creating transgenic animals which may be used as commercial end products [Figure 4.1].

4.3.3.2 DNA microinjection method

Direct microinjection of DNA into male pronucleus of fertilized oocytes is the most frequently used technique for production of transgenic mice. The number of fertilized eggs to be inoculated by microinjection is increased by stimulating

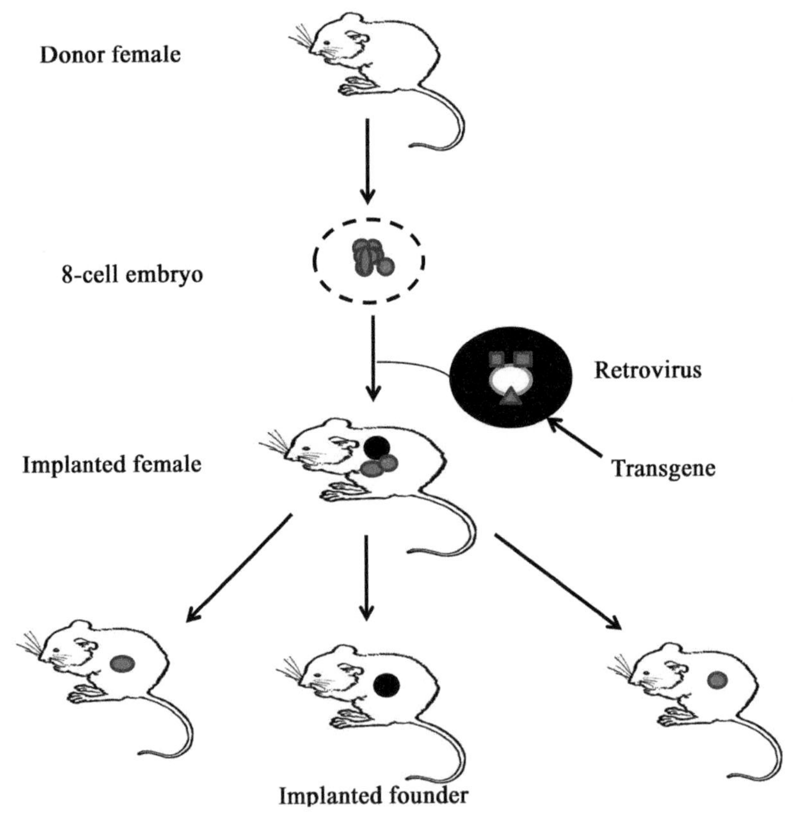

Figure 4.1 Retroviral vector method.

donor females to superovulate. Superovulation is induced by injecting females of optimum age with gonadotropins so as to obtain a sufficient number of oocytes. A superovulated mouse produces about 35 eggs instead of the normal 5 to 10. The superovulated mouse is mated and then sacrificed. The fertilized oocytes are then flushed from their oviducts 12 hours after fertilization. The male pronucleus, which tends to be larger than the female pronucleus, can be located by a dissecting microscope. DNA is microinjected in these oocytes. After inoculation 25–30 oocytes are implanted micro-surgically into the foster mother, who has been made pseudo-pregnant by mating it to a vasectomized male. In this case, because the male mate lacks sperms, none of the eggs of the foster mother are fertilized. The foster mother delivers pups from the inoculated eggs about 3 weeks from implantation. For identification of transgenic animals, DNA from a small piece of the tail can be assayed by southern blot hybridisation or polymerase chain reaction for the presence of transgene(s). Subsequently more pups can be bred with each other to form pure homozygous transgenic pups [Figure 4.2].

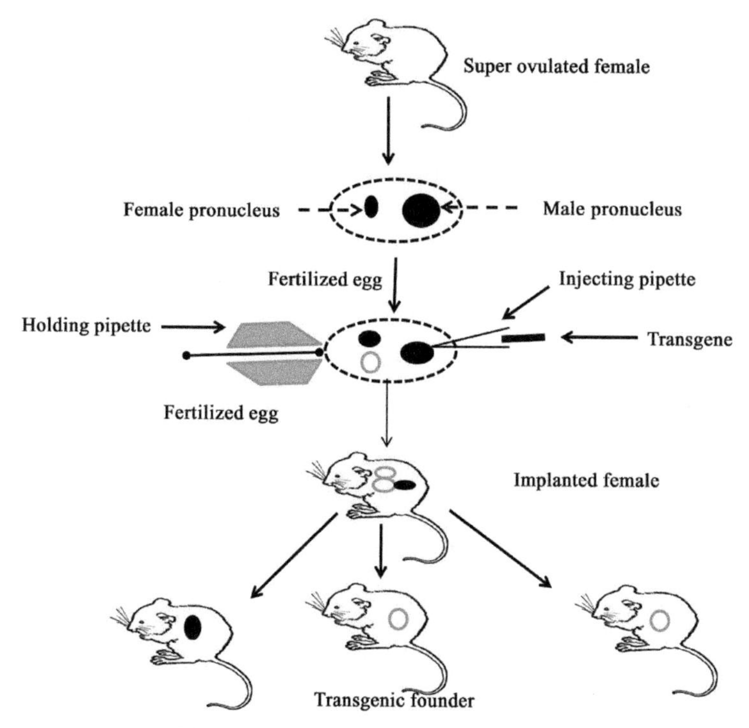

Figure 4.2 DNA microinjection method.

4.3.3.3 Nude mouse

A single gene mutation in these mice produces a hairless or 'nude' state in them. [15]. These mouse are immunologically incompetent because of absence of the thymus and a greatly reduced number of T cells. Lack of helper T cells and suppressor T cells alters the antibody response of the animals to antigen. They do not reject transplanted material. Hence, nude mice are widely used to test the tumorogenicity of cells or for testing anticancer drugs or new chemical moieties. The caging of these mice needs very strict sterile conditions and a warm environment as these mice have no body fur.

Temperature should be maintained at 26–28°C.

Some points regarding their use are:

1. Certain tumors like melanomas and colon cancer can grow very well in nude mice, whereas prostate carcinomas and most type of leukemias do not grow.
2. A large number of cells are usually required to be inoculated beneath the skin to get successful tumor rate.
3. Metastasis is seen after a long period of time (6 weeks).

4.4 Anti-cancer Drugs

Anticancer, or antineoplastic drugs are used to treat malignancies and cancerous growths. Drug therapy may be used alone, or in combination with other treatments such as surgery or radiation therapy. Here, we describe cancer chemotherapy in brief [Figure 4.3]. Several classes of drugs may be used in cancer treatment depending on the nature of the organ involved. For example, breast cancers are commonly stimulated by estrogens and may be treated with drugs which inactivate the sex hormones. Similarly, prostate cancer may be treated with drugs that inactivate androgens, the male sex hormone. However, cancer chemotherapy struggles to cause a lethal cytotoxic event or apoptosis in cancer cells that can arrest tumour progression. The anticancer drugs generally attack the DNA or metabolic sites which are essential for cell replication, e.g. purines or pyrimidine which are building blocks for DNA or RNA synthesis. Figure 4.3 shows chemotherapeutic agents affecting the availability of RNA and DNA precursors. These chemotherapeutic agents are designed to interfere only with cellular processes that are unique to malignant cells. Unfortunately most of the available anticancer agents are not specific enough to recognize neoplastic cells, but rather affect all kinds of proliferating cells both normal and abnormal. Therefore all the cancer agents have great response

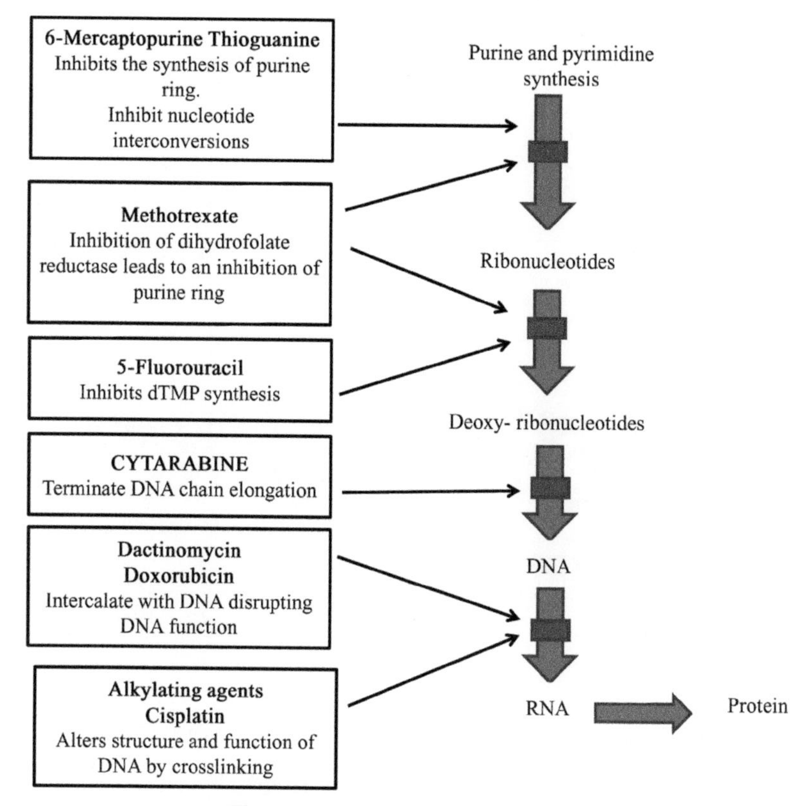

Figure 4.3 Anti cancer drug targets.

curve (that is, they start killing the cells as soon as they are injected, however, it is not known if they are killing normal cells or cancer cells) for both toxic and therapeutic effects.

Tumor susceptibility and growth cycle: Rapidly dividing cells are generally more sensitive to anticancer drugs, whereas slowly proliferating cells are less sensitive to chemotherapy. Cells in the G0 phase usually survive the toxic effect of many of these agents [16] (proliferation rate paradox). Cell cycle specificity of drugs: Both normal cell and tumor cells go through the same growth cycle. However the number of cells that are in various stages of the cycle may differ in normal and in neoplastic cells. Chemotherapeutic agents that are effective only against replicating cells are called as cell cycle specific whereas other agents are said to be cell cycle non-specific as shown in Figure 4.4. The non specific drugs have more toxicity in cycling cells and are more useful against tumors.

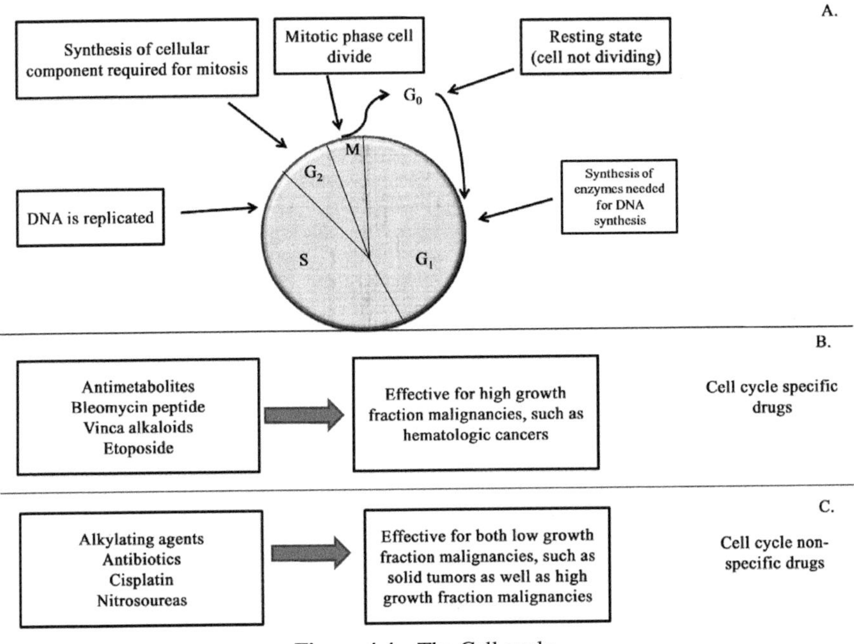

Figure 4.4 The Cell cycle.

4.5 Mouse Model of Infection

Infection is the invasion of host organism's body tissues by disease-causing organisms (bacteria/virus/fungi/parasites), their multiplication, and the reaction of host tissues to these organisms and the toxins they produce. In this section, we will briefly describe different mouse/rat/rodent models of infection and preclinical testing of new chemical drugs (antibiotics) on them.

4.5.1 Air Pouch Model

It is one of the classical methods of measuring the sub acute inflammation in the rats [17]. Wistar rats or mice (C57BL/6) can be used in this study. Groups of 5 rats or 10 mice are randomized to form the control group and test group for treatment with test or standard drugs. The back of the animal is shaved and disinfected with alcohol or disinfectants. Sterile air is taken into a syringe with the help of a sterile air catcher. The mouse is anaesthetized by ether inhalation and then subcutaneous injection with 1ml of air for three consecutive days to form air pouch. Two days later, 0.1ml of bacterial suspension containing specific quantity of bacteria $1x10^9$ cells is inoculated in the air pouch of the mouse. Starting with the formation of the air pouch the animal is treated every day with the standard and test drugs. On the 5th day animal is sacrificed under ether anesthesia. The pouch is opened and exudate is collected in the glass cylinder. The control mouse has more exudate than the ones treated with standard and test drugs. The average volume of exudates in control, standard and test drug treated rats is calculated and compared to each other for statistical analysis. This method can be used to test anti-inflammatory drugs, corticosteroids and NSAID's. (Figure 4.5).

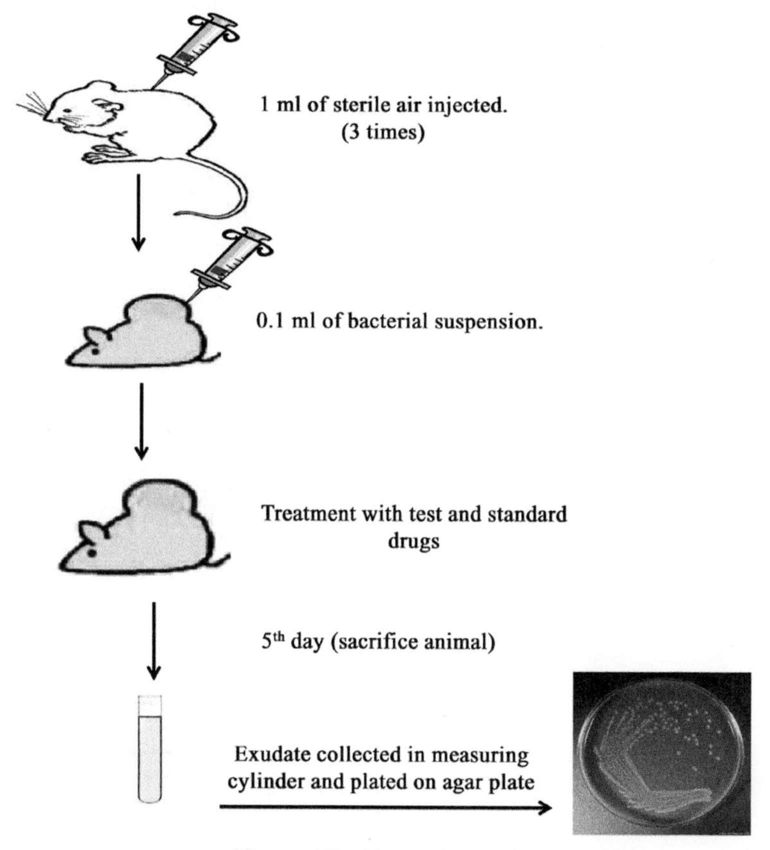

Figure 4.5 Air pouch model.

4.5.2 Dermatophytosis

Dermatophytes are keratinophilic fungi that infect tissues containing keratin such as hair, nails, and skin [18]. Dermatophytes consist of three genera: *Microsporum, Trichophyton, and Epidermophyton.*

A clinical isolate of (*M. canis)* is incubated and cultured at 25°C on Sabouraud glucose agar (SAB) with cycloheximide and chloramphenicol for 3 weeks. Fungal cells are prepared and counted by hemocytometer. Here the guinea pig is used for infection model of skin.The guinea pig is anesthetized by intramuscular administration of Zolazepam and Tiletamin. The back of the animal is shaved and disinfected. Shaving traumatizes the skin and will make it more susceptible to infection. 50μl of fungal suspension is used to inoculate four or five ring areas on each guinea pig. The animal is monitored clinically for 17 days. The clinical evaluation consists of a semi-quantitative score where the inoculated areas are evaluated and compared to the uninfected ring area on the back of the same animal. The redness is scored as follows: 0, normal; 1, pink; 2, red; and 3, violet. The lesion is scored as follows: 0, normal; 1, papule; 2, skin scales; 3, single layer of skin scales and ulcers; and 4, multiple layers skin scales and ulcers. The score values of the control, standard and test animals are compared and evaluated for statistical significance.

4.5.3 Endocarditis Model

Endocarditis is an inflammation of the inner layer of the heart, the endocardium [19]. It usually involves the heart valves (native or prosthetic valves). The bacteria *Streptococcus viridans* is usually responsible for this infection. Rabbits are generally the test animal for this type of infection model. Here, rabbits weighing about 2 kg are anaesthetized with intravenous pentobarbitone and the external jugular vein on the right side of the neck is exposed. A polyethylene catheter of external diameter 0–8 mm and internal diameter 0.4 mm, filled with sterile saline, is passed down the vein and tied in place till the tip reaches the level of right side of the heart. The cervical end of the catheter is heat sealed, and buried when the wound is closed with silk sutures. The final position of the catheter tip is usually in the right ventricle or right atrium, but occasionally in the inferior vena cava. The rabbits are not disturbed for 7–10 days, during which small sterile vegetations composed of platelets and fibrin are formed on the tricuspid valve or the endocardium at points of contact with the catheter.

Parameters to monitor:

- Bacterial counts in tissue and vegetation.
- Microscopic appearance of vegetation during the early stages of colonization.

4.5.4 Lung Infection Model

Lung infection model [20] or lung pneumonia model is the inflammatory condition of the lung affecting primarily the air sacs, known as alveoli. It is usually caused by infection with viruses or bacteria. 4 weeks old Swiss male mice are used in this study. The animals are lightly anesthetized with a mixture of ketamine and xylazine. After animal are anaesthized, the trachea of the animal is exposed by a slight incision in the neck and they are infected by intratracheal instillation with a syringe directly into the trachea with 50 μl of inoculum $\sim 1\text{x}10^9$ CFU/ mouse, and then sutured with absorbent sutures. Oral doses of standard and test drugs are given at 24 and 48 hrs post inoculation. Animals are sacrificed 72 hrs post-inoculation by cervical dislocation. The lungs are removed aseptically and homogenized in 3ml of ice-cold PBS. Colonies are counted on the plates. Average colony count are compared for control, standard and test drug treated animal lung samples.

4.5.5 Thigh Infection Model

Necrotizing soft tissue infection (NSTI) or thigh infection is a disastrous infection of the subcutaneous tissue and underlying fascial layers [21]. Even if immediate treatment is started, mortality rates are high. The underlying cause of these infections is a bacterial or viral invasion. *S. aureus* is the main cause of this types of infections. Six-week-old, pathogen free, female ICR/Swiss mice weighing 22–26 g are rendered neutropenic by injecting two doses of cyclophosphamide intraperitoneally 4days (150mg/kg) and then 1 day (100mg/kg) before the experiment. Renal impairment is induced by injecting uranyl nitrate 10mg/kg s.c 3 days before the experiment. Thigh infection with ATCC strain of *S. marcescens* 7603A is produced by injecting approximately 10^6 log-phase bacteria in 0.1 ml carboxymethylcellulose into each thigh under ether anaesthesia. Two to six mice from each group are killed 2 hrs before, at the time of injection and 2, 4, 6, 8, 12, 16, 24, 30, 36, and 48 hrs after drug injection. Treatment is started after 2 hrs of thigh infection. After 48 hrs, all the mice are killed by cervical dislocation and their thighs are dissected and homogenised in PBS. Duplicate aliquots of four to five serial 10-fold dilutions are plated on agar for CFU determination.

4.5.6 Urinary Tract Infection Model

Urinary tract infection (UTI) is the most common type of hospital-acquired infection and has been shown to occur commonly after urinary catheterization. UTI is a bacterial infection that affects any part of urinary tract which include infections of the bladder (cystitis) and kidney (pyelonephritis). It affects primarily women and is responsible for nearly 13 million annual clinician visits in the United States [22]. There are numerous other pathogens capable of infecting the urinary tract [23, 24]. These infections are most commonly caused by gram negative bacterium *Escherichia coli*, which is responsible for 80 to 85% of community-acquired UTIs.

Methodology: Female wistar rats, 5 weeks old, are used in this study. Uropathogenic *E.coli* should be maintained and prepared according to the procedure. Urinary tract infection is induced in the rat with the help of a transuretheral catheter inoculation of viable bacteria. Rats should be anaesthetized with a low dose of ketamine + xylazine. Gentle bladder massage was done to micturate the animal so as to completely empty the bladder. Papilla is grasped from one side and catheter is inserted into urethera. 0.3 ml of bacterial solution is inoculated into the bladder of the animal. Water bottle is replaced with 5% sucrose water which is given to them after 3 hrs. Animals are processed at 24, 48 and 72 hrs post infection. Bladder, ureter and kidneys are dissected and homogenised in PBS. Duplicate aliquots of four to five serial 10-fold dilutions are plated on agar for CFU determination [Figure 4.6].

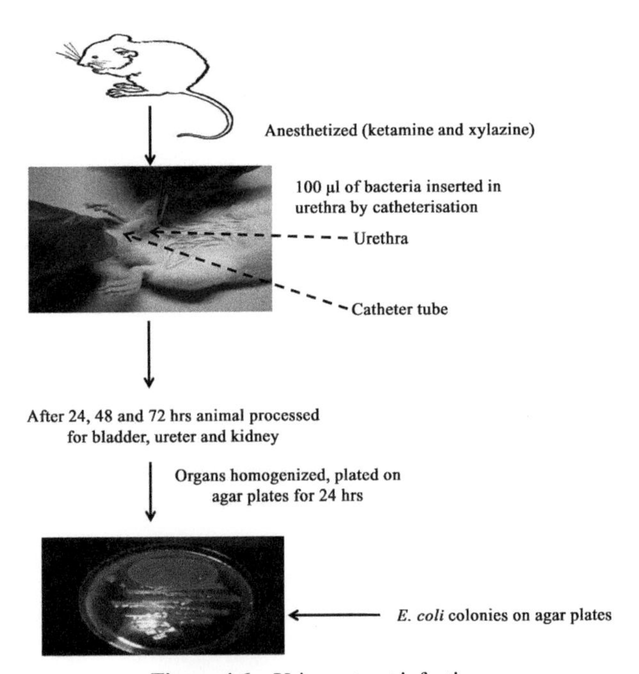

Figure 4.6 Urinary tract infection.

5

Potential Application of Natural Compounds for the Prevention and Treatment of Hepatocellular Carcinoma

Shikha Satendra Singh[1,2], Sakshi Sikka[1,2] Gautam Sethi[1,2] and Alan Prem Kumar[1,2,3,4]

[1]Department of Pharmacology, Yong Loo Lin School of Medicine,
National University of Singapore,
Singapore 117597
[2]Cancer Science Institute of Singapore,
National University of Singapore,
Singapore 117599
[3]School of Biomedical Sciences,
Faculty of Health Sciences,
Curtin University, Perth, Western Australia 6845
[4]Department of Biological Sciences, University of North Texas,
Denton, TX 76203-5017, USA

5.1 Introduction

The liver is a vital organ for xenobiotic metabolism and carries out several key biological functions to maintain homeostasis and the health of an individual. For instance, the liver is responsible for the production of proteins and hormones, detoxification of foreign chemicals, as well as glucose and lipid metabolism [1]. It is made up of many different cell types but the two main ones are the liver cells (hepatocytes) and cells lining the bile ducts (cholangiocytes). Hepatocytes constitute approximately 80% of the liver volume. Hepatocellular carcinoma (HCC) is the fifth most common cancer in the world, with an estimated incidence of half a million new cases every year around the world [2]. Moreover, primary liver cancer is the fifth largest

Post-genomic Approaches in Cancer and Nano Medicine, 101–166.

cause of cancer in men and the eighth largest in women [3]. Several risk factors are implicated as key causes for HCC such as infections due to the hepatitis B and C viruses, obesity, iron overload, both alcoholic and non-alcoholic cirrhosis, as well as dietary hepatocarcinogenes, such as aflatoxins and nitrosoamines [4, 5].

Primary liver cancer occurs due to oxidative stress and inflammation [6]. Hepatic inflammation, resulting from viral infections, chronic hepatitis, cirrhosis, as well as exposure to toxic hepatocarcinogens, represents an early malignant step with genetic and epigenetic events occurring as the result of a prolonged inflammatory process. Previous studies implicate that production of cytokines, chemokines, as well as reactive oxygen species, contribute to the process of hepatocarcinogenesis [2, 7]. Moreover, oxidative stress, due to environmental insults results in the production of superoxide anion and the hydroxyl radical which are predisposing factors to hepatocarcinogenesis [8, 9]. Besides the involvement of the oxidative stress and the inflammatory cascade, various molecular and signaling mechanisms have also been proposed in the pathogenesis of HCC [10]. Although, liver transplantation is a useful treatment for early-stage HCC, it is available only to a small number of HCC patients due to shortage of organ donors [11]. Moreover, several alternative treatment approaches, such as Yttrium-90 intra-arterial delivery, arterial chemoembolization, microwave coagulation, intra-tumor ethanol injection as well as radiofrequency ablation, are applicable only to patients with localized liver tumors, thus limiting their usefulness [12, 13]. Sorafenib, a vascular endothelial growth factor receptor and tyrosine kinase inhibitor, is currently approved in the United States for the treatment of unresectable HCC. Although sorafenib has been shown to extend the medial survival time by three months in patients with advanced HCC, it is extremely expensive and its therapeutic advantage is limited due to severe adverse effects such as risk of haemorrhage [14, 15].

Therefore, the use of dietary components, present in fruits, vegetables, nuts and spices, might be beneficial to inhibit HCC. The natural agents have demonstrated significant potential in their ability to suppress carcinogenesis in pre-clinical models and delay the occurrence of cancer in high-risk populations [16, 17]. Phytochemicals, such as dietary polyphenols have potent antioxidant as well as anti-inflammatory properties, thus providing a promising alternative towards combating HCC [18].

5.2 Risk Factors Involved in HCC

5.2.1 Viral infections

Viral infections either Hepatitis B Virus (HBV) or Hepatitis C Virus (HCV) attributes as a major risk factor in 70–80% of HCC cases. The impact of HBV or HCV infection can be reflected by strong correlation between increased incidences of HCC and prevalence of Hepatitis B surface antigen (HbsAg) or antibody to Hepatitis C virus (anti-HCV).

5.2.1.1 Hepatitis B virus (HBV)

Chronic HBV infection remains a main etiological factor of HCC worldwide with more than half of the HCC patients being chronic carriers. The risk of eventually developing HCC is inversely related to the age of acquisition of HBV. Reports have also suggested higher familial risk of first-degree relatives of patients with HBV-infected HCC [19]. This virus is a partially double stranded DNA virus belonging to the *Hepadnaviridae* family. The genome of HBV contains four overlapping open reading frames (ORFs): S, C, P, and X. The pre-S/S ORF encodes the three viral surface proteins, the pre-C/C ORF encodes the e antigen (HBeAg) and the core antigen (HBcAg), the P ORF encodes the terminal protein (TP) and the viral polymerase that possesses DNA polymerase, reverse transcriptase and RNaseH activities. The X gene encodes a small protein that is essential for virus replication but its function is less understood. Genetic mutations of the viral load have been extensively investigated for their role in hepatocarcinogenesis. Mutations in the basal core promoter region and/or of the X and preS viral genomic regions of HBV genotypes have been correlated towards liver disease progression [20]. Several genotypes of HBV are classified on the basis of sequence divergence in the entire genome into genotypes A-H with newly identified I and J [21–23]. Genotypes B and C are most prevalent in Asia [21, 24], while A and D are prevalent in Europe with A more predominant in the Northern Europe and D in the Eastern Europe and Mediterranean basin [25]. Recent studies have mentioned the role of different genotypes towards variations in the risk of developing HCC. However, although HBV genotypes influence the natural history of HBV infection, the exact role of genotypes in HCC development and anti-viral response is still inconclusive. Thus, besides the genotypes of HBV being clinically investigated for its impact on HCC patients, the "genetic mutations" in HBV contribute significantly towards HCC carcinogenesis.

5.2.1.2 Hepatitis C virus (HCV)

HCV, a *Flaviviridae* family is another important risk factor for HCC with incidence rates varying among different geographical regions. This virus causes chronic liver with eventual development of cirrhosis and HCC. Unlike HBV, HCV is a single-stranded RNA virus that does not integrate into the host genome. Many viral proteins have been implicated in the development of carcinogenesis [26]. The viral proteins like the core protein and few non-structural viral proteins like NS3 (serine protease and RNA helicase/NTPase activity)and NS4 (an essential cofactor for NS3 protease functions) have shown to have transformation potential *in vitro* [27]. Also, Human Immunodeficiency Virus (HIV) co-infection results in greater likelihood of enhanced viral replication in both HBV and HCV infections [28, 29].

5.2.2 Toxins

5.2.2.1 Chemical carcinogens

Aflatoxins are a group of approximately 20 related fungal metabolites. They are toxic metabolites contaminating major foodstuffs such as groundnuts, peanuts, maize, rice, figs and other dried foods thus, necessitating extensive quality-control measures to minimize levels in products for human consumption. Among them, Aflatoxin B1 (AFB1) is a naturally occurring most potent chemical carcinogen as well a toxic metabolite. Its role as a liver carcinogen has been supported by several experimental evidences [30]. Its biological role towards carcinogenesis of HCC has been investigated by checking its mutagenic ability tested in systems like human cell culture, as a DNA adduct profile, with the aflatoxin-N7-guanine adduct as a major adduct [31]. The adduct level in liver DNA has been quantitatively related to aflatoxin dose and to tumor yield. With development of *Aspergillusflavus* in food products under wet and humid conditions, exposure to AFB1 is commonly found in sub-tropical countries. This mycotoxin is strongly hepatocarcinogenic in experimental animal models and acts synergistically with viral factors increasing the relative risk of HCC [32].

5.2.2.2 Other carcinogens

Exogenous or endogenous exposure to varying amount of chemicals is also responsible for varying incidence rates of HCC. With increase in industrial development associated with changes in "lifestyle", exposure to chemicals like vinyl chloride monomers, hydrocarbons, processed foods, nitrites, primary metals, pesticides, and tobacco use may contribute to the causation of HCC in a sizable fraction. More than 60 known carcinogens have been detected in cigarette smoke with polycyclic aromatic hydrocarbons (PAHs), nitrosamines and aromatic amines [33].

5.2.3 Diet and Metabolic factors

Lifestyle plays a significant role in contributing towards the burden of HCC. This includes alcohol consumption, diabetes, and obesity, dietary compounds like iron and thiamine as well as chewing of betel quid. Increased consumption of alcohol primarily leads to cirrhosis with eventual HCC development. The risk of HCC development depends upon the rate of consumption of alcohol. It increases five-fold when daily alcohol consumption exceeds 80g/day for more than 10 years [33, 34]. Also, synergism between alcohol consumption and HCV infection has been observed along HCC development at an earlier age [35]. Diabetes, a part of the metabolic syndrome, is characterized by insulin resistance and is thought to predispose to non-alcoholic fatty liver disease (NAFLD) including its more severe form "non-alcoholic steatohepatitis (NASH). NASH has been identified as a cause of both "cryptogenic cirrhosis" and HCC. In addition to increasing the prevalence of chronic liver disease, diabetes is also an independent risk factor for HCC development [36]. This has been supported by another study suggesting, diabetes is associated with moderately increased risk of HCC incidence, as well as HCC mortality [37]. Similar trends have been implicated in association of increased risk of HCC and obesity [38]. Another important risk factor is excessive consumption of dietary compounds like iron and thiamine which is associated with a significant 3-fold and 2-fold increase in the risk of HCC, respectively [39].

5.2.4 Genetic Factors

The genetic susceptibility to HCC is characterized by genetic heterogeneity. Rare monogenic syndromes such as alpha1-antitrypsin deficiency, glycogen storage disease type 1, hemochromatosis, acute intermittent, and cutanea tarda porphyria, as well as hereditary tyrosinemia type 1 are associated with high risk factors [40]. There are also reports for familial aggregation of HCC owing to both environmental factors like viral infection as well as genetic factors [41]. Thus, several other diseases/ syndromes other than chronic liver diseases have been associated with HCC development.

5.2.5 Cirrhosis

Cirrhosis develops following long periods of chronic liver disease and is characterized by replacement of liver tissue by fibrosis, scar tissue and regenerative nodules leading to loss of liver function. HCC develops among 70–90% of cirrhosis patients, while only 10% of HCC patients have a non-cirrhotic liver [42], indicative of cirrhosis as a major risk factor for developing HCC. To summarize, in 90% of the HCC cases, at least one of these risk factors can be identified either alone or in combination with other factors. The presence of each risk factor among patients varies according to the geographical origin of the patients.

5.3 Types of Liver Cancers

WHO histological classification of the tumors of the liver involves:

5.3.1 Epithelial tumors (Malignant)

5.3.1.1 Hepatocholangiocarcinoma (HCC-CC)

HCC-CC is a rare form of primary liver cancer with features of both hepato-cellular and biliary epithelial differentiation. It accounts for 0.4 to 14.2% of all the primary liver carcinomas [43]. Due to it rare occurrence, its demographics and clinical features are still poorly understood, [44, 45] with its long-term prognosis not yet well-defined [46].

5.3.1.2 Hepatoblastoma

Hepatoblastoma is a rare pediatric neoplasm comprising 1% of pediatric neoplasms. The 5-year survival rate of children afflicted with this neoplasm is 70% [47]. The incidence rate of this neoplasm has been shown to increase several folds in infants afflicted with associated disorders like Beckwith-Weidemann, familial adenomatous polyposis, glycogen storage diseases I-IV, trisomy 18 or other trisomes and low-birth weight infants [48]. However, the extent of risk is difficult to determine due to the rarity of this neoplasm.

5.3.2 Non-Epithelial Tumors (Benign)

5.3.2.1 Hepatic angiomyolipoma (AML)

It is a rare benign mesenchymal tumor of the liver composed of smooth muscle cells, adipose tissue and proliferating blood vessels. Angiomyolipoma occurs more frequently in kidney with liver ranking second most frequent

Table 5.1 Types of liver cancer and pre-cancerous forms

WHO histological classification of liver cancer
EPITHELIAL TUMORS • *Malignant* **Hepatocellular carcinoma** Hepatocholangiocarcinoma Hepatoblastoma
NON- EPITHELIAL TUMORS • *Benign* HepticAngiomyolipoma Hemangioma • *Malignant* Angiosarcoma Embryonal sarcoma Rhabdomyosarcoma Hepatic Epitheloidhemangioendothelioma
MISCELLANEOUS TUMORS • *Solitaryfibroustumor* • *Teratoma* • *YorkSactumor* • *Carcinosarcoma* • *Rhabdiodtumor* • *Hepatic Mesenchymalhamartoma*
SECONDARY TUMORS

site for this tumor. Although the behavior of these tumors is benign, it has been considered precancerous with several reports revealing its ability to be malignant with evidence of recurrence or metastasis [49, 50]. It is found in both males and females with predominance in adult females. The lesions are cumbersome to diagnose, especially, if fatty component is lacking or scarce. Even the postoperative pathologic diagnosis has been easily mistaken as hepatocellular carcinoma (HCC). Many patients have been treated with surgical resection. Therefore, the proper treatment of hepatic AML has yet remained controversial [51].

5.3.2.2 Hemangioma

Hepatic hemangioma or cavernous hemangioma is a benign abnormal collection of blood vessels in the liver. It is 4 to 6 times more likely in females than males. The prevalence rate of this neoplasm ranges from 0.4 to 20% with increasing prevalence derived from autopsy studies indicating the lesions more common among general population [52]. It could be classified as "congenital" i.e. a birth defect or "infantile" hemangioma. Infants usually develop "benign infantile hemangioendothelioma" also referred to as multinodular hepatic hemangiomatosis.

5.3.3 Non-Epithelial Tumors (Malignant)

5.3.3.1 Hepatic Angiosarcoma

Angiosarcomas are tumors arising from the endothelial cells, lining the walls of the blood vessels. Hepatic angiosarcomas are very rare tumors, accounting for only 2% of the primary liver malignancy [53, 54]. Although rare, it is the third most common primary malignant neoplasm of the liver. Risk factors associated with this neoplasm include exposure to toxics like thorium dioxide, vinyl chloride and arsenic. Diagnosis for this tumor is often challenging with history towards exposure to toxic chemicals being an important aspect for diagnosis. Treatment modalities for this tumor involves surgery, chemotherapy and radiotherapy [55].

5.3.3.2 Hepatic epithelioidhemangioendothelioma (EH)

Hepatic epithelioidhemangioendothelioma (EH) is a rare, low grade, malignant vascular neoplasm. It is a soft tissue vascular tumor of endothelial origin with a clinical course intermediate between benign hemangioma and malignant angiosarcoma. This malignant neoplasm has an incidence rate of <0.1 per 100,000. The causative factors for this tumor remain unclear with women more likely to develop this tumor than men with a median age of 30 to 40 years of age [56, 57]. Treatment modalities include hepatic resection, orthotropic liver transplantation, radiotherapy, chemotherapy, and interferon-alpha treatment.

5.3.3.3 Embryonal sarcoma

Stocker *et al.*, named undifferentiated embryonal sarcoma of the liver, as an entity in 1978 on the basis of an Armed Forces Institute of Pathology (AFIP) series [58] as a rare, highly malignant hepatic neoplasm, affecting almost exclusively the pediatric population. The prognosis is poor but recent evidences has shown that long-term survival is possible after complete surgical resection with or without post-operative chemotherapy [59].

5.3.3.4 Hepatic Rhabdomyosarcoma

Hepatic rhabdomyocarsoma is a tumor typically found in children less than 8 years of age and very rarely found in adults. The tumor has "grape bunches" like appearance and has a very poor survival rate. The treatment modalities for this tumor is similar to that of undifferentiated embryonal sarcoma [60].

5.3.4 Miscellaneous tumors

5.3.4.1 Solitary fibrous tumor

Solitary fibrous tumor is an uncommon mesenchymal neoplasm consisting of cellular and collagenous components. Although most frequently involving the pleura, these lesions have been found in the liver parenchyma [61]. Surgical resection is the preferred treatment method, due to the uncertainties involved in its malignancy.

5.3.4.2 Hepatic teratomas

Teratomas are non–seminomatous germ cell tumors (NSCT) that arise from abnormal development of pluripotent and embryonal germ cells. Hepatic teratoma is a very rare entity with only 26 cases reported so far, with only six reported in adults. Most are discovered incidentally, as there is no typical clinical presentation. They are usually well-encapsulated lesions, easily resectable from the surrounding hepatic parenchyma. Complete resection remains the best treatment option [62, 63].

5.3.4.3 York Sac tumor

Primary yolk sac tumor of the liver was first reported by Hart in 1975 in a 18-month old boy who died 6.5 months later after extended hepatectomy. It is a very rare form of tumor with fewer than 20 cases reported [64]. The treatment modalities are usually lobectomy or liver transplantation and chemotherapy.

5.3.4.4 Carcinosarcoma

Primary hepatic carcinosarcoma is a rare tumor comprising of a mixture of carcinomatous and sarcomatous elements. Very few cases have been reported with most cases in men with a mean age range of 60 years [65].The clinical characteristics of this tumor include abdominal pain, fever and jaundice. Surgical resection is the treatment option for this tumor for both early stages and advanced stages due to its high recurrence rate [66].

5.3.4.5 Rhabdiod tumor

Rhabdoid tumors of the liver are rare tumors that are difficult to cure. The first evidence in literature of tumor with rhabdoid features was in 1982 by Gonzalez-Crussi *et al.*[67]. All the cases reported have been in pediatric patients with very rare report in young adults with very poor prognosis rate [68]. The aggressive and incurable nature of this tumor corroborates the need to develop a targeted and effective therapy for these tumors.

5.3.4.6 Hepatic mesenchymalhamartoma (HMH)

HMH is an uncommon benign tumor in children. It makes up approximately 8% of all pediatric tumors and is second in occurrence only to hepatoblastoma when only pediatric hepatic tumors are considered. 80% are found within the first 2 years of life and the remainder are detected by 5 years of age [69]. Very rare cases been reported in adults [70]. Pathologically, they vary greatly in size, from a few centimeters up to 30 cm [69]. Microscopically, it consists of myxomatous connective tissue containing scattered bland stellate-shaped mesenchymal cells. Although hamartomas of the liver are histologically benign, their clinical course and the complications of surgical treatment can be significant. They can often pose diagnostic dilemmas and may have a propensity for local recurrence and malignant degeneration [71].

5.3.5 Secondary tumors

Liver metastasis is a tumor that spreads to the liver from other areas of the body. The most common sites for primary tumor are breast, lung and colorectal cancer [72, 73]. The high incidence of hepatic metastasis has been attributed to two mechanisms. First, the dual blood supply of the liver from the portal and systemic circulation increases the likelihood of metastatic deposits in the liver. Second, the hepatic sinusoidal epithelium has fenestrations that enable easier penetration of metastatic cells into the liver parenchyma. Treatment for secondary liver cancer depends upon several factors like the type of primary cancer site, the extent of spread to the liver, extent of cancer spreading to other organs besides liver and patients' clinical condition. Aggressive surgical resection followed by chemotherapy is most commonly the treatment modality. If cancer has spread to liver and other organs, the whole body (systemic) chemotherapy is usually the treatment of choice [74].

5.4 Dysregulated Signaling Cascades in HCC

Hepatocellular carcinoma is one of the few cancers worldwide with increasing trends of incidence rate. One of the main reasons for the high mortality rate in patients with HCC is the lack of effective treatment options, especially for those with advanced disease. The scarcity of effective treatment modalities for advanced form of HCC has led to the need for development of new therapeutic strategies. HCC progression and initiation depends upon the progressive accumulation of several genetic and epigenetic effects which simultaneously alters an array of signaling cascades via dysregulation of signal activators and inhibitors. Understanding the dysregulation in these molecular oncogenic pathways will eventually lead to the identification of several possible therapeutic targets.

The molecular pathogenesis of HCC is very complex with a sequence of events developing from normal through chronic hepatitis/cirrhosis and dysplastic nodules to HCC [75]. Two mechanisms have been associated with deregulation of signaling pathways in HCC:

1. Initial cirrhosis is associated with hepatic regeneration due to tissue injury caused by several factors like hepatitis infection, toxin/environmental factors or metabolic factors. This regenerative ability activates the cellular repair mechanism which results in activation of several genes downstream.
2. Activation of several genes including oncogenes, could eventually lead to mutations in one or more oncogenes or tumor suppressor genes [75].

These events eventually lead to abnormalities in several critical signaling pathways that perpetuate the process of carcinogenesis. The various signaling pathways implicated in HCC carcinogenesis are described below.

5.4.1 STAT3 signaling

Signal transducer and activator of transcription 3 (STAT3) belongs to the STAT protein family of transcription factors that are found in many cell types and play different roles in normal cellular signaling. Several cytokines and growth factors like IL-6 and Epidermal Growth Factor (EGF) family members, as well as hepatocyte growth factor (HGF) mediate activation of STAT3 by phosphorylation [76, 77]. IL-6 and IL-22 are considered as potent inducers of STAT3 in hepatocytes. Initiation of STAT3 activation through ligand (IL6, IL-22)-receptor interaction results in dimerization of a signal transducer protein, gp130 in the cytoplasm [78, 79]. This is followed by induction of Janus-kinase (JAK) phosphorylation and subsequently STAT3 phosphorylation. Phosphorylated STAT3 monomers combine to form dimers and translocate into the nucleus to induce transcription of genes involved in cell survival and proliferation. Constitutive activation of STAT3 is often found in cancer cells [80–83]. STAT3 phosphorylation has been detected (i.e. activated) in approximately 60% of human HCCs, with STAT3-positive tumors being more aggressive in recent reports. Its activation due to upregulation of IL-6 and

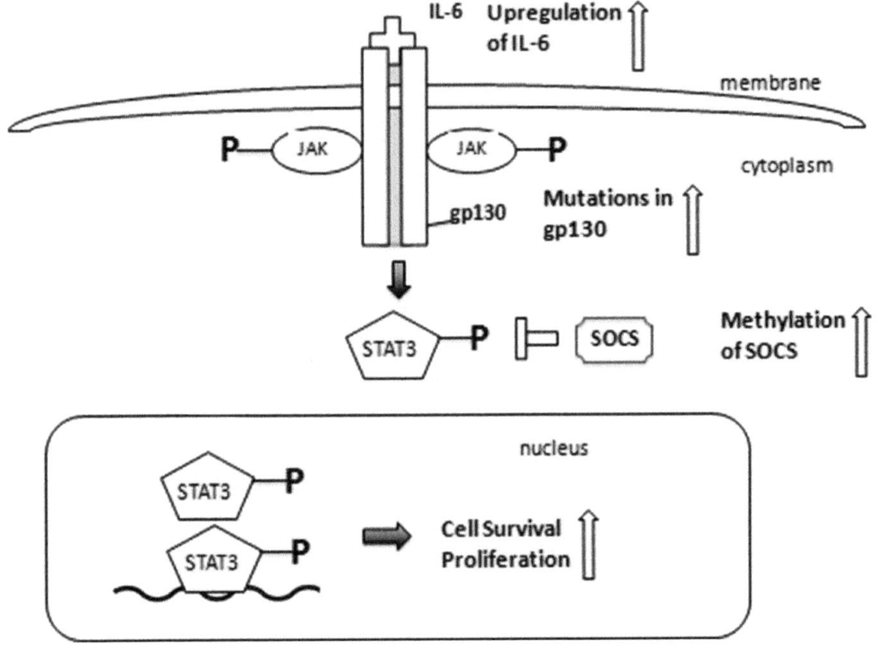

Figure 5.1 Deregulation of JAK/STAT3 pathway in Hepatocellular carcinoma: Upregulation of IL-6, Mutations in gp130, Methylation of SOCS has been reported as plausible mechanisms for deregulation of JAK/STAT3 pathway. IL6: Interleukin-6; JAK:Janus Kinase-2; STAT3:Signal Transducer and Activation of transcription 3; SOCS: Suppressor of cytokine signaling.

IL-22 has been shown to contribute towards tumorigenesis, including in HCC [84, 85]. These findings are consistent with those of other studies in which STAT3 was found to be activated in the majority of HCCs with poor prognosis and not in surrounding non-tumor tissue or normal liver [79, 86]. Also, activating mutations in the gene encoding the gp130 signaling subunit of IL-6R family has been identified in benign hepatocellular adenomas [87]. Besides activation of several genes involved in proliferation, phosphorylated STATs also activate the transcription of family of inhibitory proteins, suppressor of cytokine signaling (SOCS).

High prevalence of aberrant methylation of tumor suppressor genes (TSGs) like SOCS plays an important role towards the carcinogenesis of liver due to negative feedback mechanism [88]. Concurrently, mutations in the other signaling pathways alongside like Wnt/β-catenin and IKK/NFκB also contributes towards the occurrence of HCC [89, 90].

5.4.2 Wnt/β-catenin pathway

Wnt signaling plays an important role in normal liver function with at least eight of the Wnt ligands expressed in the hepatocytes [91]. This pathway involves three main components, the ligand-receptor complex (cell surface), the β-catenin destruction complex (cytoplasm), and the β-catenin/TCF transcription complex (nucleus) [92]. Under normal circumstances, in the absence of Wnt receptor ligand interaction, the cytosolic pool of β-catenin is degraded through the action of a multiprotein complex which includes tumor-suppressor proteins such as APC and Axin, and the Ser/Thr kinases CK1α and GSK3β. This binding of the multiprotein complex leads to phosphorylation of β-catenin, which is targeted for ubiquitination and proteolysis [93]. In the liver, this canonical signaling pathway regulates several aspects of liver specification, cell fate, and proliferation during hepatogenesis, optical hepatic homeostasis and liver metabolism [94–96]. Deregulation of the Wnt pathway has been corroborated in several studies owing to β-catenin mutations, activation, and accumulation [97]. Dominant gain-of-function mutations usually occur at the N-terminal phosphorylation sites on β-catenin, including the sites phophorylated by the GSK3β that regulates β-catenin degradation resulting in a more intrinsically stable protein which is translocated into the nucleus [98]. The nuclear β-catenin then forms a complex with TCF/LEF family of DNA binding transcription factors to activate the downstream target genes involved in cell proliferation, such as cyclin D1. β-catenin also forms core complexes with E-cadherin and α-catenin resulting in cell–cell adhesion, an important step towards development of tumor metastasis [99]. Several other mechanisms have been investigated for activation of Wnt/β-catenin signaling in HCC. Mutations in AXIN1 and AXIN2, scaffold proteins on which the complex for phosphorylation of β-catenin by GSK3β is assembled, have been detected in 5–10% and 3% of HCCs respectively [100, 101]. Upregulation of Frizzled-7 receptor (FZD7), which promotes nuclear accumulation of β-catenin and simultaneous activation of canonical Wnt signaling has also been observed in subsequent numbers of HCCs analyzed [94]. HBV infection is one of the major risk factors for development of hepatocarcinogenesis. HBV contributes towards the oncogenesis by expression of HBV-encoded X antigen (HBx). HBx transcriptionally upregulates the expression of a unique gene, URG11, which in turn transcriptionally upregulates β-catenin thereby contributing towards HCC [102]. Loss-of-function of Wnt signaling antagonist, secreted frizzled-related protein (SFRP), due to frequent methylation and thereby silencing also contributes towards constitutive action of Wnt signaling [103].With evidence of the role of Wnt/β-catenin signaling pathway towards hepatocarcinogenesis, pharmacological inhibition of β-catenin has been shown to decrease the survival of hepatoma cells [104]. Further, inactivation of β-catenin suppressor APC leads to spontaneous development of HCC in mice model, envisaging the direct contribution of activated Wnt signaling towards HCC development [105]. However, mice over-expressing a gain-of-function β-catenin mutant (exon 3 deletion) did not show spontaneous liver tumor but had increased susceptibility to HCC development only after exposure to a carcinogen diethylnitrosamine (DEN) [106]. Thus, overall, although there is little doubt about the strong correlation between aberrant Wnt signaling and HCC, due to the inherent complexity of this pathway in the liver, precise role of this pathway in the pathogenesis of liver tumor is yet to be fully understood for further implications towords therapeutic inhibition.

5.4.3 NFκB pathway

NFκB is a family of transcription factors that includes RelA (p65), NFκB1 (p50 and p105), NFκB2 (p52 and p100), c-Rel and RelB. In non-stimulated

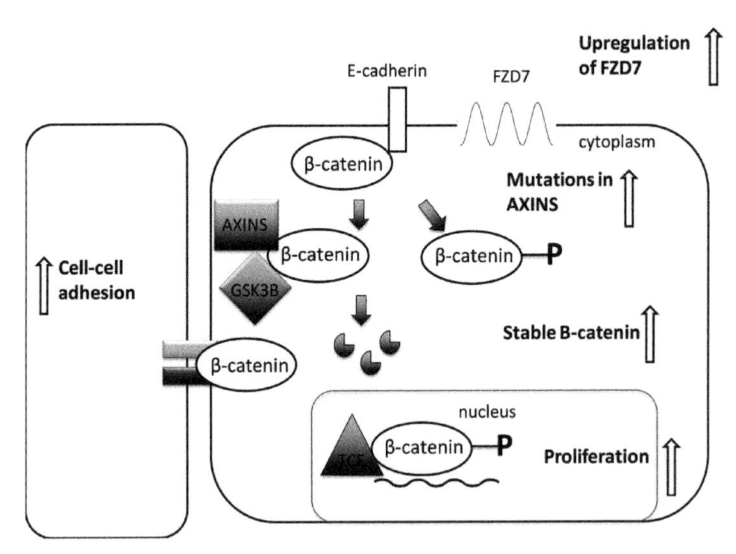

Figure 5.2 Deregulation of β-catenin in hepatocellular carcinoma: Upregulation of FZD7, mutations in AXINS, production of stable beta-catenin and increase in cell-cell adhesion have been postulated as plausible mechanisms for deregulation of beta-catenin pathway in HCC. FZD7: Frizzled-7 protein; GSK3B: Glycogen Synthase kinase 3 beta; TCF: Transcription factor.

cells, these factors are retained in the cytoplasm by IκBs that prevent NFκB activation [107], and inhibits nuclear accumulation. In liver, in response to environmental challenges that induce the production of pro-inflammatory stimuli such as tumor necrosis factor (TNF) or interleukin 1β (IL-1β) or other hepatotoxic cytokines, the IκB kinase (IKK) complex, composed of the IKKα and IKKβ catalytic subunits and the IKKγ regulatory subunit is activated, resulting in IκB phosphorylation and eventual ubiquitin-mediated degradation of IκBs which facilitates the migration of NFκB into the nucleus [108]. Here, they typically form heterodimers that bind to the promoters of many immune response genes and activate their transcription [109]. Thus, this pathway plays an important role in the hepatocytes leading to the activation of pro-survival mediators [110–112]. Constitutive activation of this pathway has been reported in HCCs, with the first evidence of the role of NFκB in HCC reported by Pikarsky *et al.* using animal models, Mdr2-knockout mice, and an

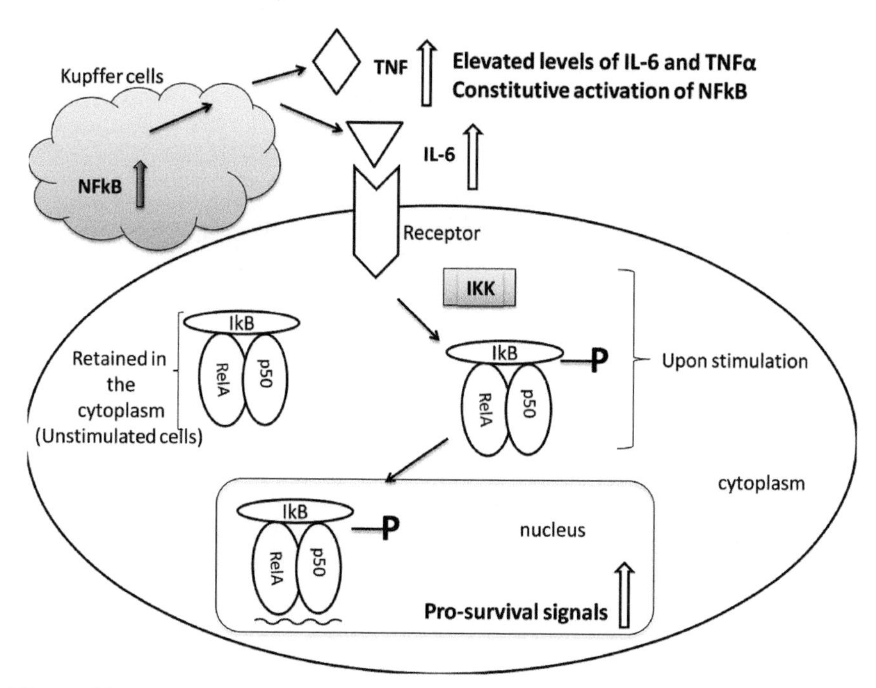

Figure 5.3 Deregulation of NFkB in hepatocellular carcinoma: Elevated levels of IL-6 and TNFα have been postulated in constitutive activation of NFkB, thereby, deregulation in hepatocellular carcinoma due to upregulation of pro-survival signals. NFkB:Nuclear factor kappa-light chain-enhancer of activated B cells; TNF: Tumor necrosis factor; IL-6: Interleukin-6; IKK: IκB kinase.

inflammation-driven model of HCC. It was shown that hepatocytes adjacent to regions of inflammation in this model had high levels of nuclear NFκB [113, 114]. Similarly, lack of NFκBsignaling strongly decreased chronic hepatitis and prevented hepatocarcinogenesis in another liver-tumor model [115]. However, another study using DEN-induced HCC model shows that hepatocyte-specific deletion of IKKβ markedly increases hepatocarcinogenesis, thus inhibition of NFκB could promote carcinogenesis [116]. Thus, *in vivo* studies have resulted in conflicting conclusions about the suggested role of NFκB. These contradictory results could emerge due to differential effects of NFkB activation between hepatocytes and liver non-parenchymal cells, including Kupffer cells. Kupffer cells produce a panel of inflammatory cytokines and growth factors in an IKK/NFκB-dependent manner. Authors for DEN model later demonstrated that deletion of IKK2 in both hepatocytes and Kupffer cells suppressed DEN-induced hepatocarcinogenesis, indicating that IKK2-deficency in Kupffer cells can suppress the enhanced liver cancer caused by the loss of IKK2 in hepatocytes [117]. Thus, NFκB activity in Kupffer cells contributes towards carcinogenesis. The mechanism by which DEN-induced administration leads to NFκB activation in Kupffer cells was found to be dependent on the release of IL-1α by necrotic hepatocytes which activates an MyD88-dependent signaling pathway upon binding to IL-1 receptor (IL-1R) on Kupffer cells [118]. One of the most important NFκB-dependent cytokines produced by activated Kupffer cells is IL-6. Elevated levels of IL-6 have been found in HCC patients [119, 120]. The precise role of elevated IL-6 levels and HCC development is, however, still elusive. Therefore, further studies are required to better understand the role of NFκB-dependent signals towards hepatocarcinogenesis.

5.4.4 PI3K/AKT/mTOR pathway

Signaling through the PI3K/Akt/mTOR pathway is initiated by mitogenic stimuli from growth factors like IGFR (Insulin-like growth factor), PDGFR (Platelet- Derived Growth Factor Receptor), EGFR (Epidermal Growth Factor Receptor) or cytokines stimulation. This results in the subsequent activation of receptor tyrosine kinases (RTKs) and G protein-coupled receptors and Phosphoinositide 3-kinases (PI3Ks) are recruited to the membrane by direct or indirect interaction with the activated receptors. PIKs then phosphorylate the hydroxyl group of phosphoinositol of the cell membrane [121]. These phosphorylated components bind to a protein called Akt. Akt, a serine-threonine kinase then becomes phosphorylated and activated [122]. Activated Akt phosphorylates multiple protein substrates and regulates a variety of critical cellular activities like cell survival, proliferation, cell cycle progression, migration and angiogenesis [122, 123]. Akt then induces the activation (phosphorylation) of mTORC1, a serine/threonine kinase. Mammalian target of rapamycin (mTOR) is one of the most important downstream effectors of Akt. Once activated, it regulates the translation of several important proliferative and angiogenic factors, such as c-myc, cyclin D1, Hypoxia Inducible Factor 1α (HIF-1α) and Vascular Endothelial Growth Factor (VEGF) [124, 125]. While this pathway is negatively regulated by Phosphatase and Tensin Homolog (PTEN), a tumor suppressor, which dephosphorylates phosphatidylinosol triphosphate [122]. Other than PTEN, there are several negative regulators, such as SH2-containing inositol phosphatase-1 (SHIP1) and PITenins (PITs) [126, 127].

Dysregulation of this pathway has been observed in several human malignancies including HCC [128–131]. In HCC, cumulative evidences have shown that several components of this signaling pathway are frequently altered, resulting in constitutive activation of this pathway. Activation of receptor kinases is believed to be one of the key mechanisms for constitutive activation of this pathway with reports indicating overexpression of c-Met, EGFR and IGF1-R [132, 133]. Activation of IGF axis has been observed in 21% of early HCCs [134]. HCV or HBV infection of hepatocytes also contributes towards aberrant activation of this pathway in HCC. The hepatitis B virus-X (HBx) protein is known as a multi-functional protein that not only activates transcription of viral and cellular genes but also coordinates the balance between proliferation and programmed cell death, by inducing or blocking apoptosis [135]. Hepatitis B virus, unlike other DNA tumor virus that blocks apoptosis in a p53-dependent manner, achieves protection from apoptosis via the HBx-PI3K-Akt-Bad pathway and by inactivating caspase-3 activity in a p53-independent way [136]. This is achieved by suppression of PTEN, a negative regulator of PI3/Akt/mTOR pathway.

Also, this PTEN suppression of HBx-mediated cell survival is PI3K pathway specific, since PTEN does not suppress the effect of HBx on protection from Fas-mediated apoptosis [136]. HCV has also been reported to increase the expression of neuroblastoma (N)-Ras in HCC cells, which then activates the PI3K pathway [137]. Dysregulation due to genetic alterations has also been indicated to play a role in hepatocarcinogenesis. Somatic loss of PTEN by gene mutation/deletion has been found to play an important role in tumorigenesis and progression of HCC by Akt activation. Thus, in HCC patients, reduced

PTEN expression has been associated with advanced tumor stage, high recurrence rate and poor survival outcome, suggesting inactivation of PTEN is involved in the pathogenesis of HCC [138]. Besides PTEN, mutations of PIK3CA, the PI3K catalytic subunit α-isoform gene encoding p110α are also found in HCC. These mutations have been shown to increase PIP3 levels, which activate Akt and induce cellular transformation [139, 140]. Studies have also shown overexpression of phosho-mTOR in 15% of liver tumors [141]. Elevated Akt phosphorylation in HCC patients has been associated with

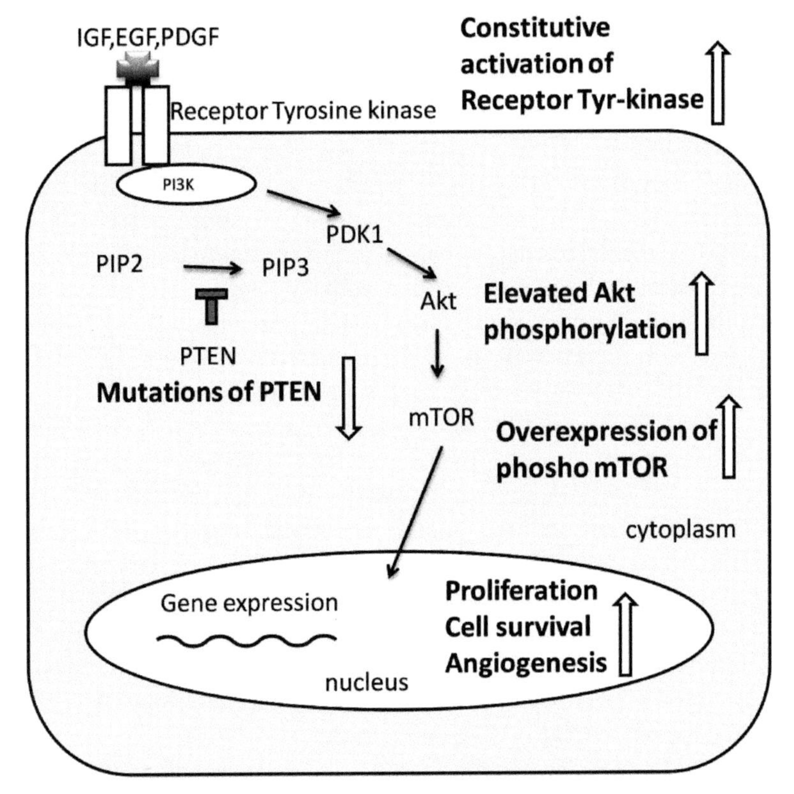

Figure 5.4 Deregulation of PI3K/Akt/mTOR pathway in hepatocellular carcinoma: Elevated levels of Akt phosphorylation, overexpression of phoshomTOR, somatic mutations of PTEN have been postulated to play a plausible role in deregulation of PI3K/Akt/mTOR pathway in hepatocellular carcinoma. PTEN : Phosphatase and tensin homolog; PI3K : Phophoinositide-3-kinase; IGF : Insulin-like growth factor; EGF : Epidermal growth factor; PDGF : Platelet derived growth factor; PDK1 : Phosphoinositide-dependent kinase-1; PIP2: Phosphophatidylinositol 4, 5-biphosphate; PIP3:Phosphophatidylinositol 3, 4, 5-triphosphate; Akt: Protein kinase B.

early recurrence and poor prognosis [142]. Hence, multiple events induce the activation of the PI3K/Akt/mTOR pathway in HCC development. Also, given the strong correlation between aberrant PI3K/Akt/mTOR signaling and HCC, pharmacological inhibition of this pathway is a viable HCC treatment strategy.

5.4.5 MAPK pathway

Mitogen-activated protein kinase (MAPK) is serine-threonine kinase that is involved in a variety of cellular activities. There are at least four subfamilies of MAPKs: extracellular signal-regulated kinase 1 and 2 (ERK1/2), Jun N-terminal kinases (JNKs), p38 MAPKs and ERK5 [143, 144]. Among these, ERK signaling has been most extensively studied for its involvement in cellular proliferation, migration and survival. MAPK signaling pathways are often deregulated in various types of human tumors, including HCC. Activation of the ERK pathway is important in HBV and HCV infection associated hepatocarcinogenesis. The viral proteins activate the MAPK signaling towards liver tumorigenesis [145]. Also, phosphorylated ERK levels have been found in significant number of human HCC samples [146]. Many genes downstream of ERK pathway activation have been shown to be upregulated in precancerous nodules and HCCs in rodents. Abundant evidences have demonstrated that the ERK pathway increases the expression of pro-survival factors, such as BCL-2, BCL-xL and MCL-1, which are largely dependent on cAMP response element-binding {CREB}-mediated transcription [147]. ERK1 and ERK2 are highly homologous serine and threonine kinases that display 88% similarity in protein sequences and share overlapping regulatory mechanisms in many aspects. According to Schmitz *et al.* [148] increase in phosphoERK1/2 (pERK1/2) and AKT expression in HCC indicates aggressive tumor behaviour. ERK signaling pathway consists of a highly conserved three-tier cascade, which typically transmits signals from the cell membrane to the nucleus. Indeed, upon activation by the MAPK kinase MEK1 and/or MEK2, the MAPK ERK1 and ERK2 are translocated into the nucleus and phosphorylate with a variety of substrates. Due to a high sequence homology, the MEK1 and MEK2 kinases, as well as ERK1 and ERK2, have been regarded as redundant isoforms. Despite the similar structures and functions between ERK1 and ERK2, studies using ERK1- and ERK2-deficient cells or mice have demonstrated different functions of these two kinases. ERK-2 deficient mice are lethal at the embryonic stage of mice development in contrast to ERK-1 deficient mice which are viable, fertile and appear normal. Thus, elucidating the specific role of the two kinases in HCC development is of importance since ERK/MEK is considered as a therapeutic target with a future potential of improved efficiency and specificity of the drug for HCC therapeutics. One of the recent studies has shown a predominant role of ERK1 over ERK2 in human hepatocellular carcinoma cells with cisplatin-induced cell death. Further studies envisage distinct roles of the two isoforms towards modulating cell survival and proliferation. The development of therapeutic tools allowing specific targeting of ERK2 could thus be beneficial in order to inhibit cell proliferation, without affecting ERK1 pro-apoptotic activities [149, 150].

c-Jun N-terminal kinase (JNK) is another MAPK signaling pathway which can be activated by two MAPK kinase (MKKs) MKK4 and MKK7 and its downstream substrates include c-Jun [144]. This pathway can be activated by various cytokines and environmental factors. There is a significant correlation between activated JNK signaling and hepatocarcinogenesis. One of the studies has shown JNK-1 over-activated in 17 out of 31 samples of Chinese HCC patients [151]. Direct evidence demonstrating the role of JNK pathway in HCC development has been shown in mouse models, wherein, JNK1 knockout mice had significant reduction in liver tumorigenesis which was chemically induced by DEN [152]. Activation of p38 pathway is induced by MKK3, 4 and 6, as well as autophosphorylation. Its substrates involve transcription factors like p53 [143]. Unlike, ERK and JNK pathways, p38 negatively regulates hepatocarcinogenesis. Direct evidence of the role of p38 in HCC came from liver-specific deletion of p38 mice models. These mice had enhanced hepatocyte proliferation and tumor development [152]. In addition, after DEN treatment, these mice developed more tumors in the liver. In summary, several studies demonstrate the role of MAPK pathways involvement during liver carcinogenesis. Identification of pharmacological inhibitors would help develop promising treatment modalities for combating HCC.

5.4.6 VEGF pathway

Angiogenesis is the fundamental event in the process of tumor growth and metastatic dissemination. Across most of the cancer, sustained angiogenesis is considered as one of the "central hallmarks of cancer" [153]. VEGF signaling pathway promotes angiogenesis via activation of the VEGFR (VEGF receptor). VEGF family consists of five members: VEGF-A (thereafter called VEGF), VEGF-B, VEGF-C, VEGF-D, and placental growth factor. Members of the VEGF family show different affinities for one of the three

VEGF tyrosine kinase receptors: VEGF receptor (VEGFR)-1, VEGFR-2, and VEGFR-3. The angiogenic effects of VEGF are achieved by binding to its receptors, VEGFR1 (Flt-1) and VEGFR2 (Flk-1), expressed on endothelial cells. Interaction between VEGF and VEGFR initiate several signaling pathways, such as Akt/PI3K/MAPK pathway, resulting in proliferation, migration, and invasion of endothelial cells [154]. In addition to endothelial cells, VEGFR expression has also been found in tumor cells, proposing an autocrine loop of VEGF/VEGFR in tumor cells [155]. Several reports suggest that VEGFR-2 is one of the major mediators of VEGF-driven responses in endothelial cells and is considered a crucial signal transducer in both physiologic and pathologic angiogenesis [156, 157]. HCC is a hyper-vascular tumor and many pro-angiogenic factors play an important role towards promoting angiogenesis. Both VEGFR-1 and VEGFR-2 play a pivotal role in augmenting HCC tumorigenesis and angiogenesis by activation of VEGF signaling cascade [158]. A higher level of VEGF expression in HCC patients is also associated with a higher proliferation index and poor encapsulation of tumors. In addition to VEGFR-1 and VEGFR-2, VEGF also interacts with the members of the neuropilin family to regulate angiogenesis [159]. These neuropilin members act together with VEGFR-1 and VEGFR-2 mediating signal transduction. High serum VEGF levels have been shown to correlate with poor response to chemotherapy and poor survival among HCC patients [123]. The VEGF promoter contains a hypoxia-responsive element, which is regulated by HIF-1, a transcription factor. HIF-1α mediated upregulation of VEGF is pivotal for hypoxia-induced angiogenesis [160]. Higher levels of HIF-1α have been associated with poor prognosis of HCC [161]. Besides upregulation of VEGF by several upstream events, VEGF can directly function on HCC since it is also expressed on the tumor cells due to the presence of autocrine loop in tumor cells which results in tumor cell proliferation [162]. The VEGF pathway is clearly important for HCC pathogenesis. The most direct evidence supporting the role of VEGF in HCC came from using anti-VEGF monoclonal antibody in patients with unresectable HCC which resulted in significant disease-stabilizing effect [163]. Also, there are several investigations towards using VEGF small molecule inhibitors for HCC treatment [123].

5.4.7 Other miscellaneous pathways

Several other signaling pathways also play a crucial role towards hepatocarcinogenesis. These include Epidermal Growth Factor Receptor (EGFR), p53,

Transforming Growth Factor (TGF)-β, apoptotic and Hedgehog pathway. The protein p53 is implicated in the control of cell cycle, apoptosis, DNA repair and angiogenesis. p53 mutations is a crucial event in the carcinogenesis of more than 50% of human cancers [164]. HCC development involves exposure to several extrinsic factors such as aflatoxins, viral infection, nutrition and alcohol intake which initiates the process of tumorigenesis. Alteration of the p53 status is an important intrinsic factor in this process. Inactivation of wild type p53 by point mutations, allelic deletion or complex formation with cellular or viral protein, plays a crucial role towards progression of liver cancer [165]. The EGFR is a receptor tyrosine kinase which is activated by ligands including epidermal growth factor (EGF) and transforming growth factor α (TGF- α) [166]. Upon activation, this pathway regulates several downstream signal transduction cascades like growth, survival, proliferation and differentiation in mammals. Aberrant EGFR signaling plays an important role in the tumorigenesis of several cancers including HCC [167]. EGFR is frequently overexpressed in HCC [168]. Higher expression of EGFR has been associated with more aggressive pathological features, increased tumor cell proliferation and reduced tumor cell apoptosis. Several anti-EGFR or EGFR blocking agents are being investigated in clinic, along with combination therapeutics to assess the effect and benefits of EGFR blocking for HCC [169]. TGF-β superfamily signaling plays a critical role in the regulation of cell growth, differentiation and development. Signaling is initiated with ligand-induced oligomerization of serine/threonine receptor kinases and thereby phosphorylation of the cytoplasmic signaling molecules Smad2 and Smad3. Activated Smad regulate diverse biological effects by partnering with transcription factors resulting in cell-state specific modulation of transcription [170]. Alterations in the TGF- β signaling pathway, including mutation or deletion of members of the signaling pathway and resistance to TGF-β mediated inhibition of proliferation are frequently observed in human cancers [171, 172]. In HCC, TGF-β regulates several steps in tumor progression, angiogenesis, production of the extracellular matrix and immune suppression [173]. It is also involved in the early stages of tumor progression and epithelial mesenchymal transition (EMT). Once the EMT is established, tumor nodes most likely arise in the liver tissues and is associated with higher TGF-β expression [173]. Furthermore, higher levels of TGF-β are present in HCC patients with aggressive and metastatic HCC in comparison with non-metastatic HCC. TGF-β1 can be considered as a hallmark of HCC because it is increased in the serum, tissue and urine of HCC patients and its increased levels correlate with tumor progression and survival [174, 175]. Hence, owing to its importance in heapatocarcinogenesis, curative therapies targeting TGF-β signaling have been investigated that blocks hepatocarcinoma growth [176]. The Hedgehog (Hh) signaling pathway plays an important role in embryonic development, tissue polarity and cell differentiation [177, 178]. This pathway is critical in the early development of the liver and contributes to differentiation between hepatic and pancreatic tissue formation, but the adult liver normally does not have detectable levels of hedgehog signaling [179, 180]. Activation of this pathway is shown to be involved in several types of gastrointestinal cancers. It has been implicated to be activated in HCC, playing an important role in HCC development and invasion [178, 181]. Blockade of Hh signaling results in decreased proliferation and migration and an increase in apoptosis in human HCC cell lines [182, 183]. With further studies required to understand the mechanism involved, it is an essential pathway for the development of hepatocarcinogenesis.

Thus, with recent identification of several key molecular pathways implicated in the pathogenesis of HCC, it has led to the development of new targeted therapies for this cancer. Combination therapy with either conventional cytotoxic drugs or another inhibitor which targets a specific molecule in a different signal transduction pathway is also a key approach for improving the effectiveness and usefulness of new molecular-targeted agents. Future work would involve identifying new molecular targets and assessing the role of targeted therapy in the adjuvant, neo-adjuvant and metastatic setting as well as determining the various combinations of treatment for hepatocellular carcinoma.

5.5 Reported Anti-Cancer Effects of Natural Compounds Against Hepatocellular Carcinoma

Hepatocellular carcinoma (HCC) is the most common form of primary hepatic carcinoma [184] and a serious medical problem especially in Asia and sub-Saharan Africa [72]. HCC has a poor prognosis with the number of deaths almost equal to the number of cases being diagnosed annually and the 5-year survival rate is below 9% [185]. The American Cancer Society [186] estimated that in 2010, more than 24,000 new cases and nearly 19,000 deaths occurred in the United States due to liver cancer. Treatment of HCC has been divided into curative and palliative. Curative treatments, such as resection, liver transplantation and percutaneous ablation, are expected to improve survival in a high proportion of patients. On the other hand, palliative treatments are not aimed to cure, but in some cases are known to have good response rates and even improve survival [187]. Phytochemicals, obtained from fruits, vegetables, nuts and spices, have drawn a great amount of attention due to their ability to selectively kill tumor cells and suppress carcinogenesis in pre-clinical animal models [188–191]. A large number of these plant-derived substances have been shown to significantly prevent the development of cancer in several high risk populations [192]. Based on various *in vitro* experiments and studies involving animal models as well as humans, it is evident that phytochemicals possess chemopreventive and therapeutic effects in liver cancer [6, 18, 193]. Since limited treatment options are currently available to patients with liver cancer, novel preventive and effective therapeutic approaches should be considered to combat this disease. In this section, we discuss few important natural compounds that have shown significant potential in both the prevention and the treatment of hepatocellular cancer as well as examine a number of studies conducted both *in vitro* and *in vivo*.

5.5.1 Curcumin

Curcumin is a constituent of the spice turmeric and possesses potent antioxidant and anti-inflammatory activities [194, 195]. Curcumin is shown to down-regulate the expression of various pro-inflammatory cytokines, tumor necrosis factor alpha (TNFα), vascular endothelial growth factor (VEGF), interleukin-1, 2, 6, 8, 12 by inactivation of the NFκB while it activates p38 mitogen-activated protein kinases (MAPK) [196]. One of the crucial findings is that curcumin suppresses NFκB, the master switch in the inflammatory cascade [197]. NFκB activation regulates several key inflammatory mediators such as cytokines, chemokines and kinases, which play a role in the pathogenesis of most chronic illnesses [198, 199]. Curcumin-mediated inhibition of the NFκB inflammatory cascade is an extremely critical mechanism of its therapeutic profile [200]. Additionally, curcumin suppresses the matrix degrading capacity of metalloproteinase family (MMPs), especially MMP-2 and MMP-9, which are involved in tumor angiogenesis. The anti-invasive and anti-migratory effects of curcumin have been shown in CBO140C12 (murine hepatocellular carcinoma cell line) cells. Ohasi and colleagues [201] presented that curcumin-mediated decrease in MMP-9 secretion was accompanied by a significant inhibition of the adhesion and migration of fibronectin and laminin. Also, Kang and coworkers [202] observed an increase in ROS levels and a decrease in histone acetyltransferase activity in Hep3B cells treated with curcumin. Furthermore, Cao *et al.* [203] studied HepG2 cells and showed that curcumin produced mitochondrial hyperpolarizaton, elevated mitochondrial membrane potential and also increased cytochrome c release. Cui *et al.* [204] demonstrated that the anti-proliferative effects of curcumin were accompanied by apoptosis and reduction of telomerase activity in SGC7901 as well as HL60 cells. Curcumin also inhibited cell survival in hepatocellular carcinoma cells and the expression of the vascular endothelial growth factor-A was observed to be down-regulated [205]. In a study by Wang *et al.*, it was found that curcumin could inhibit the growth of HepG2 cells, cause a change in the cell-surface morphology and promote cell apoptosis by inducing pro-apoptotic factors. In addition, with increasing doses of curcumin, the cell viability decreased and the 50% inhibiting concentration was $17.5 \pm 3.2 \mu M$ [206]. Cheng *et al.* treated J5 cells with various concentrations of curcumin for different time points and it was observed that curcumin inhibited the proliferation of J5 cells in a time and dose–dependent manner. Curcumin induced the GADD153 expression by causing the cleavage of caspase-12 and nuclear translocation of ATF6. Curcumin also down-regulates the expression of Mcl-1 and Bcl-2, thereby inducing mitochondrial dysfunction. In addition, curcumin induced cell cycle arrest at the G2/M phase by decreasing the Cdc2 expression. In conclusion, curcumin inhibits the proliferation of J5 cells by inducing endoplasmic reticulum stress and mitochondrial dysfunction [207]. Moreover, Ning and coworkers [208] treated HEP3B, SK-Hep-1 and SNU449 cells with curcumin which caused the down-regulation of the Notch1 signaling pathway as well as the Notch intracellular domain. Nitrosoamines and diethylnitrosoamine (DENA), either used alone or in the presence of a promoter such as phenobarbital (PB) have been a cornerstone of pre-clinical liver cancer research [209, 210]. Several studies have investigated the pharmacological effects of curcumin and its analogs in DENA-induced HCC. Chuang *et al.* [211] showed that curcumin provided protection against DENA-induced hyperplasia and HCC in rodents. A few investigators have also studied the therapeutic effects of curcumin in xenograft models of hepatocarcinogenesis. Busquets *et al.* [212] demonstrated that curcumin treatment blocked tumor growth in the Yoshida ascites hepatoma in rats [213]. Thus, it can be suggested that antioxidants like curcumin have gained immense importance for their anti-carcinogenic activities and minimum toxic manifestations in biological system. However, a drawback is that curcumin is lipophilic and thus following oral administration hardly appears in bloodstream indicating its potential therapeutic challenge in cancer therapy. Thus, nanocapsulated curcumin has been used as a drug delivery vector to focus the effectiveness of these vesicles against hepatocellular carcinoma and in combating the oxidative damage of hepatic cells [214].

5.5.2 Selenium

Dietary selenium (Se) is an essential mineral for both humans and animals. It functions as a component of selenoproteins which include glutathione peroxidases (GPx), thioredoxinreductases, iodothyroninedeiodinases, selenophosphatesynthetase, selenoprotein P, and selenoprotein W [193]. An inverse relationship exists between Se intake and cancer risk, thus, increasing the intake of Se has been suggested as a way to prevent the development of some forms of cancer in humans. Se supplementation trials have been conducted to determine whether Se is effective in reducing liver cancers in human. The first trial investigated the preventive effect of Se on primary liver cancer and discovered that Se supplementation using table salt prepared with sodium selenite resulted in a 50% decrease in the primary liver cancer cases [215]. Another study showed that salt supplementation prepared with Se reduced the rate of hepatitis, a risk factor of primary liver cancer [216]. Yu *et al.*[215] reported a significant decrease in primary liver cancer among those receiving Se yeast compared with controls. Using N-nitrosobis (2-oxopropyl) amine to induce liver tumors in hamsters, Lee *et al.*[217] found that Se injected i.p. inhibited the volume and area of tumor foci.

5.5.3 Epigallocatechin-3-gallate (EGCG)

EGCG and other polyphenols present in green tea have antioxidant activity and also prevent oxidation by chelating metal ions such as iron or copper [218]. EGCG has also been found to modulate signal transduction pathways that inhibit cell proliferation and increase apoptosis [219]. Most studies in models have demonstrated tea extracts and EGCG to inhibit liver carcinogenesis. Both green and black teas were found to inhibit the induction of hepatic tumors in mice [220]. Tong *et al.* investigated the cytotoxic effect of epigallocatechin-gallate on human hepatocellular carcinoma cell line, HepG2 cells. EGCG induced apoptosis in the HepG2 cells in a time and concentration dependent manner. Annexin V/PI assay demonstrated that early apoptosis increased upon increase in concentration, and late apoptosis also increased, when treated with a high concentration of EGCG [221]. The receptor tyrosine kinase vascular endothelial growth factor receptor (VEGFR) plays an important role in tumor angiogenesis of hepatocellular carcinoma (HCC). In a study by Shirakami *et al.*, the effects of EGCG on the activity of the VEGF-VEGFR axis in human HCC cells were examined. EGCG preferentially inhibited the growth of HuH7 cells, which express constitutive activation of the VEGF-VEGFR axis. Treatment of HuH7 cells with EGCG caused a time and dose dependent decrease in the expression of VEGFR-2 and p-VEGFR-2 proteins. Moreover, drinking EGCG significantly inhibited the growth of HuH7 xenografts in nude mice and this was associated with inhibition of the activation of VEGFR-2 and its related downstream signaling molecules such as ERK and Akt. EGCG drinking also decreased the expression of Bcl-XL protein and VEGF mRNA in the xenografts. These findings suggest that EGCG can exert its growth-inhibitive effect on HCC cells by inhibiting the VEGF-VEGFR axis [222]. Hepatitis C virus (HCV) is a major cause of liver cirrhosis and hepatocellular carcinoma. Current anti-viral therapies fail to clear infection in a large number of cases. Individuals undergoing orthotopic liver transplantation face rapid, re-infection of the graft. Therefore, there is an urgent need for anti-viral strategies that target the early stages of infection. In a study by Ciesek *et al.*, EGCG inhibited the entry of HCV into hepatoma cell lines as well as primary human hepatocytes. Green tea catechins, such as EGCG and its derivatives, epigallocatechin (EGC), epicatechingallate (ECG), and epicatechin (EC), have been previously found to exert anti-viral and anti-oncogenic properties. Furthermore, it was shown that EGCG inhibits viral attachment to the cell, thus disrupting the initial step of HCV cell entry. Thus, EGCG potently inhibits HCV entry and could be part of an anti-viral strategy aimed at the prevention of HCV re-infection after liver transplantation [223].

5.5.4 Resveratrol

Resveratrol (3,5,4'-trihydroxystilbene) is a phytoalexin found in grapes that has shown to have chemopreventive activity [224, 225]. Resveratrol can slow down the progression of a wide variety of inflammation-related diseases, including cancer. An extensive number of experimental findings reveal that resveratrol affects cellular proliferation and growth, apoptosis, inflammation, invasion, angiogenesis and metastasis [226]. A significant amount of resveratrol accumulates and is retained in the liver [227, 228] and inhibits the hepatic carcinogen-activating enzymes, such as cytochrome P450 1A1 (CYP1A1) and CYP3A/2 and induce hepatic phase 2 conjugating enzymes, namely NAD(P)H:quinineoxidoreductase, UDP-glucuronosyltransferase and glutathione S-transferase (GST) *in vitro* and *in vivo* [229, 230]. The resulting effects of these enzyme modulations by resveratrol could be the reduction of exposure of cells to carcinogens due to inhibition of carcinogen activation or elevated carcinogen detoxification and elimination. The key property of resveratrol, with regards to liver cancer, is its strong anti-inflammatory [231] and antioxidant properties [2, 232, 233]. According to a study conducted by Delmas *et al.* [234] the proliferation of human HepG2 cells were negatively impacted by the addition of resveratrol to the culture medium in both dose and time-dependent fashion. Resveratrol has the ability to prevent or delay the cells from entering mitosis thereby increasing the number of cells arrested in the S and G2/M phase. Hepatic growth factor (HGF) has been implicated in the ability of primary hepatic tumors to proliferate and invade the adjacent tissue. The effects of resveratrol on HGF-mediated invasion were determined in HepG2 cells to understand the mechanisms of resveratrol's anti-HCC property. Resveratrol was found to decrease HGF-induced scattering and invasion of liver cancer cells with simultaneous inhibition of cell proliferation [235]. Kozuki *et al.* [236] demonstrated that resveratrol inhibited both the proliferation and invasion of AH109A rat ascites hepatoma cells at higher concentrations. Moroever, Yu *et al.* [237] showed that resveratrol inhibited tumor necrosis factor-α-mediated MMP-9 expression and invasion of HepG2 cells. The inhibitory effects of resveratrol were also associated with the down-regulation of NFκB signaling pathway. Subsequent studies also confirmed the involvement of the antioxidant property of resveratrol as intake of resveratrol was found to suppress ROS-potentiated invasion of AH109A cells [238]. Upon treatment with 20–80 micromol/L resveratrol for 24 h, 48 h or 72 h, the proliferation of Hepa 1–6 cells was significantly inhibited in a time and dose-dependent manner. 20–80 micromol/L resveratrol also induced apoptosis and apoptotic morphology change in Hepa 1–6 cells accompanied with caspase-3 activation and ROS generation [239]. Another proposed mechanism of resveratrol's ability to restrict hepatoma cell invasion was studied by Zhang *et al.* [240]. By inducing hypoxia in HepG2 cell line, researchers discovered that resveratrol had an inhibitory effect on vascular endothelial growth factor gene expression via hypoxia-inducible factor-1α inhibition. Kocsis*et al.* [241] reported the effects of resveratrol on cytotoxicity, cell proliferation activity, and apoptosis in HepG2 cells in a time and concentration dependent manner. Cytotoxicity was observed at resveratrol concentration of 50 or 100 μM with treatments longer than 48 h. Moreover, cell cycle analysis showed an increase of S-phase cells at low concentrations of resveratrol (10–50 μM) and a decrease at high concentrations (100–200 μM). The ratio of apoptotic cells increased following resveratrol treatment. It has been theorized that hepatoma cell lines are more likely to undergo apoptosis rather than necrosis when exposed to resveratrol due their ability to quickly metabolize the phytochemical. Notas *et al.* [242] also found that apoptosis through cell cycle arrest is the main mechanism by which resveratrol interferes with HepG2 cell proliferation. Several studies have investigated the anti-tumor potential of resveratrol in animal models of liver cancer. A study revealed that dietary resveratrol suppressed the growth and metastasis of AH109A ascites hepatoma cells implanted into Donryu rats [243]. It was observed that tumor growth not only stopped after 12 days in rats exposed to resveratrol, but the tumors began to regress. Resveratrol was also shown to have inhibitory effects on the growth of H22 tumor cells transplanted in mice due to its effects on non-specific host immunomodulatory activity [244]. In another study, external tumors were developed by injecting H22 cells into the groin of BALB/c mice. Subsequently, tumor tissue was xenografted into the liver and allowed to grow [245]. Resveratrol treatment was found to restrain hepatic tumor growth through reduced expression of cell cycle proteins such as cyclin B1 and p34cdc2 [209]. It is known that cyclin D1 is overexpressed in liver cancers. Parekh *et al.* determined that resveratrol treatment down-regulated cyclin D1 as well as p38 MAP kinase, Akt and Pak1 expression and activity in HepG2 cells, suggesting that growth inhibitory activity of resveratrol is associated with the downregulation of cell proliferation and survival pathways. Thus, resveratrol has good potential as effective chemopreventive agent against liver cancer and requires further studies [246]. Resveratrol is rapidly absorbed following oral administration and levels are detectable in both plasma and urine with the maximum plasma concentrations being reached within an hour post-administration. Thus, it becomes apparent that resveratrol may play an important role not only in the prevention but also in the therapy of metastatic disease of the liver.

5.5.5 Ursolic Acid

Triterpenoids are a class of naturally occurring compounds that are found in a wide variety of European plants and fruits. They are popular agents for their anti-inflammatory, hepatoprotective, analgesic, anti-microbial, anti-tumor, immunomodulatory and tonic effects. Ursolic acid (3ß hydoxy urs-12 en – 28 oic acid) is a triterpenoid found in higher plants like Rosmarinusofficianalis. It is shown to lower COX-2 transcription [247] and it has been studied for its beneficial effects on liver [248] and anti-hepatoma activity in mice. In recent years, much research has focused on identifying the plant components which arrest the oxidative stress and free radical induced damage [249].

Yu *et al.* investigated the inhibitory effect and mechanisms of UA on the human hepatoma cell line, SMMC-7721. After treatment of cells with UA, the proliferation of SMMC-7721 cells was significantly inhibited in a dose and time dependent manner. UA also induced cell cycle arrest and apoptosis causing formation of apoptotic bodies and DNA fragmentation. The result of Western blotting showed the apoptotic proteins p53 and Bax to be upregulated while the anti-apoptotic proteins Bcl-2 and Survivin were down-regulated [250]. Moreover, UA was found to inhibit the proliferation of doxorubicin-resistant HepG2 cell line through apoptosis by externalization of phosphatidyl serine and loss of mitochondrial membrane potential. Furthermore, the death of HepG2 cells induced by UA was found to be mainly through the caspase-independent apoptosis-inducing factor (AIF) signaling pathway. In addition, an animal study showed that UA was effective against HepG2 cells *in vivo* with negligible body weight loss and damage of the liver, heart and spleen [251, 252].

5.5.6 Pterostilbene

Pterostilbene (*trans*-3,5-dimethoxy-4'-hydroxystilbene) is a natural dimethylated analog of resveratrol obtained from blueberries and is known to have diverse pharmacologic activities such as anti-cancer, anti-inflammation, antioxidant and analgesic activity [253]. Hepatocellular carcinoma is the most common malignant and invasive tumor in the liver, and the prognosis of patients with this type of cancer is determined by the occurrence of invading metastasis into the remaining liver parenchyma [254]. It was reported that pterostilbene significantly suppressed MMP-9 gene expression via blocking the protein kinase C/MAPK, PI3K/NFκB and AP-1 signaling pathways thereby reducing invasion and metastasis of HepG2 cells. *In vitro* invasion and migration assays were used to investigate the inhibitory effect of pterostilbene on the invasive potency of human hepatoma HepG2 cells and it was recorded that pterostilbene effectively inhibited the invasion and colony formation of HepG2 cells [255]. In another study, the effect of pterostilbene on MMP-9 activity, a key protein involved in the invasion and metastasis of hepatocellular carcinoma was examined. Pterostilbene significantly inhibited the proteolytic activity of MMP-9 in a dose-dependent manner. Treatment of HepG2 with pterostilbene, in the presence of TPA (Tissue plasminogen activator), also resulted in a decrease in the levels of MMP-9 messenger RNA. PKC acts as the major receptor in response to TPA *in vitro* and *in vivo* and [256] the activation of PKCs is correlated with the potential of tumor metastasis. It was demonstrated that pre-treatment with pterostilbene significantly reduced TPA-induced translocation of PKCα, β and γ protein from cytosol to the membrane. Thus, pterostilbene promotes a strong anti-invasive and anti-metastatic effect against TPA-mediated metastasis via down-regulation of PKC which blocks MMP-9. Therefore, inhibition of MMP-9 activity by pterostilbene has a therapeutic potential for controlling invasion and metastasis of tumors. Significant therapeutic effects were further demonstrated *in vivo* by i.p. treatment of nude mice with pterostilbene (50 and 250 mg/kg) after inoculation with HepG2 cells into the tail vein. This reveals that pterostilbene is a novel, effective, anti-metastatic agent [255].

5.5.7 Celastrol

Celastrol derived from the Chinese medicinal plant *Tripterygiumwilfordii*, has attracted great attention recently for its potent anti-cancer effects. It has been established that the constitutive activation of STAT3 plays a pivotal role in the proliferation, survival, metastasis, and angiogenesis of hepatocellular carcinoma (HCC). Thus, discovery of novel agents that can inhibit STAT3 activation have potential for prevention and treatment of HCC [257]. Celastrol was found to inhibit the constitutive activation of STAT3 in C3A cells in a dose dependent manner, with maximum inhibition occurring at 5μmol/L. Thus, celastrol abrogates the DNA-binding ability of STAT3 [258]. The effects of celastrol on constitutively active JAK1 and JAK2 were also investigated in C3A cells. It was found that celastrol suppressed the constitutive phosphorylation of JAK1. In addition, STAT3 activation has been reported to regulate the expression of various gene products involved in cell survival, proliferation, angiogenesis, and chemoresistance [259]. The expression of the cell-cycle regulator cyclin D1, the anti-apoptotic proteins Bcl-2, Bcl-XL, survivin, Mcl-1, and the angiogenic gene product VEGF was also decreased in a time dependent manner in the presence of celastrol. The anti-tumor potential – of celastrol *in vivo* was also tested via intraperitoneal administration in a subcutaneous model of human HCCs using PLC/PRF5 cells. Celastrol at doses of 1 and 2 mg/kg induced significant inhibition of tumor growth compared with the controls. The effect of celastrol on the expression of Ki67, a marker of proliferation and caspase-3 was also analyzed. The expression of Ki67 was down-regulated and that of caspase-3 was substantially increased in celastrol-treated group as compared with control group.

5.5.8 Honokiol

Honokiol is the main active compound of *Magnolia officinalis* which is widely known to inhibit microbial infection, inflammation and gastrointestinal disorder in traditional Chinese medicine and Kampo medicine in Japan [260]. The ability of honokiol to modulate constitutive STAT3 activation in HCC cells was investigated by Sethi *et al*. It was observed that honokiol inhibited the constitutive activation of STAT3 in HepG2 cells in a dose dependent manner [261]. The STAT3 was inhibited completely at around 6h. Moroever, honokiol inhibited STAT3-DNA binding activities in a time dependent manner. In HuH-7 cells, IL-6- induced STAT3 and JAK2 phosphorylation was suppressed by treatment with honokiol in a time dependent manner. It was examined whether honokiol can induce the expression of SHP-1 in HepG2 cells. SHP-1 which is involved in the suppression of a variety of cytokine signals, including STAT3 [79] and honokiol induced the expression of SHP-1 protein in HepG2 cells in a dose dependent manner which was correlated with down-regulation of constitutive STAT3 activation in HepG2 cells. The expression of cyclin D1, Bcl-2, Bcl-xL, Survivin, Mcl-1 and the angiogenic gene product VEGF decreased in a time-dependent manner. It was also noted that honokiol potentiates the apoptotic effect of doxorubicin and paclitaxel in HepG2 cells [262]. In another study, the effect of honokiol on apoptosis in human hepatocellular carcinoma SMMC-7721 cells was investigated. Upon honokiol treatment, SMMC-7721 cells displayed morphological characteristics such as cell shrinkage, detachment from the culture plate, formation of apoptotic bodies, and nuclear condensation and fragmentation. In addition, up-regulation of Bax and down-regulation of Bcl-2 were observed [263]. Honokiol was also found to induce apoptosis with a decreased expression of procaspase-3 and -9 and an increased expression of active caspase-3. In addition, exposure of HepG2 cells to honokiol resulted in the release of mitochondrial cytochrome c to the cytosol [264].

5.5.9 γ- Tocotrienol

Anti-cancer drug discovery from natural agents provides a great opportunity to improve the existing standard of care for cancers [265]. γ-tocotrienol is a member of the vitamin E superfamily derived from palm oil and rice bran that has attracted great attention for its anti-proliferative and anti-carcinogenic potential [266]. Upon treatment of HepG2 cells with different concentrations of γ-tocotrienol for 6 h inhibited the constitutive activation of STAT3 in a dose dependent manner. Also, γ-tocotrienol inhibited the translocation of STAT3 to the nucleus in HepG2 cells thereby suppressing function of transcription factors [267]. Treatment with γ-tocotrienol suppressed binding of NFkB to the DNA in HepG2 cells. In SNU-387 cells incubated with γ-tocotrienol for different times, γ-tocotrienol suppressed IL-6-induced STAT3 and JAK2 phosphorylation in a time dependent manner. γ-tocotrienol also suppressed the constitutive phosphorylation of Src kinase in a time dependent manner [79]. γ-tocotrienol induced the expression of SHP-1 protein in HepG2 cells in a dose dependent manner, with maximum expression at 25–50 mM thus causing the down-regulation of constitutive STAT3 activation in HepG2 cells. The expression of cyclin D1, Bcl-2, Bcl-xL, Survivin, Mcl-1 and VEGF were modulated by γ-tocotrienol treatment. Moreover, γ-tocotrienol inhibited the proliferation of HepG2, C3A and SNU-387 cells in a dose and time dependent manner [268, 269]. Sakai *et al.* evaluated the anti-tumor activity of tocotrienol on Hep3B cells and observed that γ-tocotrienol inhibited cell proliferation at lower concentrations and shorter treatment times than alpha-tocotrienol. In addition, γ-tocotrienol induced poly (ADP-ribose) polymerase (PARP) cleavage and caused a rise in caspase-3, 8 and 9 activities. It was also found that γ-tocotrienol induced an up-regulation of Bax and a rise in the fragments of Bid and caspase-8. These data indicate that γ-tocotrienol induces apoptosis in Hep3B cells, thus suggesting its use as a novel chemotherapeutic agent [270].

5.5.10 Butein

Butein (3,4,2,4-tetrahydroxychalcone) is derived from numerous plants including stem bark of *Semecarpusanacardium*, *Rhusverniciflua* Stokes, and the heartwood of *Dalbergiaodorifera* [271]. Butein has been traditionally used for the treatment of pain, thrombotic disease, gastritis, stomach cancer, and parasitic infections in Korea, Japan, and China and no potential toxicity of this polyphenol has been reported so far [272] thereby indicating that it is safe to be consumed by humans [273]. Previous reports indicate that butein can suppress the proliferation of different human tumor cells, including breast carcinoma, colon carcinoma, osteosarcoma, lymphoma, acute myelogenous leukemia, chronic myeloid leukemia, multiple myeloma, melanoma, and hepatic stellate cells [274–276]. Moreover, butein was recently reported to induce G_2/M phase arrest and apoptosis in HCC cells through reactive oxygen species (ROS) generation [277] suggesting that butein may have a great potential for HCC treatment. In order to investigate the ability of butein, HepG2 cells were incubated with different concentrations of butein for 6 hours and the phosphorylation of STAT3 was examined by Western blot analysis [278]. It was observed that butein inhibited the constitutive activation of STAT3 in HepG2 cells in a dose dependent manner. Because ROS generation blocks STAT3 signaling [279] and butein is known to induce ROS production in HepG2 cells [277] it was determined whether NAC, a ROS scavenger, can restore the inhibitory effect of butein on STAT3 activation. It was observed that decrease in phosphorylated STAT3 upon butein treatment was ROS dependent, as levels went back to control levels in the presence of NAC. Pre-treatment with butein suppressed JAK2 phosphorylation in a time-dependent manner. Moreover, activation of Akt has also been linked with STAT3 activation [280]. It was also examined whether butein could modulate IL-6–induced Akt activation. Treatment of SNU-387 cells with IL-6 induced phosphorylation of Akt, and treatment of cells with butein suppressed the activation in a time-dependent manner. Butein inhibited the proliferation of HepG2, SNU-387, and PLC/PRF5 cells in a dose and time-dependent manner. The anti-tumor potential of butein was investigated *in vivo* via intraperitoneal administration in a subcutaneous model of human HCC, using HCCLM3 cells. Butein at 2 mg/kg induced significant inhibition of tumor growth compared with the corn oil–treated controls. Expression of Bcl-2 was downregulated and that of caspase-3 was significantly increased in the butein-treated group as compared with the control group. Overall, the results suggest that butein suppresses HCC growth *in vitro* and *in vivo* through modulation of STAT3 activation pathway. Ma *et al.* found that butein can suppress the migration and invasion in SK-HEP-1 human hepatocarcinoma cells. The gelatin zymography assay indicated that butein inhibited the activity of MMP-2 and MMP-9. Furthermore, butein inhibited the NFκB binding activity in SK-HEP-1 cells. In conclusion, butein might be a novel anti-cancer agent for the treatment of hepatocarcinoma through inhibiting migration and invasion [281].

5.5.11 β- Escin

β-escin or aescin is a compound derived from the seeds of horse chestnut (*Aesculushippocastanum*) and has the potential to prevent and treat cancer. β-escin is a pentacyclictriterpenoid that has been reported to exhibit anti-inflammatory and anti-carcinogenic properties in various disease models [282]. HepG2 cells which were incubated with different concentrations of β-escin for 6 h, inhibited the constitutive activation of STAT3 in a dose-dependent manner, with maximum inhibition occurring at approximately 30 μM [283]. HuH-7 cells lacking constitutively active STAT3 were treated with IL-6 for different times and then phosphorylated STAT3 was examined. In

HuH-7 cells incubated with β-escin, IL-6-induced STAT3 phosphorylation was suppressed by β-escin in a time-dependent manner [284]. β-escin also suppressed the constitutive phosphorylation of JAK1. Moreover, β-escin downregulated expression of STAT3-regulated gene products, such as, cyclin D1, Bcl-2, Bcl-XL, Survivin, Mcl-1, and VEGF. Thus, the apparent pharmacologic safety of β-escin should be tested in patients for treatment of HCC and other cancers harbouring active STAT3. In another study, escin was used to analyze the anti-tumor effect in hepatocellular carcinoma *in vivo* and *in vitro*. At a dose of 2.8 mg/kg, escin had a high inhibition ratio (43.5 %) on mice H22 tumor growth *in vivo*. It was reported that escin could induce significant concentration and time dependent inhibition of HepG2 cell viability. Disruption of the G1/S phase of cell cycle progression accompanied by the induction of apoptosis was also observed in HepG2 cells following escin treatment. This study provides evidence that escin induces cell cycle checkpoint arrest and caspase-independent cell death in HepG2 cells, thus having the potential for a chemopreventive agent [285].

5.5.12 Diosgenin

Research during the past decade has shown that diosgenin can supress proliferation and induces apoptosis in a wide variety of cancer cells lines. Anti-proliferative effects of diosgenin are mediated through cell-cycle arrest, disruption of $Ca2^{+}$ homeostasis, activation of p53, release of apoptosis-inducing factor, and modulation of caspase-3 activity [286]. Diosgenin inhibited the constitutive activation of STAT3 in C3A cells in a dose-dependent manner. In HuH-7 cells incubated with diosgenin, IL-6-induced STAT3 phosphorylation was suppressed by diosgenin in a time-dependent manner [287]. The expression of cyclin D1, Bcl-2, Bcl-xL, Survivin, Mcl-1 and VEGF decreased with treatment of diosgenin in a time-dependent manner, with maximum suppression observed at 24 h. However, further *in vivo* studies with diosgenin either alone or in conjunction with existing chemotherapeutic drugs are needed to demonstrate the potential application of diosgenin for treatment of cancers. In another study, treatment of HepG2 cells with 40Umdiosgenin resulted in activation of the caspase-3, -8, -9 and cleavage of poly-ADP-ribose polymerase (PARP) and the release of cytochrome c. Diosgenin-treated cells were observed to undergo nuclear shrinkage, condensation, and fragmentation. Diosgenin also increased the expression of Bax, decreased the expression of Bid and Bcl-2, and augmented the Bax/Bcl-2 ratio. Induction of apoptosis by diosgenin was accompanied by sustained phosphorylation of JNK, p38 MAPK and apoptosis signal-regulating kinase (ASK)-1, as well as generation of the ROS [288].

5.5.13 Phyllanthusniruri

Phyllanthus, a plant genus of the family Euphorbiaceae, exhibits multiple pharmacological actions. Of these, *Phyllanthusniruri* extracts exhibit significant anti-tumor activity. The dry extract (SDEPN) reduces the viability of hepatocarcinoma cells including HepG2 and HuH-7, within a few hours. HT29 cells were observed to be more tolerable to the SDEPN [289]. Induction of apoptosis in cancer cells represents an efficient strategy for cancer therapy. In the control cells, nuclei were round in shape and appeared homogeneous, whereas in SDEPN-treated cells (for 4 hours), chromatin condensation at the nuclear periphery and nuclear fragmentation were observed. On comparing the control HepG2 and HuH-7 cells to those treated with 1 mg/mL SDEPN for eight hours revealed typical morphological characteristics for cell apoptosis, such as cell condensation, plasma membrane blebbing and formation of apoptotic bodies. This implies that the growth inhibitory effect of the SDEPN might be due to induction of apoptosis. Moreover, the SDEPN potently reduced the viability of HepG2 and HuH-7 liver cancer cells. In addition, high levels of caspase-3 expression were detected in both HepG2 and HuH-7 hepatocellular carcinoma cell lines.

5.5.14 Oleanolic Acid

OA is a triterpenoid compound, widely found in natural plants [290] and its therapeutic effect, property of liver-specific metabolism, and its wide availability make it an ideal candidate for the design of new NO-releasing compounds for the production of NO in the liver. It has been demonstrated that NO releasing derivatives of OA constitute a class of effective anti-HCC agents [291]. O^2-(2,4-dinitro-5-{[2-(12-en-28-β–D- galactopyranosyl-oleanolate-3-yl)-oxy-2-oxoethyl]amino}phenyl)1-(N-hydroxyethylmethylamino)diazen-1-ium-1, 2-diolate (NG), is a novel PABA/NO-based derivative of oleanolic acid. (OA)O^2-(2,4-dinitrophenyl) diazeniumdiolate moiety is connected to OA using naturally occurring glycine as the linking group to make NG. To evaluate the anti-tumor effect of NG *in vivo*, Liu *et al.* inoculated mice model with H22 tumor. NG inhibited tumor growth in a dose-dependent manner for 48 h [292]. NG showed significant effect on tumor weight as compared to controls. Moreover, NG had a dose-dependent inhibitory effect on HepG2, SMMC-7721, Bel-7402 and L-02 cells, especially the HepG2 cells. Furthermore, compared with control group, NG dramatically induced apoptosis in HepG2 cells in a dose-dependent manner. The percentage of apoptotic cells in the control group was 3.1%. After treatment with 2.5, 5.0, and 10 μM NG for 24 h, the percentages of apoptotic cells were 16.8%, 37.9% and 83.3%, respectively. Caspase-3 and caspase-9 were activated significantly after NG treatment for 24 h in HepG2 cells, indicating the involvement of mitochondrial pathway in NG-induced apoptosis. The release of cytochrome c and AIF from mitochondria into cytoplasm was also detected in NG treated HepG2 cells. Cytochrome c and AIF were accumulated in the cytosol and decreased in mitochondria in a concentration-dependent manner. Thus, NG is also involved in mitochondria-mediated apoptosis induced by NG.

5.5.15 Mushrooms

Mushrooms have been used all around the world as folk medicines and food for centuries and have recently been considered to possess important pharmacological activities. *Thelephoraaurantiotincta*, which belongs to the genus *Thelephora* and grows in symbiosis with pine trees, is sold as an edible mushroom [293]. Previous phytochemical investigations conducted on the genus *Thelephora* revealed that this genus is an abundant source of *p*-terphenyl derivatives [294]. It was discovered that a *T. aurantiotincta* ethanol extract (TAE) decreased cell viability in human hepatocellular carcinoma cells (HepG2). In addition, a new *p*-terphenyl derivative, thelephantin O, and a known compound, vialinin A, was isolated as the principal bioactive components of TAE [295]. The inhibitory activity of vialinin A and thelephantin O against the cell viability of HepG2, Caco2, and non-cancerous human hepatocytes was examined. These compounds showed potent inhibitory activity against both HepG2 and Caco2 in a dose-dependent manner. This suggests the potential availability of vialinin A and thelephantin O for anticancer chemoprevention. Also, Lentinulaedodes mycelia (L.E.M.) is a dried powder extracted from shiitake mushrooms (Lentinulaedodes) which is demonstrated to have immunomodulatory effects. The direct cytotoxic effects of the polysaccharide-rich fraction of L.E.M. (L.E.M. ethanol precipitate; LEP) on HepG2 cells were investigated. LEP directly killed the HepG2 cells efficaciously, but had only minor effects on normal rat hepatocytes and normal mouse dermal cells. The morphological changes associated with apoptosis such as shrinkage, rounding as well as chromatin condensation was observed. The caspase-3 and -8 death receptor pathway was found largely responsible for the apoptotic death of HepG2 cells treated with LEP. In conclusion, LEP can directly induce apoptosis of HepG2 cells, and thus may have potential chemotherapeutic applications for the treatment of HCC [296]. Youn *et al.* investigated the anti-proliferative and apoptotic effects of Chaga mushroom (Inonotusobliquus) extract on HepG2 and Hep3B cells. HepG2 cells were more sensitive to Chaga extract than Hep3B cells. Chaga extract inhibited the cell growth in a dose-dependent manner, accompanied with G0/G1-phase arrest and apoptotic cell death. In addition, G0/G1 arrest in the cell cycle was closely associated with down-regulation of p53, pRb, p27, cyclin D1, D2, E, cyclin-dependent kinase (Cdk) 2, Cdk4, and Cdk6 expression. Thus, Chaga mushroom may provide a new therapeutic option, as a potential anti-cancer agent, in the treatment of hepatoma [297].

5.5.16 JuglansMandshurica

Juglansmandshurica Maxim (Juglandaceae) is a rare medicinal herb, which is widely distributed in Korea and northeast China. Its roots, leaves, and seeds have been used as traditional medicine for cancer therapy in Korea and China. In recent years, a number of natural compounds, such as phenolics, naphthoquinones, tetralones, flavonoids, naphthalenylglucosides, diarylheptanoid and galloyl glycosides, have been isolated from Juglans species [298]. In particular, Juglanthraquinone C (JC), extracted from J. mandshurica has shown strong cytotoxicity in hepatocellular cancer cell lines [299]. JC treatment reduced cell viability in time and dose dependent manner in HepG2 cells. Also, JC showed strong cytotoxicity in HepG2 cells in time and dose dependent manners, but no cytotoxicity in L02 cells. These results suggest that JC may be a promising anti-cancer drug for the treatment of human liver cancer. JC treatment significantly inhibited the expression of proliferation marker Ki-67. The percentages of cells accumulated at the S phase were significantly increased by 14 and 22 % after exposure to JC. In addition, the G2/M population was decreased from 13.4 to 2.47 %, and the sub-G1 population was increased from 9.8 to 18.4 % after treatment with JC, thus indicating the involvement of apoptosis-associated chromatin degradation [300]. HepG2 cells treated with JC showed chromatin condensation in a time-dependent manner as compared to the controls. The level of cleaved caspase-3 increased in a dose-dependent manner and Western blotting also indicated that the cleaved caspase- 3 level increased in a time-dependent manner when cells were treated with JC. Upon treatment with JC, Bcl-2 levels were markedly decreased and Bax levels were increased in HepG2 cells. Thus, these results suggest that the increase in Bax/Bcl-2 ratios may be involved in JC-induced mitochondrion-dependent apoptosis [301] since the Bax/Bcl-2 ratio in cells can regulate the susceptibility of cells to apoptosis [302].

5.5.17 Quercetin

The flavone quercetin (3,3',4',5,7-pentahydroxyflavone) is found in a broad range of fruits, vegetables, and beverages such as tea and wine [303] and is known to block NFκB activation [304]. Flavonoids, and specifically flavonols such as quercetin, have been reported to exhibit a broad spectrum of biological activities [305] including anti-carcinogenic and anti-inflammatory actions. In a study performed by Mario *et al.* [306] it was observed that quercetin caused a significant arrest of HepG2 cell growth. Moreover, quercetin concentrations over 10 μM totally inhibited replication of HepG2 cells. Leakage of intra-cellular lactate dehydrogenase (LDH) into the culture medium is used as an indicator of cytotoxicity. A trend of increasing LDH leakage was observed with low doses (0.1–5 μM) of quercetin whereas higher quercetin doses (50–100 μM) were cytotoxic, causing a 4- to 5-fold increase in LDH liberation into the culture medium. In another study, the regulatory effect of quercetin on NFκB and AP-1 related to survival/proliferation pathways in HepG2 cells was investigated. Quercetin induced a significant time-dependent inactivation of the NFκB pathway along with a down-regulation of the NFκB binding activity. A time-dependent activation of the AP-1/JNK pathway was also recorded, which played an important role in the control of the cell death induced by the flavonoid and contributed to the regulation of survival/proliferation (AKT, ERK) and death (caspase-3, p38, unbalance of Bcl-2 pro-apoptotic and anti-apoptotic proteins) signals. The results suggest that NFκB and AP-1 play an important role in the regulation of survival/proliferation pathways exerted by quercetin and that the sustained JNK/AP-1 activation and inhibition of NFκ B provoked by the flavonoid induces HepG2 death [307].

5.5.18 Ginger

Ginger (Zingiberofficinale) is a natural dietary rhizome that is widely used as a flavoring agent and occasionally used as a traditional medicinal herb due to

Table 5.2 List of natural compounds and their mode of action in HCC

Compound Name	Mode of action
Curcumin	DownregulatesNFκB, TNFα, VEGF, IL-6, MMP-2 and 9. Induces apoptosis and telomerase activity in HL60, SGC7901 cells. Increases cytochrome c release and downregluates Notch1 signaling pathway. Blocks tumor growth in Yoshida ascites hepatoma in rats
Selenium	Se injected i.p. inhibits volume and area of tumor foci in hamsters
EGCG	Inhibits growth and migration of HepG2 cells. Inhibits HCV entry. Induces apoptosis in HepG2 cells. Inhibits growth of HuH7 xenograts in nude mice.
Resveratrol	Decreases invasion of liver cancer cells. Inhibits proliferation and invasion of AH109A rat ascites hepatoma cells. Inhibits TNFα and MMP9 expression in HepG2 cells.
Ursolic Acid	Inhibits proliferation of SMMC-7221 cells. Decreases Bcl-2 and survivin expression thus inducing apoptosis.
Pterostilbene	Suppresses MMP9 expression by blocking AP-1 signaling pathway.
Celastrol	Inhibits constitutive activation of STAT3 and JAK1.
Honokiol	Suppresses IL-6 induced STAT3 and JAK2 phosphorylation. Induces SHP-1 in HepG2 cells. Downregulates Bcl-2 and Bcl-XL expression and releases cytochrome c to cytosol .
Γ-tocotrienol	Inhibits NFκB binding to DNA. Suppresses IL-6 induced STAT3 and JAK2 phosphorylation.
Butein	Induces G2/M phase arrest and apoptosis in HCC cells through ROS generation. Inhibits constitutive STAT3 activation. Suppresses migration and invasion in SK-Hep1 cells.
β-escin	Inhibits constitutive STAT3 activation. Downregulates expression of cyclin D1, Bcl-2, Bcl-XL
Diosgenin	Inhibits IL-6 induced STAT3 activation. Releases cytochrome c and activated caspase 3,8, 9.
Quercetin	Inhibits NFκB and AP-1 in HepG2 cells.
JuglansMandshlerica	Reduced HepG2 cell viability decreases Bcl-2 levels and increases BAX expression in HepG2 cells.

its anti-oxidative, anti-inflammatory, and anti-carcinogenic properties [308, 309]. Several pungent compounds, such as gingerols, shogaols, paradols, and gingerdiols, have been identified in ginger. Among these compounds, shogaols and gingerols are two phenolic substances that provide ginger its characteristic odor and flavor [310]. Phenolic substances present in fruits and vegetables, have been found to protect against cancer both *in vitro* and *in vivo* [311, 312]. 6-Shogaol [1-(4- hydroxy-3-methoxyphenyl)-4-decen-3-one] is a lipid-soluble organic compound that is known to exhibit significant anti-hepatotoxic effect against galactosamine-induced cytotoxicity in primary cultured rat hepatocytes [313]. 6-Gingerol [5 Hydroxy-1-(40-hydroxy-30-methoxyphenyl)-3-decanone] is an abundant constituent of ginger and also possesses anti-oxidative and anti-inflammatory activities [314]. Hepatocellular carcinoma is a highly metastatic cancer, representing 83% of all liver cancer cases. Thus, the anti-invasion activity of MMP-2 and MMP-9 in HepG2 cells upon treatment with 6-shogaol and 6-gingerol was investigated [315]. The activity of MMP-9 was suppressed in a dose-dependent manner, but the MMP-2 activity was not significantly changed when HepG2 and Hep3B cells were treated with 6-shogaol or 6-gingerol. The results suggest that 6-shogaol and 6- gingerol could possess potential anti-invasive activity against hepatoma cells and the anti-invasion mechanisms for the compounds might be mediated through the inhibition of MMP-9 and the induction of TIMP-1. Furthermore, it was demonstrated that 6-shogaol induces apoptosis in human hepatocellular carcinoma cells in relation to caspase activation and endoplasmic reticulum (ER) stress signaling. 6-shogaol was found to affect the ER stress signaling by regulating unfolded protein response (UPR) sensor PERK and its downstream target eIF2α. In prolonged ER stress, 6-shogaol inhibited the phosphorylation of eIF2α and triggered apoptosis in SMMC-7721 cells. Overexpression of eIF2α prevented 6-shogaol-mediated apoptosis in SMMC-7721 cells, whereas inhibition of eIF2α by small interfering RNA markedly enhanced 6-shogaol-mediated cell death. Furthermore, 6-shogaol caused inhibition of tumor growth of mouse SMMC-7721 xenograft by inducing apoptosis, activating caspase-3, and inactivating eIF2α. Altogether the results indicate that the PERK/eIF2α pathway plays an important role in 6-shogaol-mediated ER stress and apoptosis in SMMC-7721 cells *in vitro* and *in vivo*[316]. In another study, the inhibition of matrix metalloproteinases (MMPs) and urokinase-type plasminogen activator (uPA) in Hep3B cells as well as the anti-angiogenic activity of 6-gingerol and 6-shogaol was investigated. 6-Gingerol and 6-shogaol inhibited the phosphorylation of mitogen-activated protein kinase (MAPK) and PI3K/Aktsignaling, the activation of NFκB, and the translocation of NFκB and STAT3. Incubation of 6-gingerol or 6-shogaol with human umbilical vein endothelial cells or rat aortas significantly attenuated tube formation. Thus, 6-Shogaol and 6-gingerol effectively inhibit invasion and metastasis of hepatocellular carcinoma through diverse molecular mechanisms [315]. 6-gingerol, a major component of ginger, has antioxidant, anti-apoptotic, and anti-inflammatory activities. The genotoxic effects of 6-gingerol was evaluated using HepG2 cells. Exposure of the cells to 6-gingerol caused significant increase in DNA migration thus indicating that 6-gingerol causes DNA strand breaks and chromosome damage. The intracellular generation of reactive oxygen species (ROS) and reduced glutathione (GSH) was also studied and results showed that GSH and ROS levels were significantly increased after treatment for 60 min. Thus, 6-gingerol induces genotoxicity probably by oxidative stress [317]. Zerumbone (ZER), found in the subtropical ginger (Zingiberzerumbet Smith), is known to possess anti-proliferative properties in several cancer cells lines, including the cervical, skin and colon cancers. Thus, the anti-tumorigenic effects of ZER were assessed in rats induced to develop liver cancer with intraperitoneal injection of diethylnitrosamine (DEN, 200 mg/kg). The rats also received intraperitoneal ZER injections. Histopathological evaluations showed that ZER protects the rat liver from the carcinogenic effects of DEN. There was also significant reduction in the hepatic tissue glutathione (GSH) concentrations. There were significantly higher numbers of apoptotic cells in DEN/AAF rats treated with ZER than those untreated. Zerumbone treatment also increased Bax and decreased Bcl-2 protein expression in the livers of DEN/AAF rats, which suggests increased apoptosis. This study suggests that ZER reduces oxidative stress, inhibits proliferation, induces mitochondria-regulated apoptosis, thus minimizing DEN/AAF-induced carcinogenesis in rat liver. Therefore, ZER has great potential in the treatment of liver cancers [318].

5.6 Conclusions

Liver cancer remains the third leading cause of cancer mortality throughout the world with few therapeutic options to combat this disease. Natural compounds obtained from different sources such as fruits, vegetables, herbs and even fungi, open up a novel and exciting prospective on liver cancer prevention and treatment. Among the various secondary metabolites produced by plants and other organisms, terpenoids have emerged as a promising group of phytochemicals. Terpenoids selectively kill liver cancer cells with a pleiotropic mode of action while showing no effect on normal cells. However, there still remains a lot to be determined about these agents' molecular mechanisms, including their effects on genes that are overexpressed or mutated in hepatocarcinogenesis, and subsequent studies may increase our knowledge as to whether any of these antioxidants have the capability to prevent liver cancer. These future studies should focus on molecular aspects of the antioxidant agents. Animal and cell culture models in which oncogenes are overexpressed or tumor suppressor genes are silenced will provide important information about antioxidant actions. Transgenic and knockout models that target the metabolism of the antioxidants will also be important in understanding the role that specific antioxidants play in the prevention of hepatocarcinogenesis. Thus, there is a great need for the development of novel and effective cancer therapies.

References

[1] Subramaniam, A., et al., Potential role of signal transducer and activator of transcription (STAT)3 signaling pathway in inflammation, survival, proliferation and invasion of hepatocellular carcinoma. Biochim Biophys Acta, 2013. **1835**(1): p. 46–60.

[2] Berasain, C., et al., Inflammation and liver cancer: new molecular links.Ann N Y Acad Sci, 2009. **1155**: p. 206–21.

[3] El-Serag, H. B. and K. L. Rudolph, Hepatocellular carcinoma: epidemiology and molecular carcinogenesis.Gastroenterology, 2007. **132**(7): p. 2557–76.

[4] Schutte, K., J. Bornschein, and P. Malfertheiner, Hepatocellular carcinoma–epidemiological trends and risk factors.Dig Dis, 2009. **27**(2): p. 80–92.

[5] Ribes, J., et al., The influence of alcohol consumption and hepatitis B and C infections on the risk of liver cancer in Europe.J Hepatol, 2008. **49**(2): p. 233–42.

[6] Bishayee, A., T. Politis, and A. S. Darvesh, Resveratrol in the chemoprevention and treatment of hepatocellular carcinoma. Cancer Treat Rev, 2010. **36**(1): p. 43–53.

[7] Kundu, J. K. and Y. J. Surh, Inflammation: gearing the journey to cancer. Mutat Res, 2008. **659**(1–2): p. 15–30.

[8] Kawanishi, S., et al., Oxidative and nitrative DNA damage in animals and patients with inflammatory diseases in relation to inflammation-related carcinogenesis.Biol Chem, 2006. **387**(4): p. 365–72.

[9] Gius, D. and D. R. Spitz, Redox signaling in cancer biology.Antioxid Redox Signal, 2006. **8**(7–8): p. 1249–52.

[10] Ng, I. O., Chimerism in transplant allografts.Br J Surg, 2005. **92**(6): p. 661–2.

[11] Okuda, K., Hepatocellular carcinoma.J Hepatol, 2000. **32**(1 Suppl): break p. 225–37.

[12] Rossi, S., et al., Percutaneous radiofrequency interstitial thermal ablation in the treatment of small hepatocellular carcinoma.Cancer J Sci Am, 1995. **1**(1): p. 73–81.

[13] Sato, M., et al., Microwave coagulation therapy for hepatocellular carcinoma.Gastroenterology, 1996. **110**(5): p. 1507–14.
[14] Je, Y., F. A. Schutz, and T. K. Choueiri, Risk of bleeding with vascular endothelial growth factor receptor tyrosine-kinase inhibitors sunitinib and sorafenib: a systematic review and meta-analysis of clinical trials.Lancet Oncol, 2009. **10**(10): p. 967–74.
[15] Lu, S. C., Where are we in the chemoprevention of hepatocellular carcinoma? Hepatology, 2010. **51**(3): p. 734–6.
[16] Khan, N., F. Afaq, and H. Mukhtar, Cancer chemoprevention through dietary antioxidants: progress and promise.Antioxid Redox Signal, 2008. **10**(3): p. 475–510.
[17] Stan, S. D., et al., Bioactive food components and cancer risk reduction.J Cell Biochem, 2008. **104**(1): p. 339–56.
[18] Darvesh, A. S., B. B. Aggarwal, and A. Bishayee, Curcumin and liver cancer: a review.Curr Pharm Biotechnol, 2012. **13**(1): p. 218–28.
[19] Sherlock, S., Hepatic adenomas and oral contraceptives.Gut, 1975. **16**(9): p. 753–6.
[20] Labrune, P., et al., Hepatocellular adenomas in glycogen storage disease type I and III: a series of 43 patients and review of the literature. J Pediatr Gastroenterol Nutr, 1997. **24**(3): p. 276–9.
[21] Rebouissou, S., P. Bioulac-Sage, and J. Zucman-Rossi, Molecular pathogenesis of focal nodular hyperplasia and hepatocellular adenoma. J Hepatol, 2008. **48**(1): p. 163–70.
[22] Wanless, I. R., C. Mawdsley, and R. Adams, On the pathogenesis of focal nodular hyperplasia of the liver. Hepatology, 1985. **5**(6): p. 1194–200.
[23] Shortell, C. K. and S. I. Schwartz, Hepatic adenoma and focal nodular hyperplasia.Surg Gynecol Obstet, 1991. **173**(5): p. 426–31.
[24] Di Bisceglie, A. M., et al., NIH conference. Hepatocellular carcinoma. Ann Intern Med, 1988. **108**(3): p. 390–401.
[25] Parikh, S. and D. Hyman, Hepatocellular cancer: a guide for the internist. Am J Med, 2007. **120**(3): p. 194–202.
[26] Tangkijvanich, P., et al., Gender difference in clinicopathologic features and survival of patients with hepatocellular carcinoma. World J Gastroenterol, 2004. **10**(11): p. 1547–50.
[27] Sherman, M., Hepatocellular carcinoma: epidemiology, risk factors, and screening. Semin Liver Dis, 2005. **25**(2): p. 143–54.
[28] Bosch, F. X., et al., Primary liver cancer: worldwide incidence and trends. Gastroenterology, 2004. **127**(5 Suppl 1): p. S5-S16.

[29] Srivatanakul, P., H. Sriplung, and S. Deerasamee, Epidemiology of liver cancer: an overview.Asian Pac J Cancer Prev, 2004. **5**(2): p. 118–25.
[30] Yu, M. W., et al., Familial risk of hepatocellular carcinoma among chronic hepatitis B carriers and their relatives. J Natl Cancer Inst, 2000. **92**(14): p. 1159–64.
[31] Lin, C. L., et al., Basal core-promoter mutant of hepatitis B virus and progression of liver disease in hepatitis Be antigen-negative chronic hepatitis B. Liver Int, 2005. **25**(3): p. 564–70.
[32] Sumi, H., et al., Influence of hepatitis B virus genotypes on the progression of chronic type B liver disease. Hepatology, 2003. **37**(1): p. 19–26.
[33] Olinger, C. M., et al.,Possible new hepatitis B virus genotype, southeast Asia. Emerg Infect Dis, 2008. **14**(11): p. 1777–80.
[34] Tatematsu, K., et al., A genetic variant of hepatitis B virus divergent from known human and ape genotypes isolated from a Japanese patient and provisionally assigned to new genotype J. J Virol, 2009. **83**(20): p. 10538–47.
[35] Zhu, B., K. Luo, and Z. Hu, [Establishment of a method for classification of HBV genome and it's application]. Zhonghua Shi Yan He Lin Chuang Bing Du Xue Za Zhi, 1999. **13**(4): p. 309–13.
[36] Schaefer, S., Hepatitis B virus genotypes in Europe.Hepatol Res, 2007. **37**(s1): p. S20–6.
[37] You, L. R., C. M. Chen, and Y. H. Lee, Hepatitis C virus core protein enhances NF-kappaB signal pathway triggering by lymphotoxin-beta receptor ligand and tumor necrosis factor alpha. J Virol, 1999. **73**(2): p. 1672–81.
[38] Steinkuhler, C., L. Tomei, and R. De Francesco, In vitro activity of hepatitis C virus protease NS3 purified from recombinant Baculovirus-infected Sf9 cells. J Biol Chem, 1996. **271**(11): p. 6367–73.
[39] Viriyavejakul, P., et al., Does HIV infection accelerate the development of hepatocellular carcinoma? A case report in a young man. Southeast Asian J Trop Med Public Health, 2001. **32**(3): p. 504–6.
[40] Garcia-Samaniego, J., et al., Hepatocellular carcinoma in HIV-infected patients with chronic hepatitis C. Am J Gastroenterol, 2001. **96**(1): p. 179–83.
[41] Kolars, J. C., Aflatoxin and hepatocellular carcinoma: a useful paradigm for environmentally induced carcinogenesis. Hepatology, 1992. **16**(3): p. 848–51.

[42] Groopman, J. D., et al., Molecular dosimetry of urinary aflatoxin-N7-guanine and serum aflatoxin-albumin adducts predicts chemoprotection by 1,2-dithiole-3-thione in rats. Carcinogenesis, 1992. **13**(1): p. 101–6.
[43] Wu, H. C. and R. Santella, The role of aflatoxins in hepatocellular carcinoma. Hepat Mon, 2012. **12**(10 HCC): p. e7238.
[44] Wogan, G. N., et al., Environmental and chemical carcinogenesis. Semin Cancer Biol, 2004. **14**(6): p. 473–86.
[45] Morgan, T. R., S. Mandayam, and M. M. Jamal, Alcohol and hepatocellular carcinoma.Gastroenterology, 2004. **127**(5 Suppl 1): p. S87–96.
[46] Montalto, G., et al., Epidemiology, risk factors, and natural history of hepatocellular carcinoma. Ann N Y Acad Sci, 2002. **963**: p. 13–20.
[47] El-Serag, H. B., H. Hampel, and F. Javadi, The association between diabetes and hepatocellular carcinoma: a systematic review of epidemiologic evidence. Clin Gastroenterol Hepatol, 2006. **4**(3): p. 369–80.
[48] Wang, P., et al., Diabetes mellitus and risk of hepatocellular carcinoma: a systematic review and meta-analysis.Diabetes Metab Res Rev, 2012. **28**(2): p. 109–22.
[49] Caldwell, S. H., et al., Obesity and hepatocellular carcinoma. Gastroenterology, 2004. **127**(5 Suppl 1): p. S97–103.
[50] Dragani, T. A., Risk of HCC: genetic heterogeneity and complex genetics. J Hepatol, 2010. **52**(2): p. 252–7.
[51] Gao, Y., et al., HBV infection and familial aggregation of liver cancer: an analysis of case-control family study. Cancer Causes Control, 2004. **15**(8): p. 845–50.
[52] Fattovich, G., et al., Hepatocellular carcinoma in cirrhosis: incidence and risk factors.Gastroenterology, 2004. **127**(5 Suppl 1): p. S35–50.
[53] Gonciarz, Z., et al., [Hepatocholangiocarcinoma]. Wiad Lek, 1993. **46**(19–20): p. 782–5.
[54] Ng, I. O., et al., Combined hepatocellular-cholangiocarcinoma: a clinicopathological study. J Gastroenterol Hepatol, 1998. **13**(1): p. 34–40.
[55] Goodman, Z. D., et al., Combined hepatocellular-cholangiocarcinoma. A histologic and immunohistochemical study. Cancer, 1985. **55**(1): p. 124–35.
[56] Chok, K. S., et al., An update on long-term outcome of curative hepatic resection for hepatocholangiocarcinoma. World J Surg, 2009. **33**(9): p. 1916–21.
[57] Roebuck, D. J. and G. Perilongo, Hepatoblastoma: an oncological review. Pediatr Radiol, 2006. **36**(3): p. 183–6.

[58] Fukuzawa, R., et al., Beckwith-Wiedemann syndrome-associated hepatoblastoma: wnt signal activation occurs later in tumorigenesis in patients with 11p15.5 uniparental disomy. Pediatr Dev Pathol, 2003. **6**(4): p. 299–306.

[59] Flemming, P., et al., Common and epithelioid variants of hepatic angiomyolipoma exhibit clonal growth and share a distinctive immunophenotype. Hepatology, 2000. **32**(2): p. 213–7.

[60] Croquet, V., et al., Late recurrence of a hepatic angiomyolipoma. Eur J Gastroenterol Hepatol, 2000. **12**(5): p. 579–82.

[61] Yang, C. Y., et al., Management of hepatic angiomyolipoma. J Gastrointest Surg, 2007. **11**(4): p. 452–7.

[62] Karhunen, P. J., Benign hepatic tumors and tumor like conditions in men. J Clin Pathol, 1986. **39**(2): p. 183–8.

[63] Molina, E. and A. Hernandez, Clinical manifestations of primary hepatic angiosarcoma.Dig Dis Sci, 2003. **48**(4): p. 677–82.

[64] Locker, G. Y., et al., The clinical features of hepatic angiosarcoma: a report of four cases and a review of the English literature. Medicine (Baltimore), 1979. **58**(1): p. 48–64.

[65] Kim, H. R., et al., Clinical features and treatment outcomes of advanced stage primary hepatic angiosarcoma. Ann Oncol, 2009. **20**(4): p. 780–7.

[66] Lyburn, I. D., et al., Hepatic epithelioid hemangioendothelioma: sonographic, CT, and MR imaging appearances.AJR Am J Roentgenol, 2003. **180**(5): p. 1359–64.

[67] Demuynck, F., et al., [Hepatic Epithelioid hemangioendothelioma: a rare liver tumor]. J Radiol, 2009. **90**(7–8 Pt 1): p. 845–8.

[68] Stocker, J. T. and K. G. Ishak, Undifferentiated (embryonal) sarcoma of the liver: report of 31 cases. Cancer, 1978. **42**(1): p. 336–48.

[69] Lightfoot, N. and M. Nikfarjam, Embryonal sarcoma of the liver in an adult patient. Case Rep Surg, 2012. **2012**: p. 382723.

[70] Schoofs, G., et al., Hepatic rhabdomyosarcoma in an adult: a rare primary malignant liver tumor. Case report and literature review. Acta Gastroenterol Belg, 2011. **74**(4): p. 576–81.

[71] Chan, J. K., Solitary fibrous tumor–everywhere, and a diagnosis in vogue. Histopathology, 1997. **31**(6): p. 568–76.

[72] Martin, L. C., et al., Best cases from the AFIP: liver teratoma. Radiographics, 2004. **24**(5): p. 1467–71.

[73] Winter, T. C., 3rd and P. Freeny, Hepatic teratoma in an adult. Case report with a review of the literature. J Clin Gastroenterol, 1993. **17**(4): p. 308–10.

[74] Wong, N. A., et al., Primary yolk sac tumor of the liver in adulthood. J Clin Pathol, 1998. **51**(12): p. 939–40.

[75] Fayyazi, A., et al., Carcinosarcoma of the liver. Histopathology, 1998. **32**(4): p. 385–7.

[76] Garcez-Silva, M. H., et al., Carcinosarcoma of the liver: a case report. Transplant Proc, 2006. **38**(6): p. 1918–9.

[77] Gonzalez-Crussi, F., et al., Infantile sarcoma with intracytoplasmic filamentous inclusions: distinctive tumor of possible histiocytic origin. Cancer, 1982. **49**(11): p. 2365–75.

[78] Woehrer, A., et al., Incidence of atypical teratoid/rhabdoid tumors in children: a population-based study by the Austrian Brain Tumor Registry, 1996–2006. Cancer, 2010. **116**(24): p. 5725–32.

[79] Yen, J. B., M. S. Kong, and J. N. Lin, Hepatic mesenchymal hamartoma. J Paediatr Child Health, 2003. **39**(8): p. 632–4.

[80] Klaassen, Z., P. R. Paragi, and R. S. Chamberlain, Adult Mesenchymal Hamartoma of the Liver: Case Report and Literature Review. Case Rep Gastroenterol, 2010. **4**(1): p. 84–92.

[81] Karpelowsky, J. S., et al., Difficulties in the management of mesenchymal hamartomas.Pediatr Surg Int, 2008. **24**(10): p. 1171–5.

[82] Herrine, S. K., Schiff's diseases of the liver. Gastroenterology, 1999. **116**(6): p. 1501–2.

[83] Adam, R., E. Hoti, and L. C. Bredt, [Oncosurgical strategies for metastatic liver cancer]. Cir Esp, 2011. **89**(1): p. 10–9.

[84] Thorgeirsson, S. S. and J. W. Grisham, Molecular pathogenesis of human hepatocellular carcinoma. Nat Genet, 2002. **31**(4): p. 339–46.

[85] Takeda, K. and S. Akira, STAT family of transcription factors in cytokine-mediated biological responses. Cytokine Growth Factor Rev, 2000. **11**(3): p. 199–207.

[86] Hirano, T., K. Ishihara, and M. Hibi, Roles of STAT3 in mediating the cell growth, differentiation and survival signals relayed through the IL-6 family of cytokine receptors. Oncogene, 2000. **19**(21): p. 2548–56.

[87] Wang, H., et al., Signal transducer and activator of transcription 3 in liver diseases: a novel therapeutic target. Int J Biol Sci, 2011. **7**(5): p. 536–50.

[88] Calvisi, D. F., et al., Ubiquitous activation of Ras and Jak/Stat pathways in human HCC. Gastroenterology, 2006. **130**(4): p. 1117–28.

[89] Weerasinghe, P., et al., T40214/PEI complex: a potent therapeutics for prostate cancer that targets STAT3 signaling. Prostate, 2008. **68**(13): p. 1430–42.

[90] de Araujo, V. C., et al., STAT3 expression in salivary gland tumors. Oral Oncol, 2008. **44**(5): p. 439–45.

[91] Chen, S. H., et al., Activated STAT3 is a mediator and biomarker of VEGF endothelial activation. Cancer Biol Ther, 2008. **7**(12): p. 1994–2003.

[92] Chen, R. J., et al., Rapid activation of Stat3 and ERK1/2 by nicotine modulates cell proliferation in human bladder cancer cells. Toxicol Sci, 2008. **104**(2): p. 283–93.

[93] Kishimoto, T., IL-6: from its discovery to clinical applications. Int Immunol, 2010. **22**(5): p. 347–52.

[94] Bromberg, J. and T. C. Wang, Inflammation and cancer: IL-6 and STAT3 complete the link. Cancer Cell, 2009. **15**(2): p. 79–80.

[95] He, G., et al., Hepatocyte IKKbeta/NF-kappaB inhibits tumor promotion and progression by preventing oxidative stress-driven STAT3 activation. Cancer Cell, 2010. **17**(3): p. 286–97.

[96] Spannbauer, M. M. and C. Trautwein, Frequent in-frame somatic deletions activate gp130 in inflammatory hepatocellular tumors. Hepatology, 2009. **49**(4): p. 1387–9.

[97] Yang, B., et al., Aberrant promoter methylation profiles of tumor suppressor genes in hepatocellular carcinoma. Am J Pathol, 2003. **163**(3): p. 1101–7.

[98] Anson, M., et al., Oncogenic beta-catenin triggers an inflammatory response that determines the aggressiveness of hepatocellular carcinoma in mice. J Clin Invest, 2012. **122**(2): p. 586–99.

[99] He, G. and M. Karin, NF-kappaB and STAT3 - key players in liver inflammation and cancer. Cell Res, 2011. **21**(1): p. 159–68.

[100] Thompson, M. D. and S. P. Monga, WNT/beta-catenin signaling in liver health and disease. Hepatology, 2007. **45**(5): p. 1298–305.

[101] Giles, R. H., J. H. van Es, and H. Clevers, Caught up in a Wnt storm: Wnt signaling in cancer. Biochim Biophys Acta, 2003. **1653**(1): p. 1–24.

[102] Gordon, M. D. and R. Nusse, Wnt signaling: multiple pathways, multiple receptors, and multiple transcription factors. J Biol Chem, 2006. **281**(32): p. 22429–33.

[103] Merle, P., et al., Functional consequences of frizzled-7 receptor overexpression in human hepatocellular carcinoma. Gastroenterology, 2004. **127**(4): p. 1110–22.

[104] Cadoret, A., et al., New targets of beta-catenin signaling in the liver are involved in the glutamine metabolism. Oncogene, 2002. **21**(54): p. 8293–301.

[105] Sekine, S., et al., Liver-specific loss of beta-catenin results in delayed hepatocyte proliferation after partial hepatectomy. Hepatology, 2007. **45**(2): p. 361–8.

[106] Cieply, B., et al., Unique phenotype of hepatocellular cancers with exon-3 mutations in beta-catenin gene. Hepatology, 2009. **49**(3): p. 821–31.

[107] Takigawa, Y. and A. M. Brown, Wnt signaling in liver cancer. Curr Drug Targets, 2008. **9**(11): p. 1013–24.

[108] Nelson, W. J. and R. Nusse, Convergence of Wnt, beta-catenin, and cadherin pathways. Science, 2004. **303**(5663): p. 1483–7.

[109] Taniguchi, K., et al., Mutational spectrum of beta-catenin, AXIN1, and AXIN2 in hepatocellular carcinomas and hepatoblastomas. Oncogene, 2002. **21**(31): p. 4863–71.

[110] Satoh, S., et al., AXIN1 mutations in hepatocellular carcinomas, and growth suppression in cancer cells by virus-mediated transfer of AXIN1. Nat Genet, 2000. **24**(3): p. 245–50.

[111] Lian, Z., et al., Enhanced cell survival of Hep3B cells by the hepatitis B x antigen effector, URG11, is associated with upregulation of beta-catenin. Hepatology, 2006. **43**(3): p. 415–24.

[112] Takagi, H., et al., Frequent epigenetic inactivation of SFRP genes in hepatocellular carcinoma. J Gastroenterol, 2008. **43**(5): p. 378–89.

[113] Zeng, G., et al., siRNA-mediated beta-catenin knockdown in human hepatoma cells results in decreased growth and survival. Neoplasia, 2007. **9**(11): p. 951–9.

[114] Colnot, S., et al., Liver-targeted disruption of Apc in mice activates beta-catenin signaling and leads to hepatocellular carcinomas. Proc Natl Acad Sci U S A, 2004. **101**(49): p. 17216–21.

[115] Harada, N., et al., Lack of tumorigenesis in the mouse liver after adenovirus-mediated expression of a dominant stable mutant of beta-catenin. Cancer Res, 2002. **62**(7): p. 1971–7.

[116] Tergaonkar, V., et al., Distinct roles of IkappaB proteins in regulating constitutive NF-kappaB activity. Nat Cell Biol, 2005. **7**(9): p. 921–3.

[117] Karin, M. and Y. Ben-Neriah, Phosphorylation meets ubiquitination: the control of NF-[kappa]B activity. Annu Rev Immunol, 2000. **18**: p. 621–63.

[118] Ghosh, S., M. J. May, and E. B. Kopp, NF-kappa B and Rel proteins: evolutionarily conserved mediators of immune responses. Annu Rev Immunol, 1998. **16**: p. 225–60.

[119] Beg, A. A., et al., Embryonic lethality and liver degeneration in mice lacking the RelA component of NF-kappa B. Nature, 1995. **376**(6536): p. 167–70.

[120] Doi, T. S., et al., Absence of tumor necrosis factor rescues RelA-deficient mice from embryonic lethality. Proc Natl Acad Sci U S A, 1999. **96**(6): p. 2994–9.

[121] Rudolph, D., et al., Severe liver degeneration and lack of NF-kappaB activation in NEMO/IKKgamma-deficient mice. Genes Dev, 2000. **14**(7): p. 854–62.

[122] Qiao, L., et al., Constitutive activation of NF-kappaB in human hepatocellular carcinoma: evidence of a cytoprotective role. Hum Gene Ther, 2006. **17**(3): p. 280–90.

[123] Tai, D. I., et al., Constitutive activation of nuclear factor kappaB in hepatocellular carcinoma. Cancer, 2000. **89**(11): p. 2274–81.

[124] Haybaeck, J., et al., A lymphotoxin-driven pathway to hepatocellular carcinoma. Cancer Cell, 2009. **16**(4): p. 295–308.

[125] Maeda, S., et al., IKKbeta couples hepatocyte death to cytokine-driven compensatory proliferation that promotes chemical hepatocarcinogenesis. Cell, 2005. **121**(7): p. 977–90.

[126] Sakurai, T., et al., Loss of hepatic NF-kappa B activity enhances chemical hepatocarcinogenesis through sustained c-Jun N-terminal kinase 1 activation. Proc Natl Acad Sci U S A, 2006. **103**(28): p. 10544–51.

[127] Sakurai, T., et al., Hepatocyte necrosis induced by oxidative stress and IL-1 alpha release mediate carcinogen-induced compensatory proliferation and liver tumorigenesis.Cancer Cell, 2008. **14**(2): p. 156–65.

[128] Tilg, H., et al., Serum levels of cytokines in chronic liver diseases. Gastroenterology, 1992. **103**(1): p. 264–74.

[129] Trikha, M., et al., Targeted anti-interleukin-6 monoclonal antibody therapy for cancer: a review of the rationale and clinical evidence. Clin Cancer Res, 2003. **9**(13): p. 4653–65.

[130] Vanhaesebroeck, B. and M. D. Waterfield, Signaling by distinct classes of phosphoinositide 3-kinases.Exp Cell Res, 1999. **253**(1): p. 239–54.

[131] Paez, J. and W. R. Sellers, PI3K/PTEN/AKT pathway. A critical mediator of oncogenic signaling. Cancer Treat Res, 2003. **115**: p. 145–67.

[132] Whittaker, S., R. Marais, and A. X. Zhu, The role of signaling pathways in the development and treatment of hepatocellular carcinoma. Oncogene, 2010. **29**(36): p. 4989–5005.

[133] Mamane, Y., et al., mTOR, translation initiation and cancer.Oncogene, 2006. **25**(48): p. 6416–22.

[134] Dunlop, E. A. and A. R. Tee, Mammalian target of rapamycin complex 1: signaling inputs, substrates and feedback mechanisms. Cell Signal, 2009. **21**(6): p. 827–35.

[135] Miao, B., et al., Small molecule inhibition of phosphatidylinositol-3,4,5-triphosphate (PIP3) binding to pleckstrin homology domains. Proc Natl Acad Sci U S A, 2010. **107**(46): p. 20126–31.

[136] Freeburn, R. W., et al., Evidence that SHIP-1 contributes to phosphatidylinositol 3,4,5-trisphosphate metabolism in T lymphocytes and can regulate novel phosphoinositide 3-kinase effectors. J Immunol, 2002. **169**(10): p. 5441–50.

[137] Ding, L., et al., Somatic mutations affect key pathways in lung adenocarcinoma. Nature, 2008. **455**(7216): p. 1069–75.

[138] Thomas, R. K., et al., High-throughput oncogene mutation profiling in human cancer. Nat Genet, 2007. **39**(3): p. 347–51.

[139] Samuels, Y., et al., High frequency of mutations of the PIK3CA gene in human cancers.Science, 2004. **304**(5670): p. 554.

[140] Wood, L. D., et al., The genomic landscapes of human breast and colorectal cancers. Science, 2007. **318**(5853): p. 1108–13.

[141] Tavian, D., et al., u-PA and c-MET mRNA expression is co-ordinately enhanced while hepatocyte growth factor mRNA is down-regulated in human hepatocellular carcinoma.Int J Cancer, 2000. **87**(5): p. 644–9.

[142] Sakata, H., et al., Hepatocyte growth factor/scatter factor overexpression induces growth, abnormal development, and tumor formation in transgenic mouse livers. Cell Growth Differ, 1996. **7**(11): p. 1513–23.

[143] Tovar, V., et al., IGF activation in a molecular subclass of hepatocellular carcinoma and pre-clinical efficacy of IGF-1R blockage. J Hepatol, 2010. **52**(4): p. 550–9.

[144] Shih, W. L., et al., Hepatitis B virus X protein inhibits transforming growth factor-beta -induced apoptosis through the activation of phosphatidylinositol 3-kinase pathway. J Biol Chem, 2000. **275**(33): p. 25858–64.

[145] Kang-Park, S., et al., PTEN modulates hepatitis B virus-X protein induced survival signaling in Chang liver cells. Virus Res, 2006. **122**(1–2): p. 53–60.

[146] Mannova, P. and L. Beretta, Activation of the N-Ras-PI3K-Akt-mTOR pathway by hepatitis C virus: control of cell survival and viral replication. J Virol, 2005. **79**(14): p. 8742–9.

[147] Hu, T. H., et al., Expression and prognostic role of tumor suppressor gene PTEN/MMAC1/TEP1 in hepatocellular carcinoma. Cancer, 2003. **97**(8): p. 1929–40.

[148] Samuels, Y., et al., Mutant PIK3CA promotes cell growth and invasion of human cancer cells. Cancer Cell, 2005. **7**(6): p. 561–73.

[149] Bader, A. G., et al., Oncogenic PI3K deregulates transcription and translation. NatRev Cancer, 2005. **5**(12): p. 921–9.

[150] Sahin, F., et al., mTOR and P70 S6 kinase expression in primary liver neoplasms. Clin Cancer Res, 2004. **10**(24): p. 8421–5.

[151] LoPiccolo, J., et al., Targeting Akt in cancer therapy. Anticancer Drugs, 2007. **18**(8): p. 861–74.

[152] Min, L., B. He, and L. Hui, Mitogen-activated protein kinases in hepatocellular carcinoma development. Semin Cancer Biol, 2011. **21**(1): p. 10–20.

[153] Keshet, Y. and R. Seger, The MAP kinase signaling cascades: a system of hundreds of components regulates a diverse array of physiological functions. Methods Mol Biol, 2010. **661**: p. 3–38.

[154] Zhao, L. J., et al., Hepatitis C virus E2 protein promotes human hepatoma cell proliferation through the MAPK/ERK signaling pathway via cellular receptors. Exp Cell Res, 2005. **305**(1): p. 23–32.

[155] Lee, H. C., et al., Loss of Raf kinase inhibitor protein promotes cell proliferation and migration of human hepatoma cells. Gastroenterology, 2006. **131**(4): p. 1208–17.

[156] Feo, F., et al., Genetic and epigenetic control of molecular alterations in hepatocellular carcinoma. Exp Biol Med (Maywood), 2009. **234**(7): p. 726–36.

[157] Schmitz, K. J., et al., Activation of the ERK and AKT signaling pathway predicts poor prognosis in hepatocellular carcinoma and ERK activation in cancer tissue is associated with hepatitis C virus infection.J Hepatol, 2008. **48**(1): p. 83–90.

[158] Guegan, J. P., et al., MAPK signaling in cisplatin-induced death: predominant role of ERK1 over ERK2 in human hepatocellular carcinoma cells. Carcinogenesis, 2013. **34**(1): p. 38–47.

[159] Gailhouste, L., et al., RNAi-mediated MEK1 knock-down prevents ERK1/2 activation and abolishes human hepatocarcinoma growth in vitro and in vivo. Int J Cancer, 2010. **126**(6): p. 1367–77.

[160] Chang, Q., et al., Sustained JNK1 activation is associated with altered histone H3 methylations in human liver cancer. J Hepatol, 2009. **50**(2): p. 323–33.

[161] Hui, L., et al., p38alpha suppresses normal and cancer cell proliferation by antagonizing the JNK-c-Jun pathway. Nat Genet, 2007. **39**(6): p. 741–9.

[162] Hanahan, D. and R. A. Weinberg, Hallmarks of cancer: the next generation.Cell, 2011. **144**(5): p. 646–74.

[163] Vogel, C., et al., Flt-1, but not Flk-1 mediates hyperpermeability through activation of the PI3-K/Akt pathway. J Cell Physiol, 2007. **212**(1): p. 236–43.

[164] Price, D. J., et al., Role of vascular endothelial growth factor in the stimulation of cellular invasion and signaling of breast cancer cells. Cell Growth Differ, 2001. **12**(3): p. 129–35.

[165] Holmqvist, K., et al., The adaptor protein shb binds to tyrosine 1175 in vascular endothelial growth factor (VEGF) receptor-2 and regulates VEGF-dependent cellular migration. J Biol Chem, 2004. **279**(21): p. 22267–75.

[166] Takahashi, T., et al., A single autophosphorylation site on KDR/Flk-1 is essential for VEGF-A-dependent activation of PLC-gamma and DNA synthesis in vascular endothelial cells. EMBO J, 2001. **20**(11): p. 2768–78.

[167] Yoshiji, H., et al., Different cascades in the signaling pathway of two vascular endothelial growth factor (VEGF) receptors for the VEGF-mediated murine hepatocellular carcinoma development. Oncol Rep, 2005. **13**(5): p. 853–7.

[168] Staton, C. A., et al., Neuropilins in physiological and pathological angiogenesis. J Pathol, 2007. **212**(3): p. 237–48.

[169] von Marschall, Z., et al., Dual mechanism of vascular endothelial growth factor upregulation by hypoxia in human hepatocellular carcinoma. Gut, 2001. **48**(1): p. 87–96.

[170] Huang, G. W., L. Y. Yang, and W. Q. Lu, Expression of hypoxia-inducible factor 1alpha and vascular endothelial growth factor in hepatocellular carcinoma: Impact on neovascularization and survival. World J Gastroenterol, 2005. **11**(11): p. 1705–8.

[171] Liu, C., et al., Dual-specificity phosphatase DUSP1 protects overactivation of hypoxia-inducible factor 1 through inactivating ERK MAPK. Exp Cell Res, 2005. **309**(2): p. 410–8.

[172] Siegel, A. B., et al., Phase II trial evaluating the clinical and biologic effects of bevacizumab in unresectable hepatocellular carcinoma. J Clin Oncol, 2008. **26**(18): p. 2992–8.

[173] Yu, Q., Restoring p53-mediated apoptosis in cancer cells: new opportunities for cancer therapy. Drug Resist Updat, 2006. **9**(1–2): p. 19–25.

[174] Zhang, Y. J., et al., Aflatoxin B1 and polycyclic aromatic hydrocarbon adducts, p53 mutations and p16 methylation in liver tissue and plasma of hepatocellular carcinoma patients. Int J Cancer, 2006. **119**(5): p. 985–91.

[175] Herbst, R. S., Review of epidermal growth factor receptor biology. Int J Radiat Oncol Biol Phys, 2004. **59**(2 Suppl): p. 21–6.

[176] Zhang, X., et al., Inhibition of the EGF receptor by binding of MIG6 to an activating kinase domain interface.Nature, 2007. **450**(7170): p. 741–4.

[177] Buckley, A. F., et al., Epidermal growth factor receptor expression and gene copy number in conventional hepatocellular carcinoma. Am J Clin Pathol, 2008. **129**(2): p. 245–51.

[178] Zhu, A. X., et al., Phase 2 study of cetuximab in patients with advanced hepatocellular carcinoma. Cancer, 2007. **110**(3): p. 581–9.

[179] Gold, L. I., The role for transforming growth factor-beta (TGF-beta) in human cancer. Crit Rev Oncog, 1999. **10**(4): p. 303–60.

[180] Katakura, Y., et al., Transforming growth factor beta triggers two independent-senescence programs in cancer cells. Biochem Biophys Res Commun, 1999. **255**(1): p. 110–5.

[181] Desruisseau, S., et al., Divergent effect of TGFbeta1 on growth and proteolytic modulation of human prostatic-cancer cell lines. Int J Cancer, 1996. **66**(6): p. 796–801.

[182] Giannelli, G., et al., Inhibiting TGF-beta signaling in hepatocellular carcinoma. Biochim Biophys Acta, 2011. **1815**(2): p. 214–23.

[183] Tsai, J. F., et al., Clinical relevance of transforming growth factor-beta 1 in the urine of patients with hepatocellular carcinoma. Medicine (Baltimore), 1997. **76**(3): p. 213–26.

[184] Bedossa, P., et al., Transforming growth factor-beta 1 (TGF-beta 1) and TGF-beta 1 receptors in normal, cirrhotic, and neoplastic human livers. Hepatology, 1995. **21**(3): p. 760–6.

[185] Mazzocca, A., et al., Inhibition of transforming growth factor beta receptor I kinase blocks hepatocellular carcinoma growth through neo-angiogenesis regulation.Hepatology, 2009. **50**(4): p. 1140–51.

[186] Scales, S. J. and F. J. de Sauvage, Mechanisms of Hedgehog pathway activation in cancer and implications for therapy. Trends Pharmacol Sci, 2009. **30**(6): p. 303–12.

[187] Rubin, L. L. and F. J. de Sauvage, Targeting the Hedgehog pathway in cancer. Nat Rev Drug Discov, 2006. **5**(12): p. 1026–33.

[188] Berman, D. M., et al., Widespread requirement for Hedgehog ligand stimulation in growth of digestive tract tumors. Nature, 2003. **425**(6960): p. 846–51.

[189] Deutsch, G., et al., A bipotential precursor population for pancreas and liver within the embryonic endoderm. Development, 2001. **128**(6): p. 871–81.

[190] Cheng, W. T., et al., Role of Hedgehog signaling pathway in proliferation and invasiveness of hepatocellular carcinoma cells. Int J Oncol, 2009. **34**(3): p. 829–36.

[191] Chen, X., et al., Epithelial mesenchymal transition and hedgehog signaling activation are associated with chemoresistance and invasion of hepatoma subpopulations. J Hepatol, 2011. **55**(4): p. 838–45.

[192] Chen, X. L., et al., Gli-1 siRNA induced apoptosis in Huh7 cells. World J Gastroenterol, 2008. **14**(4): p. 582–9.

[193] Nakashima, T., et al., Pathology of hepatocellular carcinoma in Japan. 232 Consecutive cases autopsied in ten years. Cancer, 1983. **51**(5): p. 863–77.

[194] Jemal, A., et al., Cancer statistics, 2010. CA Cancer J Clin, 2010. **60**(5): p. 277–300.

[195] Llovet, J. M., A. Burroughs, and J. Bruix, Hepatocellular carcinoma.Lancet, 2003. **362**(9399): p. 1907–17.

[196] Naithani, R., et al., Comprehensive review of cancer chemopreventive agents evaluated in experimental carcinogenesis models and clinical trials. Curr Med Chem, 2008. **15**(11): p. 1044–71.

[197] Kaefer, C. M. and J. A. Milner, The role of herbs and spices in cancer prevention. JNutr Biochem, 2008. **19**(6): p. 347–61.

[198] Moiseeva, E. P. and M. M. Manson, Dietary chemopreventive phytochemicals: too little or too much? Cancer Prev Res (Phila), 2009. **2**(7): p. 611–6.

[199] Huang, W. Y., Y. Z. Cai, and Y. Zhang, Natural phenolic compounds from medicinal herbs and dietary plants: potential use for cancer prevention. Nutr Cancer, 2010. **62**(1): p. 1–20.

[200] Kruger, J., M. M. Yore, and H. W. Kohl, 3rd, Physical activity levels and weight control status by body mass index, among adults–National Health and Nutrition Examination Survey 1999–2004. Int J Behav Nutr Phys Act, 2008. **5**: p. 25.

[201] Glauert, H. P., et al., Dietary antioxidants in the prevention of hepatocarcinogenesis: a review. Mol Nutr Food Res, 2010. **54**(7): p. 875–96.

[202] Plummer, S. M., et al., Inhibition of cyclo-oxygenase 2 expression in colon cells by the chemopreventive agent curcumin involves inhibition of NF-kappaB activation via the NIK/IKK signaling complex. Oncogene, 1999. **18**(44): p. 6013–20.

[203] Shishodia, S., G. Sethi, and B. B. Aggarwal, Curcumin: getting back to the roots. Ann N Y Acad Sci, 2005. **1056**: p. 206–17.

[204] Shehzad, A., F. Wahid, and Y. S. Lee, Curcumin in cancer chemoprevention: molecular targets, pharmacokinetics, bioavailability, and clinical trials. Arch Pharm (Weinheim), 2010. **343**(9): p. 489–99.

[205] Singh, S. and B. B. Aggarwal, Activation of transcription factor NF-kappa B is suppressed by curcumin (diferuloylmethane) [corrected]. J Biol Chem, 1995. **270**(42): p. 24995–5000.

[206] Libby, P., Inflammatory mechanisms: the molecular basis of inflammation and disease. Nutr Rev, 2007. **65**(12 Pt 2): p. S140–6.

[207] Ralhan, R., M. K. Pandey, and B. B. Aggarwal, Nuclear factor-kappa B links carcinogenic and chemopreventive agents. Front Biosci (Schol Ed), 2009. **1**: p. 45–60.

[208] Aggarwal, B. B., et al., Potential of spice-derived phytochemicals for cancer prevention. Planta Med, 2008. **74**(13): p. 1560–9.

[209] Ohashi, Y., et al., Prevention of intrahepatic metastasis by curcumin in an orthotopic implantation model. Oncology, 2003. **65**(3): p. 250–8.

[210] Kang, J., et al., Curcumin-induced histone hypoacetylation: the role of reactive oxygen species. Biochem Pharmacol, 2005. **69**(8): p. 1205–13.

[211] Cao, J., et al., Curcumin induces apoptosis through mitochondrial hyperpolarization and mtDNA damage in human hepatoma G2 cells. Free Radic Biol Med, 2007. **43**(6): p. 968–75.

[212] Cui, S. X., et al., Curcumin inhibits telomerase activity in human cancer cell lines. Int J Mol Med, 2006. **18**(2): p. 227–31.

[213] Jia, L., et al., CD147 regulates vascular endothelial growth factor-A expression, tumorigenicity, and chemosensitivity to curcumin in hepatocellular carcinoma. IUBMB Life, 2008. **60**(1): p. 57–63.

[214] Wang, M., et al., Curcumin induced HepG2 cell apoptosis-associated mitochondrial membrane potential and intracellular free Ca(2+) concentration. Eur J Pharmacol, 2011. **650**(1): p. 41–7.

[215] Cheng, C. Y., Y. H. Lin, and C. C. Su, Curcumin inhibits the proliferation of human hepatocellular carcinoma J5 cells by inducing endoplasmic reticulum stress and mitochondrial dysfunction.Int J Mol Med, 2010. **26**(5): p. 673–8.

[216] Ning, L., et al., Down-regulation of Notch1 signaling inhibits tumor growth in human hepatocellular carcinoma. Am J Transl Res, 2009. **1**(4): p. 358–66.

[217] Bishayee, A. and N. Dhir, Resveratrol-mediated chemoprevention of diethylnitrosamine-initiated hepatocarcinogenesis: inhibition of cell proliferation and induction of apoptosis. Chem Biol Interact, 2009. **179**(2–3): p. 131–44.

[218] Bishayee, A., A. Sarkar, and M. Chatterjee, Further evidence for chemopreventive potential of beta-carotene against experimental carcinogenesis: diethylnitrosamine-initiated and phenobarbital-promoted hepatocarcinogenesis is prevented more effectively by beta-carotene than by retinoic acid. Nutr Cancer, 2000. **37**(1): p. 89–98.

[219] Thapliyal, R., et al., Inhibition of nitrosodiethylamine-induced hepatocarcinogenesis by dietary turmeric in rats. Toxicol Lett, 2003. **139**(1): p. 45–54.

[220] Busquets, S., et al., Curcumin, a natural product present in turmeric, decreases tumor growth but does not behave as an anticachectic compound in a rat model. Cancer Lett, 2001. **167**(1): p. 33–8.

[221] Tharappel, J. C., et al., Effect of antioxidant phytochemicals on the hepatic tumor promoting activity of 3,3',4,4'-tetrachlorobiphenyl (PCB-77). Food Chem Toxicol, 2008. **46**(11): p. 3467–74.

[222] Ghosh, D., et al., Nanocapsulated curcumin: oral chemopreventive formulation against diethylnitrosamine induced hepatocellular carcinoma in rat. Chem Biol Interact, 2012. **195**(3): p. 206–14.

[223] Yu, S. Y., Y. J. Zhu, and W. G. Li, Protective role of selenium against hepatitis B virus and primary liver cancer in Qidong. Biol Trace Elem Res, 1997. **56**(1): p. 117–24.

[224] Yu, M. W., et al., Plasma selenium levels and risk of hepatocellular carcinoma among men with chronic hepatitis virus infection. Am J Epidemiol, 1999. **150**(4): p. 367–74.

[225] Lee, C. Y., et al., Chemopreventive effect of selenium and Chinese medicinal herbs on N-nitrosobis(2-oxopropyl)amine-induced hepatocellular carcinoma in Syrian hamsters. Liver Int, 2008. **28**(6): p. 841–55.

[226] Cabrera, C., R. Artacho, and R. Gimenez, Beneficial effects of green tea–a review. J Am Coll Nutr, 2006. **25**(2): p. 79–99.

[227] Khan, N., et al., Targeting multiple signaling pathways by green tea polyphenol (-)-epigallocatechin-3-gallate. Cancer Res, 2006. **66**(5): p. 2500–5.

[228] Cao, J., et al., Chemopreventive effects of green and black tea on pulmonary and hepatic carcinogenesis. Fundam Appl Toxicol, 1996. **29**(2): p. 244–50.

[229] Tong, J. L., et al., [Epigallocatechin gallate induces apoptosis in human hepatocellular carcinoma HepG2 cells via TGF/Smad signaling pathway]. Zhonghua Zhong Liu Za Zhi, 2009. **31**(9): p. 646–50.

[230] Shirakami, Y., et al., (-)-Epigallocatechin gallate suppresses the growth of human hepatocellular carcinoma cells by inhibiting activation of the vascular endothelial growth factor-vascular endothelial growth factor receptor axis. Cancer Sci, 2009. **100**(10): p. 1957–62.

[231] Ciesek, S., et al., The green tea polyphenol, epigallocatechin-3-gallate, inhibits hepatitis C virus entry. Hepatology, 2011. **54**(6): p. 1947–55.

[232] Liang, Y. C., et al., Resveratrol-induced G< sub< 2</sub< arrest through the inhibition of CDK7 and p34< sup< CDC2</sup< kinases in colon carcinoma HT29 cells. Biochemical pharmacology, 2003. **65**(7): p. 1053–1060.

[233] Whitlock, N. C. and S. J. Baek, The anticancer effects of resveratrol: modulation of transcription factors. Nutr Cancer, 2012. **64**(4): p. 493–502.

[234] Athar, M., et al., Multiple molecular targets of resveratrol: Anti-carcinogenic mechanisms. Arch Biochem Biophys, 2009. **486**(2): p. 95–102.

[235] Gester, S., et al., Synthesis and biodistribution of an 18F-labelled resveratrol derivative for small animal positron emission tomography. Amino Acids, 2005. **29**(4): p. 415–28.

[236] Abd El-Mohsen, M., et al., Distribution of [3H]trans-resveratrol in rat tissues following oral administration. Br J Nutr, 2006. **96**(1): p. 62–70.

[237] Hebbar, V., et al., Toxicogenomics of resveratrol in rat liver. Life Sci, 2005. **76**(20): p. 2299–314.

[238] Canistro, D., et al., Alteration of xenobiotic metabolizing enzymes by resveratrol in liver and lung of CD1 mice. Food Chem Toxicol, 2009. **47**(2): p. 454–61.

[239] Das, S. and D. K. Das, Anti-inflammatory responses of resveratrol. Inflamm Allergy Drug Targets, 2007. **6**(3): p. 168–73.

[240] Rubiolo, J. A., G. Mithieux, and F. V. Vega, Resveratrol protects primary rat hepatocytes against oxidative stress damage: activation of the Nrf2 transcription factor and augmented activities of antioxidant enzymes. Eur J Pharmacol, 2008. **591**(1–3): p. 66–72.

[241] Mantovani, A., et al., Cancer-related inflammation.Nature, 2008. **454**(7203): p. 436–44.

[242] Delmas, D., et al., Inhibitory effect of resveratrol on the proliferation of human and rat hepatic derived cell lines. Oncol Rep, 2000. **7**(4): p. 847–52.

[243] De Ledinghen, V., et al., Trans-resveratrol, a grapevine-derived polyphenol, blocks hepatocyte growth factor-induced invasion of hepatocellular carcinoma cells. Int J Oncol, 2001. **19**(1): p. 83–8.

[244] Kozuki, Y., Y. Miura, and K. Yagasaki, Resveratrol suppresses hepatoma cell invasion independently of its anti-proliferative action.Cancer Lett, 2001. **167**(2): p. 151–6.

[245] Yu, H., et al., Resveratrol inhibits tumor necrosis factor-alpha-mediated matrix metalloproteinase-9 expression and invasion of human hepatocellular carcinoma cells. Biomed Pharmacother, 2008. **62**(6): p. 366–72.

[246] Miura, D., Y. Miura, and K. Yagasaki, Resveratrol inhibits hepatoma cell invasion by suppressing gene expression of hepatocyte growth factor via its reactive oxygen species-scavenging property. Clin Exp Metastasis, 2004. **21**(5): p. 445–51.

[247] Du, Q., et al., [Effects of resveratrol on apoptosis and ROS production in Hepa 1–6 hepatocarcinoma cells]. Zhong Yao Cai, 2012. **35**(3): p. 443–8.

[248] Zhang, Q., et al., Resveratrol inhibits hypoxia-induced accumulation of hypoxia-inducible factor-1alpha and VEGF expression in human tongue squamous cell carcinoma and hepatoma cells. Mol Cancer Ther, 2005. **4**(10): p. 1465–74.

[249] Kocsis, Z., et al., Chemopreventive properties of trans-resveratrol against the cytotoxicity of chloroacetanilide herbicides in vitro. Int J Hyg Environ Health, 2005. **208**(3): p. 211–8.

[250] Notas, G., et al., Resveratrol exerts its antiproliferative effect on HepG2 hepatocellular carcinoma cells, by inducing cell cycle arrest, and NOS activation. Biochim Biophys Acta, 2006. **1760**(11): p. 1657–66.

[251] Miura, D., Y. Miura, and K. Yagasaki, Hypolipidemic action of dietary resveratrol, a phytoalexin in grapes and red wine, in hepatoma-bearing rats. Life Sci, 2003. **73**(11): p. 1393–400.

[252] Liu, H. S., et al., Antitumor and immunomodulatory activity of resveratrol on experimentally implanted tumor of H22 in Balb/c mice. World J Gastroenterol, 2003. **9**(7): p. 1474–6.

[253] Yu, L., et al., Effect of resveratrol on cell cycle proteins in murine transplantable liver cancer. World J Gastroenterol, 2003. **9**(10): p. 2341–3.

[254] Parekh, P., et al., Downregulation of cyclin D1 is associated with decreased levels of p38 MAP kinases, Akt/PKB and Pak1 during chemopreventive effects of resveratrol in liver cancer cells. Exp Toxicol Pathol, 2011. **63**(1–2): p. 167–73.

[255] Subbaramaiah, K., et al., Ursolic acid inhibits cyclooxygenase-2 transcription in human mammary epithelial cells. Cancer Res, 2000. **60**(9): p. 2399–404.

[256] Martin-Aragon, S., et al., Pharmacological modification of endogenous antioxidant enzymes by ursolic acid on tetrachloride-induced liver damage in rats and primary cultures of rat hepatocytes. Exp Toxicol Pathol, 2001. **53**(2–3): p. 199–206.

[257] Gayathri, R., et al., Ursolic acid attenuates oxidative stress-mediated hepatocellular carcinoma induction by diethylnitrosamine in male Wistar rats. Asian Pac J Cancer Prev, 2009. **10**(5): p. 933–8.

[258] Yu, Y. X., et al., Ursolic acid induces human hepatoma cell line SMMC-7721 apoptosis via p53-dependent pathway. Chin Med J (Engl), 2010. **123**(14): p. 1915–23.

[259] Yang, L., et al., Ursolic acid induces doxorubicin-resistant HepG2 cell death via the release of apoptosis-inducing factor. Cancer Lett, 2010. **298**(1): p. 128–38.

[260] Tang, C., et al., Downregulation of survivin and activation of caspase-3 through the PI3K/Akt pathway in ursolic acid-induced HepG2 cell apoptosis. Anticancer Drugs, 2009. **20**(4): p. 249–58.

[261] Remsberg, C. M., et al., Pharmacometrics of pterostilbene: preclinical pharmacokinetics and metabolism, anticancer, antiinflammatory, antioxidant and analgesic activity. Phytother Res, 2008. **22**(2): p. 169–79.

[262] Bergsland, E. K. and A. P. Venook, Hepatocellular carcinoma. Curr Opin Oncol, 2000. **12**(4): p. 357–61.

[263] Pan, M. H., et al., Pterostilbene inhibited tumor invasion via suppressing multiple signal transduction pathways in human hepatocellular carcinoma cells. Carcinogenesis, 2009. **30**(7): p. 1234–42.

[264] Guo, K., et al., Role of PKCbeta in hepatocellular carcinoma cells migration and invasion in vitro: a potential therapeutic target. Clin Exp Metastasis, 2009. **26**(3): p. 189–95.

[265] Rajendran, P., et al., Celastrol suppresses growth and induces apoptosis of human hepatocellular carcinoma through the modulation of

STAT3/JAK2 signaling cascade in vitro and in vivo.Cancer Prev Res (Phila), 2012. **5**(4): p. 631–43.

[266] Zauberman, A., et al., Stress activated protein kinase p38 is involved in IL-6 induced transcriptional activation of STAT3. Oncogene, 1999. **18**(26): p. 3886–93.

[267] Aggarwal, B. B., et al., Signal transducer and activator of transcription-3, inflammation, and cancer: how intimate is the relationship? Ann N Y Acad Sci, 2009. **1171**: p. 59–76.

[268] Arora, S., et al., Honokiol: a novel natural agent for cancer prevention and therapy. Curr Mol Med, 2012.

[269] Rajendran, P., et al., Honokiol inhibits signal transducer and activator of transcription-3 signaling, proliferation, and survival of hepatocellular carcinoma cells via the protein tyrosine phosphatase SHP-1. J Cell Physiol, 2012. **227**(5): p. 2184–95.

[270] Jin, C., et al., Combination chemotherapy of doxorubicin and paclitaxel for hepatocellular carcinoma in vitro and in vivo. J Cancer Res Clin Oncol, 2010. **136**(2): p. 267–74.

[271] Han, L. L., et al., Reactive oxygen species production and Bax/Bcl-2 regulation in honokiol-induced apoptosis in human hepatocellular carcinoma SMMC-7721 cells. Environ Toxicol Pharmacol, 2009. **28**(1): p. 97–103.

[272] Deng, J., et al., Involvement of p38 mitogen-activated protein kinase pathway in honokiol-induced apoptosis in a human hepatoma cell line (hepG2). Liver Int, 2008. **28**(10): p. 1458–64.

[273] Newman, D. J., Natural products as leads to potential drugs: an old process or the new hope for drug discovery? J Med Chem, 2008. **51**(9): p. 2589–99.

[274] Aggarwal, B. B., et al., Tocotrienols, the vitamin E of the 21st century: its potential against cancer and other chronic diseases. Biochem Pharmacol, 2010. **80**(11): p. 1613–31.

[275] Rajendran, P., et al., gamma-Tocotrienol is a novel inhibitor of constitutive and inducible STAT3 signaling pathway in human hepatocellular carcinoma: potential role as an antiproliferative, pro-apoptotic and chemosensitizing agent. BrJ Pharmacol, 2011. **163**(2): p. 283–98.

[276] Seitz, S. J., et al., Chemotherapy-induced apoptosis in hepatocellular carcinoma involves the p53 family and is mediated via the extrinsic and the intrinsic pathway. Int J Cancer, 2010. **126**(9): p. 2049–66.

[277] Weng-Yew, W., et al., Suppression of tumor growth by palm tocotrienols via the attenuation of angiogenesis. Nutr Cancer, 2009. **61**(3): p. 367–73.

[278] Sakai, M., et al., Apoptosis induction by gamma-tocotrienol in human hepatoma Hep3B cells.J Nutr Biochem, 2006. **17**(10): p. 672–6.

[279] Pandey, M. K., et al., Butein suppresses constitutive and inducible signal transducer and activator of transcription (STAT) 3 activation and STAT3-regulated gene products through the induction of a protein tyrosine phosphatase SHP-1. Mol Pharmacol, 2009. **75**(3): p. 525–33.

[280] Samoszuk, M., J. Tan, and G. Chorn, The chalcone butein from Rhus verniciflua Stokes inhibits clonogenic growth of human breast cancer cells co-cultured with fibroblasts. BMC Complement Altern Med, 2005. **5**: p. 5.

[281] Lee, S. H., et al., 2',4',6'-Tris(methoxymethoxy) chalcone attenuates hepatic stellate cell proliferation by a heme oxygenase-dependent pathway. Biochem Pharmacol, 2006. **72**(10): p. 1322–33.

[282] Wang, Y., et al., The plant polyphenol butein inhibits testosterone-induced proliferation in breast cancer cells expressing aromatase. Life Sci, 2005. **77**(1): p. 39–51.

[283] Jang, H. S., et al., Flavonoids purified from Rhus verniciflua Stokes actively inhibit cell growth and induce apoptosis in human osteosarcoma cells. Biochim Biophys Acta, 2005. **1726**(3): p. 309–16.

[284] Kang, H. M., et al., Relationship between flavonoid structure and inhibition of farnesyl protein transferase. Nat Prod Res, 2004. **18**(4): p. 349–56.

[285] Moon, D. O., et al., Butein induces G(2)/M phase arrest and apoptosis in human hepatoma cancer cells through ROS generation. Cancer Lett, 2010. **288**(2): p. 204–13.

[286] Rajendran, P., et al., Suppression of signal transducer and activator of transcription 3 activation by butein inhibits growth of human hepatocellular carcinoma in vivo. Clin Cancer Res, 2011. **17**(6): p. 1425–39.

[287] Kurdi, M. and G. W. Booz, Evidence that IL-6-type cytokine signaling in cardiomyocytes is inhibited by oxidative stress: parthenolide targets JAK1 activation by generating ROS. J Cell Physiol, 2007. **212**(2): p. 424–31.

[288] Chen, R. H., et al., Interleukin-6 inhibits transforming growth factor-beta-induced apoptosis through the phosphatidylinositol 3-kinase/Akt and signal transducers and activators of transcription 3 pathways. J Biol Chem, 1999. **274**(33): p. 23013–9.

[289] Ma, C. Y., et al., Butein inhibits the migration and invasion of SK-HEP-1 human hepatocarcinoma cells through suppressing the ERK, JNK, p38,

and uPA signaling multiple pathways. J Agric Food Chem, 2011. **59**(16): p. 9032–8.

[290] Sirtori, C. R., Aescin: pharmacology, pharmacokinetics and therapeutic profile. Pharmacol Res, 2001. **44**(3): p. 183–93.

[291] Moran, D. M., et al., Interleukin-6 mediates G(0)/G(1) growth arrest in hepatocellular carcinoma through a STAT 3-dependent pathway. J Surg Res, 2008. **147**(1): p. 23–33.

[292] Ihle, J. N., STATs: signal transducers and activators of transcription. Cell, 1996. **84**(3): p. 331–4.

[293] Zhou, X. Y., et al., Escin, a natural mixture of triterpene saponins, exhibits antitumor activity against hepatocellular carcinoma. Planta Med, 2009. **75**(15): p. 1580–5.

[294] Sung, B., et al., Cancer cell signaling pathways targeted by spice-derived nutraceuticals. Nutr Cancer, 2012. **64**(2): p. 173–97.

[295] Li, F., et al., Diosgenin, a steroidal saponin, inhibits STAT3 signaling pathway leading to suppression of proliferation and chemosensitization of human hepatocellular carcinoma cells. Cancer Lett, 2010. **292**(2): p. 197–207.

[296] Kim, D. S., et al., Diosgenin Induces Apoptosis in HepG2 Cells through Generation of Reactive Oxygen Species and Mitochondrial Pathway. Evid Based Complement Alternat Med, 2012. **2012**: p. 981675.

[297] de Araujo Junior, R. F., et al., A dry extract of Phyllanthus niruri protects normal cells and induces apoptosis in human liver carcinoma cells. Exp Biol Med (Maywood), 2012. **237**(11): p. 1281–8.

[298] Liu, J., Pharmacology of oleanolic acid and ursolic acid. J Ethnopharmacol, 1995. **49**(2): p. 57–68.

[299] Huang, Z., et al., Synthesis and anti-human hepatocellular carcinoma activity of new nitric oxide-releasing glycosyl derivatives of oleanolic acid. Org Biomol Chem, 2010. **8**(3): p. 632–9.

[300] Liu, L., et al., NG, a novel PABA/NO-based oleanolic acid derivative, induces human hepatoma cell apoptosis via a ROS/MAPK-dependent mitochondrial pathway. Eur J Pharmacol, 2012. **691**(1–3): p. 61–8.

[301] Hu, L. and J. K. Liu, p-Terphenyls from the basidiomycete Thelephora aurantiotincta. Z Naturforsch C, 2003. **58**(5–6): p. 452–4.

[302] Quang, D. N., T. Hashimoto, and Y. Asakawa, Inedible mushrooms: a good source of biologically active substances. Chem Rec, 2006. **6**(2): p. 79–99.

[303] Norikura, T., et al., Anticancer activities of thelephantin O and vialinin A isolated from Thelephora aurantiotincta. J Agric Food Chem, 2011. **59**(13): p. 6974–9.

[304] Yukawa, H., et al., Direct cytotoxicity of Lentinula edodes mycelia extract on human hepatocellular carcinoma cell line. Biol Pharm Bull, 2012. **35**(7): p. 1014–21.

[305] Youn, M. J., et al., Chaga mushroom (Inonotus obliquus) induces G0/G1 arrest and apoptosis in human hepatoma HepG2 cells. World J Gastroenterol, 2008. **14**(4): p. 511–7.

[306] Min, B. S., et al., Anti-complement activity of constituents from the stem-bark of Juglans mandshurica. Biol Pharm Bull, 2003. **26**(7): p. 1042–4.

[307] Yao, Y., et al., Juglanthraquinone C, a novel natural compound derived from Juglans mandshurica Maxim, induces S phase arrest and apoptosis in HepG2 cells. Apoptosis, 2012. **17**(8): p. 832–41.

[308] Tounekti, O., J. Belehradek, Jr., and L. M. Mir, Relationships between DNA fragmentation, chromatin condensation, and changes in flow cytometry profiles detected during apoptosis. Exp Cell Res, 1995. **217**(2): p. 506–16.

[309] Antonsson, B. and J. C. Martinou, The Bcl-2 protein family. Exp Cell Res, 2000. **256**(1): p. 50–7.

[310] Vander Heiden, M. G. and C. B. Thompson, Bcl-2 proteins: regulators of apoptosis or of mitochondrial homeostasis? Nat Cell Biol, 1999. **1**(8): p. E209–16.

[311] Wang, P., et al., Effects of quercetin on the apoptosis of the human gastric carcinoma cells. Toxicol In Vitro, 2012. **26**(2): p. 221–8.

[312] Murakami, A., H. Ashida, and J. Terao, Multitargeted cancer prevention by quercetin. Cancer Lett, 2008. **269**(2): p. 315–25.

[313] Bravo, L., Polyphenols: chemistry, dietary sources, metabolism, and nutritional significance. Nutr Rev, 1998. **56**(11): p. 317–33.

[314] Alia, M., et al., Influence of quercetin and rutin on growth and antioxidant defense system of a human hepatoma cell line (HepG2). Eur J Nutr, 2006. **45**(1): p. 19–28.

[315] Granado-Serrano, A. B., et al., Quercetin modulates NF-kappa B and AP-1/JNK pathways to induce cell death in human hepatoma cells. Nutr Cancer, 2010. **62**(3): p. 390–401.

[316] Manju, V. and N. Nalini, Chemopreventive efficacy of ginger, a naturally occurring anticarcinogen during the initiation, post-initiation stages of 1,2 dimethylhydrazine-induced colon cancer. Clin Chim Acta, 2005. **358**(1–2): p. 60–7.

[317] Lantz, R. C., et al., The effect of extracts from ginger rhizome on inflammatory mediator production. Phytomedicine, 2007. **14**(2–3): p. 123–8.

[318] O'Hara, M., et al., A review of 12 commonly used medicinal herbs. Arch Fam Med, 1998. **7**(6): p. 523–36.

[319] Surh, Y., Molecular mechanisms of chemopreventive effects of selected dietary and medicinal phenolic substances. Mutat Res, 1999. **428**(1–2): p. 305–27.

[320] Mahmoud, N. N., et al., Plant phenolics decrease intestinal tumors in an animal model of familial adenomatous polyposis. Carcinogenesis, 2000. **21**(5): p. 921–7.

[321] Kyung, K. S., et al., 6-Shogaol, a natural product, reduces cell death and restores motor function in rat spinal cord injury. Eur J Neurosci, 2006. **24**(4): p. 1042–52.

[322] Wang, C. C., et al., Effects of 6-gingerol, an antioxidant from ginger, on inducing apoptosis in human leukemic HL-60 cells. In Vivo, 2003. **17**(6): p. 641–5.

[323] Weng, C. J., et al., Anti-invasion effects of 6-shogaol and 6-gingerol, two active components in ginger, on human hepatocarcinoma cells. Mol Nutr Food Res, 2010. **54**(11): p. 1618–27.

[324] Hu, R., et al., 6-Shogaol induces apoptosis in human hepatocellular carcinoma cells and exhibits anti-tumor activity in vivo through endoplasmic reticulum stress. PLoS One, 2012. **7**(6): p. e39664.

[325] Yang, G., et al., Genotoxic effect of 6-gingerol on human hepatoma G2 cells. Chem Biol Interact, 2010. **185**(1): p. 12–7.

[326] Taha, M. M., et al., Potential chemoprevention of diethylnitrosamine-initiated and 2-acetylaminofluorene-promoted hepatocarcinogenesis by zerumbone from the rhizomes of the subtropical ginger (Zingiber zerumbet). Chem Biol Interact, 2010. **186**(3): p. 295–305.

6

Nanomaterials: A Ray of Hope in Infectious Disease Treatment

Rashmi M. Bhande and C. N. Khobragade

School of Life Sciences, Swami Ramanand Teerth Marathwada University, Nanded -431 606 (MS) India

6.1 Introduction

Since the dawn of civilization and especially in the early phase of 20th century, infectious diseases are said to be leading cause of death throughout the world. These are clinically evident diseases resulting from the presence of pathogenic agent such as virus, bacteria, fungi or a parasites. Upon invasion of the epithelial surfaces, infectious microorganisms spread throughout the body *via* circulatory system and are removed from the blood by macrophages. After being phagocytosed by macrophages, these infectious microorganisms are trapped in phagosomes, which then fuse with lysosomal granules inside cell cytoplasm forming phagolysosomes. Subsequently, oxygen- dependent or oxygen-independent bacterial killing mechanisms induced by enzymes inside the phagolysosomes occur to digest the trapped microorganisms. However, many microorganisms are able to evade the macrophage digestion by escaping from the phagosomes, inhibiting the phagosome-lysosome fusion, by escaping withstanding the lysosomal enzymes, or resisting oxidative and non-oxidative killing mechanisms. These bacterial defense mechanisms make intracellular infections difficult to eradicate resulting in infectious diseases that range from *Staphylococcus* to Tuberculosis infections [1].

Infectious diseases are also called communicable diseases due to their ability to get transmitted from one person to another and sometimes from one species to another. Infectious diseases can be broadly classified as:

1. Known diseases which are persistently there (e.g., Dengue, Malaria, Tuberculosis);

Post-genomic Approaches in Cancer and Nano Medicine, 167–198.

2. New, previously unknown diseases (e.g., Severe Acute Respiratory Syndrome) and
3. Diseases which threaten to increase in the near future (e.g., Avian Influenza).

Diagnosis of an infectious disease involves the detection of the infecting agent, either directly or indirectly. Microbial culture, Microscopy, Biochemical tests, Immunoassays and Molecular diagnostics are some common methods used for the diagnosis of infectious diseases independently or in combination with each other but some of these methods do not provide the strain specific information. Therefore, there is a need to find methods to overcome the pressing demand of novel strategies to identify new antimicrobial agents from natural and inorganic substances. In 1960s many antimicrobial drugs have been introduced so as to decrease the morbidity and mortality caused due to invading microbe however resistance to antibiotics has reached a critical level, invalidating major antimicrobial drugs that are currently used in clinical settings [2, 3].

The bacterial resistance to antimicrobial drugs has been attempted to be resolved by discovering new antibiotics and chemically modifying existing antimicrobial drugs. Despite these facts there is no assurance that the development of new antimicrobial drugs can catch up to the microbial pathogen's fast and overcome resistance in a timely manner. Now a days drug resistant infections in hospitals and in the communities caused by both Gram-positive and Gram-negative bacterial pathogens are growing [4], and the continued evolution of antimicrobial resistance threatens human health seriously compromising our ability to treat serious infections [5]. Thus, the treatment demands a long-term solution to this ever-growing and foreseeable problem [6].

One of the recent efforts in addressing this challenge lies in exploring antimicrobial nanomaterials with novel nano sized platforms for efficient antibiotic(s) delivery. Recently it has been observed that some metal nano constructors are known to possess antimicrobial activities and therefore could be employed in treatment of infectious diseases [7–9].

Antimicrobial nanoparticles (NPs) offer many distinctive advantages such as reduced toxicity, potential to overcome resistance and lowercost over conventional antibiotics [10–12]. Various nano sized drug carriers are also available to efficiently administer antibiotics by improving pharmacokinetics and pharmacodynamics by reducing the adverse effects of antibiotics. Theoretically, NPs are retained much longer in the body than small molecule antibiotics, rendering sustained therapeutic effects.

Over the last few decades, the applications of nanotechnology in medicine have been extensively explored in many medical areas, especially in drug delivery. Nanotechnology concerns the understanding and control of matters in the 1–100 nm range, at which scale materials have unique physicochemical properties including ultra small size, large surface to mass ratio, high reactivity and unique interactions with biological systems [13–15]. Advantages of nanoparticle-based drug delivery include improving serum solubility of the drugs, prolonging the systemic circulation lifetime, releasing drugs in a sustained and controlled manner, preferentially delivering drugs to the tissues and cells of interest, and concurrently delivering multiple therapeutic agents to the same cells for combination therapy [16–18]. Moreover, drug-loaded nanoparticles can enter host cells through endocytosis and then release drug payloads to treat microbes-induced intracellular infections. As a result, a number of nanoparticle-based drug delivery systems have been approved for clinical uses to treat a variety of diseases and many other therapeutic nanoparticle formulations are currently under various stages of clinical tests [19]. A few types of nanoparticles including liposome, polymeric nanoparticles, solid lipid nanoparticles and dendrimers have been widely investigated as antimicrobial drug delivery platforms, of which several products have been introduced into pharmaceutical market [20–21].

Currently, the metallic nanoparticles are thoroughly being explored and investigated as potential antimicrobials. Metal nanoparticles with antimicrobial activity when embedded and coated on to surfaces can find immense applications in water treatment, synthetic textiles, biomedical and surgical devices, food processing and packaging [22]. Moreover, the composites prepared using metal nanoparticles and polymers can find better utilization due to the enhanced antimicrobial activity [23].

Nanodiagnostics involve the use of nanotechnology in clinical diagnosis to meet the demands for increased sensitivity and early detection in less time. The large surface area of nonmaterial enables attachment of large number of target-specific molecules of interest for ultra-sensitive detection. With such capability, diagnosis at the molecular and single cell level is possible. Because of high sensitivity, nanotechnology enables detection of a few microorganisms or target molecular specific to pathogens whereas the conventional methods are limited in achieve this ultra-sensitivity.

In addition, unique properties of nanomaterial could allow rapid (few minutes duration) and real-time detection of the pathogens. Moreover, relatively small sample volumes are required for the tests. All these attractive features would have positive impact on the cost-effectiveness for implementing nanotechnology for diagnosis of infectious diseases [24].

Till date, nanoparticles (NPs) of various types have been primarily studied and have shown great promise for nanodiagnostics of infectious diseases. Nanoparticle technology based on fluorescent NPs (e.g. dye-loaded NPs, quantum dots (Qdots)), magnetic NPs and metallic NPs (e.g. gold and silver NPs) has been successfully used to image, track and detect various infectious microorganisms[25].

This chapter introduces employing nanotechnology as a new paradigm in controlling infectious diseases, especially in overcoming antimicrobial drug resistance, in the context of research and clinical prospective of this novel and promising strategy.

6.2 Systemic Applications of Nanoparticles

Metal Nanoparticles, such as silver and metal oxides, are promising agents for antibacterial applications having some general mechanism of toxicity toward bacteria that mammalian cells do not have. Nanoparticles bind to bacterial cell walls, causes membrane disruption through direct interactions or through free radical production [26]. Mammalian cells are able to phagocytes nanoparticles and subsequently degrade them by lysozomal fusion, reducing toxicity and free radical damage [27]. This may allow for selectivity of the same nanoparticle to promote tissue-forming cell functions, while also inhibiting bacterial functions that lead to infection. The antibacterial activity of silver and metal oxides has long been known, but by decreasing particle size into the nanometer range, surface area gets increased, and the antibacterial activity of the material increases [28]. The use of colloidal silver for minimizing infection has been investigated for over 50 years and now with recent advances in chemical, biological, and material characterization techniques, silver is being more widely adopted in the medical community.

An appropriate target mechanism directs the antibacterial treatment to the site of infection by selectively targeting bacteria and all virulent strains over other cells by which the therapeutic effectiveness of the drug can be ascertained. The selective targeting of nanoparticles to an infection site minimizes uptake by surrounding tissues as well as decreases exposure of nonpathogenic bacterial flora (altering the balance of natural flora that would exacerbate virulent bacterial growth). Chemical targeting is highly specific and requires identification of an epitope (a molecule or protein) in the bacterial biofilm for nanoparticle delivery [29]. Finally, therapeutic feedback provides effectiveness of the treatment, location of infection, and delivery efficiency. Now a day's Modular nanopharmaceutical systems are designed to address all of these multifunctional capabilities for the ideal bacterial treatment. To achieve such multifunctionality nanoparticles are made to tailor through surface

conjugation where in the final chemical modification of nanoparticles by various mechanisms such as:

1. Photo catalytic production of reactive oxygen species (ROS) that damage cellular and viral components,
2. Compromising the bacterial cell wall/membrane,
3. Interruption of energy transduction and
4. Inhibition of enzyme activity and DNA synthesis.

6.2.1 Antimicrobial Nanotechnology Based Drug Delivery

6.2.1.1 NPs for efficient antimicrobial drug delivery

Despite the well-established efficacy of antimicrobial drugs, a sub optimized therapeutic index and local/systemic adverse reactions limits the use in order to obtain maximized therapeutic effects. In addition, intracellular infections and acquired resistance of infectious microbes are also key challenges for many antimicrobial drugs. Novel NPs by virtue of their unique physicochemical properties (e.g., ultra small and controllable size, large surface area to mass ratio, high interactions with microorganisms and host cells, and structural/functional versatility) are a promising platform to overcome those limitations [30]. The advantages of NP-based antimicrobial drug delivery include improved solubility of poorly water-soluble drugs, prolonged drug half-life and systemic circulation time, sustained and stimuli-responsive drug release, which eventually lowers administration frequency and dose. Moreover, minimized systemic side effects via targeted delivery of antimicrobial drugs as well as combined, synergistic, and resistance-overcoming effects via co-delivery of multiple antimicrobial drugs can be achieved using NP carriers [31].

6.2.1.2 Liposome for antimicrobial drug delivery

Liposomes are nano- to micro-sized vesicles comprising of a phospholipid bilayer with an aqueous core. Liposomes are known to be promising clinically acceptable delivery carriers of enzymes, proteins, and drugs as their lipid bilayer structure mimics the cell membrane and can readily fuse with infectious microbes. In addition, both hydrophilic and hydrophobic antimicrobial drugs can be easily encapsulated and retained, without chemical modifications, in aqueous core and in the phospholipid bilayer, respectively. Upon administration, liposomes are rapidly cleared from the blood by mononuclear phagocytic system (MPS). For example, incorporation of certain glycolipids (e.g., monosialoganglioside and phosphatidylinositol) in the liposomes resulted in prolonged circulation time and reduced uptake by the MPS in the liver and the spleen. Conjugating "stealth" material (e.g., polyethylene glycol, PEG) on the surface of liposome not only resulted in enhanced *in vivo* stability (i.e., long-circulation) but also enabled targeted delivery of antimicrobial drugs after treated with various targeting ligands like antibody, antibody segments, aptamers, peptides. Aminoglycosides and Polymixin B when formulated with Liposomes reduce Nephrotoxicity, Cytotoxicity and Neuromuscular blockage etc. and successfully were employed to treat *P.aeruginosa* infections by affecting efflux pump and suppressing the drug resistance of microbes. Lauric acid loaded in liposome is reported to be safe, and effective formulation for treating *Acne vulgaris* and other *Propioni bacterium acne*-associated diseases. The drug stability and antimicrobial activity against *Micrococcus luteus* were shown to be greatly enhanced when Ampicillin was loaded in liposomes, in comparison with free drugs. Completely inhibited growth of *S.aureus* strain by benzyl penicillin-encapsulating cationic liposome was reported at lower drug concentrations for shorter exposure times in comparison with free drug effect. Ciprofloxacin in liposomal formulation was found to be rapidly cleared from the blood which suggests that liposomal Ciprofloxacin can be an effective therapy for systemic *Salmonella* infections. Successful treatment of *Mycobacterium avium* infected mice by liposomal streptomycin was also demonstrated. Encapsulation of Vancomycin and Teicoplanin in liposome resulted in significantly improved elimination of intracellular MRSA infection [32].

6.2.1.3 Solid lipid (SL) NPs

SLNPs offer combined advantages of traditional solid NPs and liposome by avoiding some of their disadvantages. Improved bioavailability and targeted delivery of antimicrobial drug using SLNPs have been investigated via parenteral, topical, ocular, oral, and pulmonary administration routes. The SLNPs in various formulations for oral administration (e.g., tablets, capsules, and pellets) can also be used for antimicrobial drug delivery. When applied onto the skin, SLNPs tend to adhere to the surface and form a dense hydrophobic film that is occlusive and affords a long residence time on the stratum corneum. In addition, increased transdermal diffusion of water-insoluble Azole antifungal drugs (e.g., Clotrimazole, Miconazole, Econazole, Oxiconazole, and Ticonazole) are reported by encapsulating them in SLNPs. Tobramycin-loaded SLNPs easily crosses the blood-brain barrier, provide significantly higher bioavailability in the aqueous humor than standard eye drops and may replace the advantages of subconjunctival injections for *Pseudomonal keratitis* and preoperative prophylaxis. SLNPs are also a promising means for prolonged Ciprofloxacin release, particularly in ocular and skin infections via local delivery.

Unlike liposome and polymeric NPs, inhalable SLNPs are stable, have a high drug incorporation capability, and offer a significantly reduced risk of retaining residual organic solvents. SLNPs are assumed to be phagocyted by alveolar macrophages in the lungs, and subsequently transported to the lymphoid tissues, thereby taking premium period to obtain equivalent therapeutic effects over the free drugs.

6.2.1.4 Polymeric NPs

The first polymer-based delivery of macromolecules (e.g., albumin and peptide hormones using poly[ethylene vinyl acetate] polymer) was demonstrated in 1976 and later on controlled drug release using biocompatible and biodegradable polymers further emerged in the 1980s which has been extensively investigated in clinic for enhanced intracellular drug delivery and reduced rapid clearance by Reticulo Endothelial System (RES).

Antimicrobial drug delivery using polymeric NPs offers several advantages such as 1) structural stability in biological fluids and under harsh and various conditions for preparation (e.g., spray drying and ultrafine milling) and storage, 2) precisely tunable properties (e.g., size, zeta-potentials, and drug release profiles) by manipulating polymer lengths, surfactants, and organic solvents used for NP preparation, 3) facile and versatile surface functionalization for conjugating drugs and targeting ligands. Linear polymers (e.g., polyalkyl acrylates and polymethyl methacrylate) and Amphiphilic block copolymers are two major types of polymeric NPs have been explored for antimicrobial drug delivery. Majority of polymeric NPs prepared with linear polymers are either nanocapsules or solid nanospheres. In polymeric nanocapsules, a polymeric membrane that controls the release rate surrounds the drugs are solubilized either in aqueous or oily solvents. Amphiphilic block copolymers spontaneously self-assemble micellar NPs with the drug encapsulating hydrophobic core and the hydrophilic corona shielding the core from opsonization and degradation. A library of biodegradable polymers, including poly(lactic acid) (PLA), poly(glycolic acid)(PGA), poly(lactide-co-glycolide) (PLGA), poly (ϵ-carprolactone) (PCL), and poly(cyanoacrylate)(PCA), has been used as hydrophobic segments (forming drug-encapsulating core for controlled drug release) of the amphiphilic copolymers, whereas PEG has been most commonly used as a hydrophilic segment. Often targeting ligands (e.g., Aptamers, Apt) are conjugated on the termini of PEG (e.g., PLGA-b-PEGb-Apt) for selective delivery.

Polymeric NPs have been explored to deliver various antimicrobial agents and greatly enhance the therapeutic efficacy in treating many types of infectious diseases [33].

6.2.1.5 Dendrimers

Dendrimers are hyper branched polymers with precise nano architecture and low polydispersity, which are synthesized in a layer-by-layer fashion around a core unit, resulting in a high level control of size, branching points (drug conjugation capability), and surface functionality. The highly branched nature of dendrimers provides bigger surface area to size ratios that generate great reactivity to microorganisms *in vivo*. The highly dense surface of functional groups allows the synthesis of dendrimers with specific and high binding affinities to a wide variety of viral and bacterial receptors. Both hydrophobic and hydrophilic drugs can be loaded/conjugated/adsorbed inside empty internal cavities in the core and on the multivalent surfaces of dendrimers, respective. In addition, the dendrimers functionalized with quaternary ammonium, which is known as antimicrobials, on the surface at a high density displayed greater antibacterial activity than free antibiotics. Directly destroying the cell membrane of microorganisms or disrupting multivalent binding interactions between microorganism and host cell are the primary mechanisms of antimicrobial action of dendrimer biocides. PAMAM dendrimer is a promising drug delivery carrier but its cytotoxicity due to amine-terminated nature has been a limiting factor for clinical use. Carboxylic- or hydroxyl-terminated PAMAM dendrimers, which appear to be more biocompatible and less toxic than unmodified ones, can be easily conjugated with antimicrobial agents via abundant functional groups. Sulfamethoxazole (SMZ)-encapsulating PAMAM dendrimers led to sustained release of the drug *in vitro* and 4–8 folds increased antibacterial activity against *E.coli*, compared to free SMZ. Aqueous insoluble quinolones, which prevents liquid formulations and restricts their use in topical application, were loaded in PAMAM dendrimers, generating not only excellent solubility but also similar or increased antibacterial activity. Solubilization and controlled delivery of a hydrophobic anti-malarial drug was achieved using PEGylated lysine-based dendrimers. Many other antimicrobial drugs have been successfully incorporated into dendrimer NPs for improved solubility and, hence prove to be therapeutically effective [34].

6.2.1.6 Drug-infused nanoparticles

In order to enhance the delivery and efficacy of antibiotics, nanoparticles and liposomes have been investigated as potential drug carriers because of their ability to be endocytosed and releasing phagocytic cells carrying intracellular pathogens [35].

Many studies have reported the use of nano sized vehicles to deliver antibiotics, including β-lactams such as Penicillin, Ampicillin, and Cephalosporin, as well as Macrolides (Azithromycin), Aminoglycosides and Fluoroquinolones by enhancing microbial killing [36–37].

Antifungals encapsulated into nanoparticles usually lead to enhanced efficacy against molds and yeasts. Activation of nano-carrier drug delivery by cellular activity, such as pH change or oxidative burst, can aid targeted therapy. Sometime extrinsic modes of targeting and activation, such as magnetic guidance and radio frequency mediated drug release can be used for effective localized delivery of nano-drugs [38].

6.2.1.7 Chitosan Nanoparticle

Chitosan is a natural polysaccharide biopolymer derived from chitin, which is the principal structural component of the crustacean exoskeleton. The antimicrobial properties of chitosan result from its polycationic character, which favors interaction with negatively-charged microbial cell wall and cytoplasmic membrane, resulting in decreased osmotic stability, membrane disruption and eventual leakage of intracellular elements. Chitosan is able to enter the nuclei of bacteria and fungi and inhibit mRNA and protein synthesis by binding to microbial DNA. Nano-scaled chitosan has a higher surface-to-volume ratio, translating into higher surface charge density, increased affinity to bacteria and fungi and greater antimicrobial activity. Several studies have demonstrated the efficacy of chitosan nanoparticles against a variety of pathogens, including Gram-negative *E.coli* and Gram-positive *S.aureus*. Chitosan nanoparticles were found to be more effective against these bacteria than chitosan alone.

In addition, chitosan's polycationic nature and high affinity to metal allow it to be used as a carrier system and platform stabilizer for a variety of other nanoparticles including silver- and copper-containing nanoparticles, nitric oxide-releasing nanoparticles, and drug-containing nanoparticles that allows for targeted delivery of various medications. Chitosan platforms also augment the antimicrobial properties of these nanoparticles. For example, the antimicrobial efficacy of silver-loaded membranes is enhanced up to 70% with increasing chitosan contents, resulting in larger zones of inhibition against both *S.aureus* and *E.coli* [39].

6.2.1.8 Silver Nanoparticle

It has been proposed that silver and silver ions (such as $AgNO_3$) penetrate into bacterial cell wall and membranes via interaction with sulfur-containing proteins or thiol groups. Once inside the cell, $AgNO_3$ targets and damages bacterial DNA and respiratory enzyme, leading to loss of the cell's replicating abilities and ultimately cell death. AgNPs have been shown to be effective against a variety of pathogens, including viruses, fungi, and many bacterial species including *E.coli*, *S.aureus*, *B.subtilis* and *S.typhi*. The stronge interaction of AgNPs with microbial surfaces might also allow for the use of lower drug concentrations as compared to current silver agents, and may limit silver's toxicity. AgNPs have also been found to augment the efficacy of other antimicrobial agents. In particular, the antibacterial activities of Penicillin G, Amoxicillin, Erythromycin, Clindamycin and Vancomycin increased against *S.aureus* and *E.coli* when mixed with AgNPs, with erythromycin having the greatest synergy with silver. Microbes are also less likely to develop resistance against silver and AgNPs, their broad range of targets would require multiple and simultaneous compensatory mutations. As a result, silver can be used to arrest bacterial resistance to antibiotics and enhance their efficacy. Given these properties, there are many diverse applications for AgNPs, including coatings on medical devices to prevent microbial colonization, wound dressings and augmentation of antibiotics [40].

6.2.1.9 Copper Nanoparticle

Copper oxide (CuO) is cheaper and more easily mixed with polymers as compared to Ag. The use of CuO nanoparticles (CuO-NPs) as a novel antimicrobial agent has recently been investigated. As compared to AgNPs, CuO-NPs are less effective against *E.coli* and Methicillin-resistant *S.aureus* but more effective against *B.subtilis*. These variations may be due to copper's greater interaction with amine and carboxyl groups on the cell surface of pathogens. Even though Ag may be a stronger antibacterial agent, Cu potentially has a broader range of activity, especially against fungi. The antimicrobial capacity of copper nanoparticle loaded polymer thin films (CuNP), which uniformly release copper, has been demonstrated against *S.cerevisiae* yeast, molds and bacteria including *E.coli*, *S.aureus* and *Listeria monocytogenes*. Therefore, despite a possibly weaker antibacterial activity CuNPs are effective agent against a wide range of bacteria and fungi and have many potential applications, including the prevention of microbial surface colonization [41].

6.2.1.10 Titanium Nanoparticle

Titanium dioxide (TiO_2) forms active oxygen species when exposed to ultraviolet light, a process called photo catalysis. These oxygen species, including hydrogen peroxide and hydroxyl radicals, obliterate bacterial cell membranes resulting in cell death. This antimicrobial property has been utilized in water and air purification and recently it has been investigated against pathogenic and opportunistic microorganisms. TiO_2nanoparticle-infused thin film composite (TFC) membranes have been shown to reduce *E.coli* biofilm formation by disrupting the bacterial membrane and thus inhibiting bacterial attachment to the membrane surface. TiO_2 has also been combined with silver to create TiO_2-Ag nanoparticles (TiO_2-AgNPs) that were tested against Gram-negative Gram-positive bacteria and various fungi that are responsible for opportunistic infection and colonization of medical devices. Based on these results, it might be possible to combine metal nanoparticles to augment the antimicrobial activity of each alone [42].

6.2.1.11 Magnesium Nanoparticles

Halogens such as chlorine, bromine and fluorine are well known for their bactericidal capabilities, but their significant toxicity limits their direct use. The antimicrobial activity of halogens is mediated by the formation of covalent metal-halogen complexes that interact with and inhibit specific cellular enzymes. Mg-halogen nanoparticle have been shown to damage the microbial cell envelope mediated by lipid peroxidation of the various metals. Magnesium oxide (MgO) is unique in its ability to absorb and retain halogens, and this capacity is increased up to five times in nanoparticle formulations (MgO-NPs). Combination with MgO-NPs not only increases the antimicrobial capability of the respective halogen, but also converts the halogen into an easy to handle powder form. This enhanced antimicrobial activity was demonstrated when nanoparticulate formulations of MgO, chloride and bromide were tested against *E.coli*, *B.megaterium* and *B.subtilis* endospores. *E.coli* and *B.megaterium* were extremely susceptible to the MgO-halogen nanoparticles, as both species were completely killed (100%) in as little as 20 min. The *B.subtilis* endospores were more resistant, with only 36% of bacteria killed in 20 min. The antibacterial effects of the nanoparticulate formulations were much stronger than the halogens alone, which resulted in 68% killing in 20 min. Similar results were found using magnesium fluoride nanoparticles (MgF_2-NPs), which reduced the growth of *E.coli* and *S.aureus* in a dose-dependent manner [43].

6.2.1.12 Zinc Nanoparticle

Another metal oxide of interest is zinc oxide (ZnO), a compound that is approved by the FDA as a result of its antimicrobial properties and safety profile. ZnO nanoparticles (ZnO-NPs) have received considerable attention by the food industry given their demonstrated efficacy against food-borne pathogens such as *E. coli* O157:H7, *Listeria monocytogenes* and *Salmonella* spp. ZnO-NPs have been shown to inhibit the growth of *E.coli* O157:H7 in a dose-dependent manner, with increasing inhibitory effects as the concentration of ZnO rises. These antimicrobial properties are mediated by the strong adherence of ZnO-NPs to bacterial cell membranes, which results in destruction of membrane lipids and proteins, altered membrane permeability and leakage of intracellular contents, much like the mechanism of other metal oxide nanoparticles. In addition, ZnO-NPs are thought to be strong inducers of reactive oxygen species that are harmful to bacterial cells [44].

6.2.1.13 Nitric oxide-releasing Nanoparticles

Nitric oxide (NO) has many roles in the human body that encompass virtually every physiological system, including host defense. When stimulated, phagocytic cells such as macrophages upregulate the production and release of NO through the transcription of inducible nitric oxide synthesis (iNOS). NO is then able to exert its antimicrobial effects through several mechanisms, including direct microbial DNA damage through the generation of peroxynitrite as well as interference of cellular respiration by inactivation of zinc metalloproteins. In addition, NO can stimulate several innate antimicrobial pathways, enhancing the host's own immune response.

One such delivery system capitalizes on the benefits of nanotechnology. Utilizing NO-releasing nanoparticles (NO-NPs) housing NO within a dry matrix, the system allows release of gaseous NO free radicals only upon exposure to moisture. This delivery system is ideal for the topical treatment of wounds and infections, as NO can be easily stored and applied to the skin, and provides sustained delivery to the affected areas over a prolonged period of time. These NO-NPs have been tested against a variety of pathogens including *S.aureus* and *Acinetobacter* with considerable efficacy. In a murine wound model, MRSA-infected full thickness wounds treated with NO-NPs demonstrate accelerated wound closure and less bacterial burden when compared to wounds that were untreated or treated with control nanoparticles (with no nitric oxide release) [45].

6.2.1.14 Immunomodulatory effects of nanotechnology-based drug delivery systems

The immune system acts as the body's major defense mechanism against foreign pathogens, and is comprised of the innate and adaptive systems. The innate or nonspecific immune system recognizes foreign pathogens, once they have crossed the body's physical barriers by recognizing microbial characteristics called pathogen associated molecular patterns (PAMPs). PAMPs are recognized by pattern-recognition receptors (PRPs) that reside on a variety of host cells, initiating antigen uptake by antigen presenting cells (APCs). APCs subsequently present the antigen to cells of the adaptive or specific, immune system for the induction of specific memory T and B cells, which leads to the downstream activation of CD8 and CD4 lymphocytes. This entire process is carefully coordinated by the interactions of various cytokines, which often dictate the type of response that is generated for a given pathogen.

For example, interferon-gamma (IFNγ) promotes a T helper type-1 (Th1) response that mediates antibody-independent immune responses, while interleukin-4 (IL-4) and IL-5 result in Th2 response, essential for antibody production. Thus, the harmonized interactions of APCs, T cells, B cells and inflammatory cytokines are imperative to an effective immune system [46–49].

6.2.1.15 Nanotechnology-based vaccines and immunostimulatory adjuvant

Vaccines are largely responsible for the reduction in mortalities caused due to infectious diseases. To be effective, a vaccine must induce both an innate and adaptive immune response in a manner that is safe and beneficial to the patients. There are various existing vaccines like live attenuated and subunit vaccines that display different degrees of immunogenicity and safety. Nanoparticles have the ability to modulate various aspects of the immune response and can improve the efficiency of all sorts of vaccines because of increased exposure time and uptake of antigens by APCs, improved immunogenicity of viral, bacterial components and modulation of the cytokine response. [50–51].

6.2.1.16 Synthetic polymers

Polymeric-based nanoparticles can act as carriers for a wide range of materials including protein or DNA vaccines. The characteristics of the nanoparticle *in vivo* is largely dependent on the type of polymer used, and therefore may be manipulated to fit the requirements of the particular vaccine. For example, to overcome compartmental differences of the digestive tract for oral vaccination, poly (ϵ-caprolactone) (PCL) polymers were used to create multi-component particles. These particles are made up of nanoparticle-encapsulated DNA surrounded by a PCL microparticle, known as the nanoparticle in- microsphere hybrid oral delivery system (NiMOS). Although synthetic polymers facilitate vaccine delivery, they do not seem to be immune stimulatory on their own. Thus, synthetic polymers cannot be used as immune stimulatory adjuvant alone, but they can enhance immunization via delivery of antigens across mucosal barriers. Polymethyl methyl methacrylate (PMMA) nanoparticles have demonstrated adjuvant immune stimulatory properties.

Nanoparticles modified with Toll-like receptor agonists have been shown to enhance immune responsiveness [52, 53].

6.2.1.17 Nanoemulsions

Nanoemulsions(NEs) encapsulate lipophilic or hydrophilic substances in a dispersed phase are manufactured as water-in-oil (W/O) or oil-in-water (O/W) systems. These carrier systems can be used in the development of mucosal vaccines, as they can be endocytosed by cells on the mucosal surface (epithelial or M cells) and subsequently delivered to APCs. In addition, NEs are themselves immune stimulatory and can be used to boost the immune response of a mucosal vaccine. The current hepatitis B vaccine require an intramuscular (IM) injection and consist of recombinant hepatitis B surface antigen (HBsAg) formulated with an aluminum salt (alum) adjuvant. NE-based vaccines have also been shown to be effective mucosal adjuvant for inactivated *Influenza* and *Vaccinia* viruses among other pathogens [54].

6.2.1.18 Immune-stimulating complexes

Immune-stimulating complexes (ISCOMs) are nanosized spherical micelles that act as carriers and immune- stimulatory adjuvants due to their saponin derived components such as Quil A, which originates from the bark of the Quillaja tree. [55].

6.2.1.19 Cytidine-phosphate-guanosine (CpG) motifs

Bacterial DNA sequences, specifically oligodeoxynucleotides containing unmethylatedCpG motifs have been found to be immune stimulatory and therefore have been used as vaccine adjuvant. When incorporated into nanoparticle formulations, CpG motifs induce a strong and sustained immune response. For example, a multiple nanoemulsion system (W/O/W) formulated with CpG and inactivated influenza virus (PELC/CpG) was found to induce a higher antigen- specific serum antibody response after only one dose when compared to two doses of a similar but non-adjuvant vaccine.

6.2.1.20 Fullerenes (C60) and fullerene-derivatives

Fullerenes antimicrobial properties are a very recent finding. Although native fullerenes are nearly aqueous-insoluble, they can be dispersed in water. In particular, numerous techniques for creating the stable colloidal C60 aggregates (nC60) in water are noted for their potent and broad antibacterial activity. The antibacterial mechanism for nC60 includes photo catalytic ROS production in eukaryotic cells. Antibacterial activity of nC60 to prokaryotic cells is mediated by lipid peroxidation in the cell membrane. The antimicrobial activity of carboxyfullerene is mediated by insertion into the cell wall, followed by disruption of the cell membrane structure. Fullerol can also be used as a drug carrier that bypasses the blood ocular barriers [56–58].

6.2.1.21 Carbon nanotubes (CNTs)

CNTs are cylindrical nanostructures made of pure carbon atoms covalently bonded in hexagonal arrays, and their unique optical, electrical, mechanical, and thermal properties have been of great interest. Single-walled nanotubes (SWNTs) are a single pipe with a diameter in the range of 1–5 nm, while multi-walled tubes (MWNTs) have several nested tubes with lengths varying from 100 nm up to several tens of micrometers.

Among various carbon-based nonmaterial, which are cytotoxic in general, SWNTs exhibit the strongest antimicrobial activity via combination of membrane and oxidative stress, possibly in a synergic way. The detailed antimicrobial mechanisms of SWNTs in three-steps: initial SWNT bacteria contact, membrane perturbation, and membrane oxidation in an electronic structure (i.e., metallic vs. semiconducting)-dependent manner. CNTs can also be used for antimicrobial photothermal therapy by delivering CNT nanoclusters to an infected area, followed by spontaneous bacterial adsorption to the clusters and selective destruction of drug-resistance microorganisms upon near infrared irradiation [59–60].

6.2.1.22 Surfactant-based nanoemulsions

Nanoemulsions are found to be thermodynamically stable and either transparent or translucent and show antimicrobial properties. Bactericidal properties of soybean oil-based nanoemulsion against Gram-positive, but not against enteric Gram-negative species are reported. Stable and antimicrobial O/W micro emulsions with various compositions of Tween 80, pentanol, and ethyl oleate were also obtained. These micro emulsions were found to be effective in killing *S.aureus* and resistant *P.aeruginosa* as well as biofilms of *P.aeruginosa*. NB-401, a mixture of BCTP and P10 liposome (Tween 60, Soybean oil, Glycerol mono oleate, Refined soya Sterols, and the Cationic Cetylpyridinium Chloride), is a fast-working antimicrobial agent against bacteria that are planktonically grown as a biofilm or in the sputum [61].

6.3 Nanocarriers

By using drug discovery system the pharmaceutical properties of drugs can be improved with particulate nanocarriers which are composed of lipids and polymers. Properties of nanopartilce such as potency in addition to stability, solubility, size and charge are important. Although carrier toxicity, metabolism and elimination or biodegradability is the problems which are yet to be solved [62].

6.3.1 Types of nanocarriers

6.3.1.1 Liposome

The attractive property of liposome is its biocompatibility and entrapping hydrophilic pharmaceutical agents in their internal compartment so that the incorporated drugs are protected from the inactivation and deliver drug in to the cells. Liposomes are more suitable than polymeric nanoparticles for the encapsulation of hydrophilic small drugs. Polyethylene glycol (PEG) – grafted liposomes (70 to 200 nm) are having extended circulation of half-lives [63].

6.3.1.2 Polymeric micelles

Micelles represent colloidal dispersions with particle size normally 5–100 nm range and polymeric micelles are more stable in comparison to micelles prepared from conventional detergents. Micelles posses an excellent ability to solubilize poorly water soluble drugs within the cores whereas polar molecule get adsorbed on the micelle surface and the substances within intermediated polarity distributed along surfactants within intermediated position. High loading capacity, controlled release profile and good compatibility between drug and core block, reduction of toxicity and other adverse effects are the characteristics. They enhance permeability to pathological areas with compromised leaky vasculature due to their small size [64].

6.3.1.3 Polymer blended nanoparticles

Among the polymeric nanoparticles for controlled drug delivery, biodegradable and biocompatible poly (D, Lactic acid/Poly (D, lactic/glycolic acid)/ (PLA/PLGA) based nanopartilces have been investigated as carriers for therapeutic bioactive molecules. The main advantage of PEG nanoparticles are their ability to control the release of encapsulated compound, since PEG-PLA helps to stabilize inner core, reduce droplet size, and encapsulate drugs [65–67].

6.3.1.4 Fluorescent Nanoparticles

Fluorescent NPs are considered as a new class of photostable highly-sensitive fluorescent tags for labeling various biological specimens such as cells and tissue samples. These fluorescent tags serve as new tools for biologists to perform sensitive bioimaging and sensing studies in real time. Fluorescent Qdots, NPs have been successfully used for imaging and sensing of various infectious diseases. Qdots are semiconductor nanocrystals, typically in the size range between 1 nm and 10 nm, composed of groups II–VI (e.g., CdSe) or II–V (e.g., InP) elements of the periodic table. Qdots are extremely bright, photo stable and possess high quantum yield. These properties are suitable for real-time sensitive imaging and sensing applications. Qdots exhibit size and composition tunable fluorescence properties that are suitable for multiplexed imaging using single excitation wavelength. Moreover, broad absorption spectra and narrow emission spectra allow for simultaneous excitation and detection, respectively of multi colorQdots. [68, 69].

6.4 Synergism of Antibiotics with Zinc Oxide Nanoparticles: A Study of Urinary Tract Infections

Bacterial urinary tract infections (UTIs) are common in outpatients and nosocomial settings. Commercially available antibiotics (III rd, IV th and Vth generation) are ineffective to overcome the infection caused by bacteria specially of *Enterobacteriaceae* family as they produced Extended spectrum β-lactamases. Nanoparticles in combination with β-lactam antibiotics damage the cell membrane of bacteria and finally lead to bacterial death. Recently Zinc Oxide is frequently used in several areas of Nanotechnology. Zinc Oxide nanoparticles are used in various commercial products such as Cosmetics, Sunscreens and are known for their antibacterial activity. In the present study electronic properties of Zinc Oxide nanoparticles are determined by their chemical composition which can be modulated by crystalline structures, defect, size, shape and morphology. All these properties of Zinc Oxide nanoparticles are characterized by X-ray Diffraction (XRD), Scanning Electron Microscopy (SEM), High Resolution Transmission Electron –Microscopy (HR-TEM), Selective Area Electron Diffraction (SAED), X-ray Photoelectron Spectroscopy (XPS) and UV-Visible spectroscopy. The characterized Zinc Oxide nanoparticles in combination with antibiotics revealed enhanced bactericidal activity.

6.4.1 Structural and Morphological Evaluation of Synthesized ZnO NPS

The XRD analysis (Figure 6.1) clearly revealed the sharp and distinct peaks. These peaks were analyzed with JCPDS data file 05–0664, and confirmed the presence of pure crystalline ZnO. Absence of zinc metal peaks indicates that

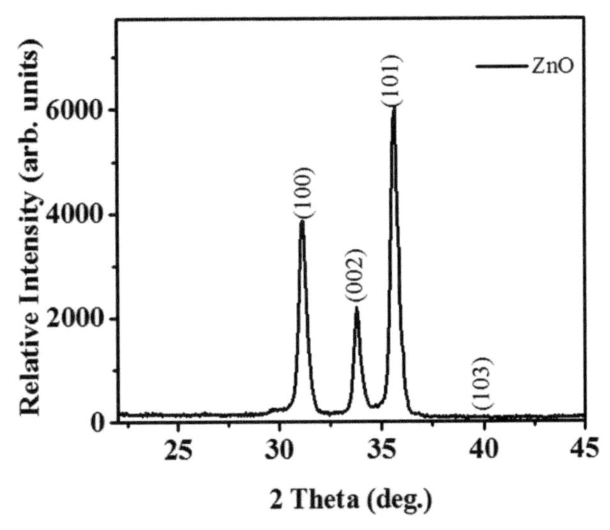

Figure 6.1 The XRD spectra of ZnO nanoparticles.

the deposited ZnOnanoparticles are highly pure in phase. The obtained four peaks corresponds to (100), (002), (101), and (103) reflections. The average grain size calculated from first three peaks is ~15 nm.

Surface morphological study was carried out using field-emission scanning electron microscopy (SEM) image. The synthesized ZnO nanoparticles were dispersed in water with 50 wt. % and ultrasonic treatment was given for 3 hrs to remove the agglomeration, and then spin-coated on to a cleaned glass slide for the scanning electron microscopy wherein, a uniform growth of ZnO spherical nanoparticles were obtained. Moreover, the ZnO nanoparticle film was free from the agglomeration effect indicating high dispersion rate in water (Figure 6.2).

6.4.2 Surface Analysis of Synthesized ZnO NPS

X-ray photoelectron spectroscopy (XPS) was used for contribution of Zn and O elements. A peak with a binding energy peak position at about 531 eV corresponding to O (1s).The peak on the low binding energy side of the O 1s spectrum to the O^{2-}ions in the wurtzite structure of the hexagonal Zn^{2+}ion array was surrounded by zinc atoms with the full supplement of

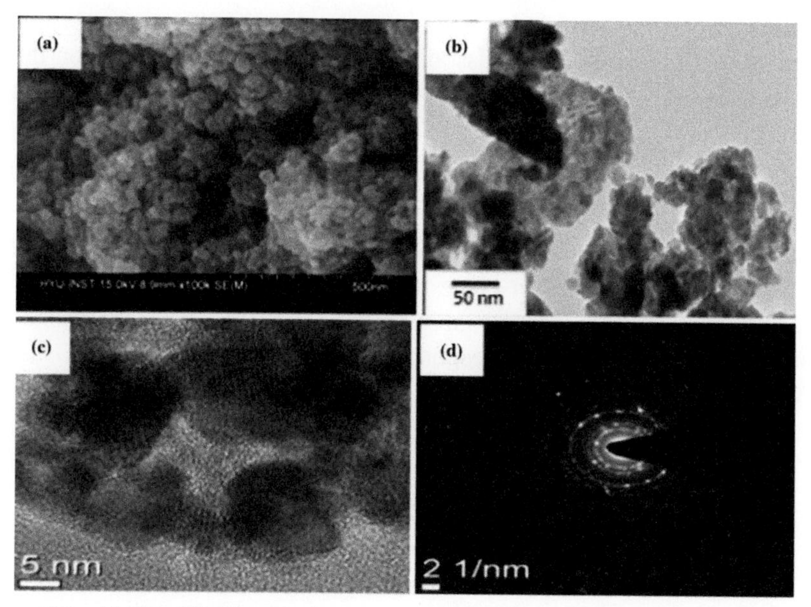

Figure 6.2 (a) The SEM image of synthesized ZnO NPs. (b) TEM images of ZnO NPs. (c) HR-TEM of ZnO NPs (d) SAED pattern.

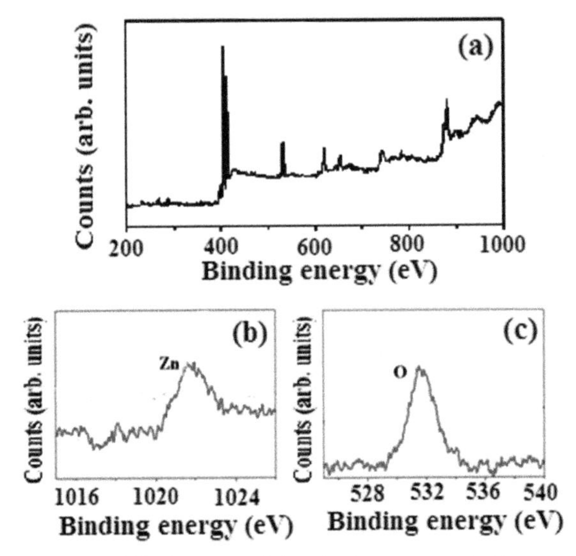

Figure 6.3 XPS spectra of ZnO nanoparticles. (a, b, c represents the scan over wide range and magnified band structure at Zn and O level).

nearest neighbor O^{2-}ions. Accordingly, this peak of the O 1s spectrum can be attributed to the Zn–O bonds. The higher binding energy at 531.96 eV is attributed to chemisorbed and dissociated oxygen and OH species on the surface of the ZnO thin film (such as adsorbed H_2O or O_2).The Zn 2p at 1022 eV is noticed (Figure 6.3).

6.4.3 Optical Analysis

The optical absorption measurement of ZnO nanoparticles film was recorded within 300–900 nm wavelength range by UV-Visible spectrum. Plot shows that ZnO nanoparticles are transparent in the visible range. Sharp absorbance edge at 390 nm wavelength, corresponds to band gap energy of ZnO i.e. 3.17 eV (Figure 6.4).

6.4.4 TIME–KILL ASSAY

The enhanced synergistic bioactivity of Zinc Oxide nanoparticles with standard β-lactam antibiotics were tested by Time-Kill curve. The dynamics of time–kill assay was monitored by broth dilution method against *E.coli, P. aeruginosa, S. paucimobilis*, and *K.pneumoniae*. All pathogens delayed the exponential phase when treated with antibiotic and ZnO NPs separately than

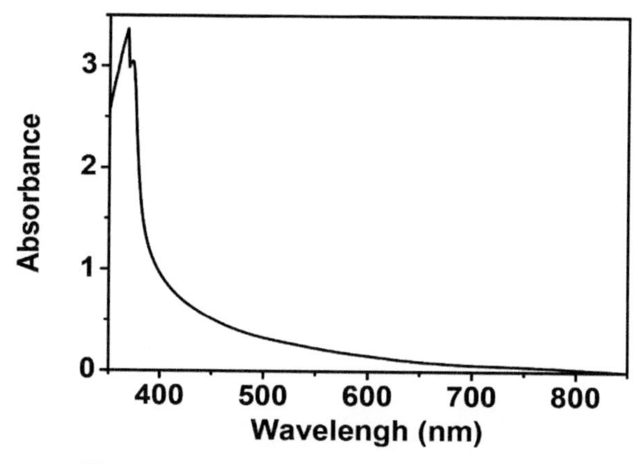

Figure 6.4 UV-Visible spectrum of ZnO NPs.

the normal exponential growth of pathogens (without nanoparticles and antibiotics). When *E.coli* (5×10^5cfu/ml) was treated with Cefotaxime, Ampicillin, Ceftriaxone and Cefepime alone, the exponential phase gets delayed similar to that of ZnO NPs but when the same *E.coli* were treated with combination of Cefotaxime + ZnO NPs, Ampicillin + ZnO NPs, Ceftriaxone + ZnO NPs and Cefepime + ZnO NPs a sudden decrease in exponential phase duration and growth transition was observed from the exponential to a very low stationary phase (Figure 6.5).

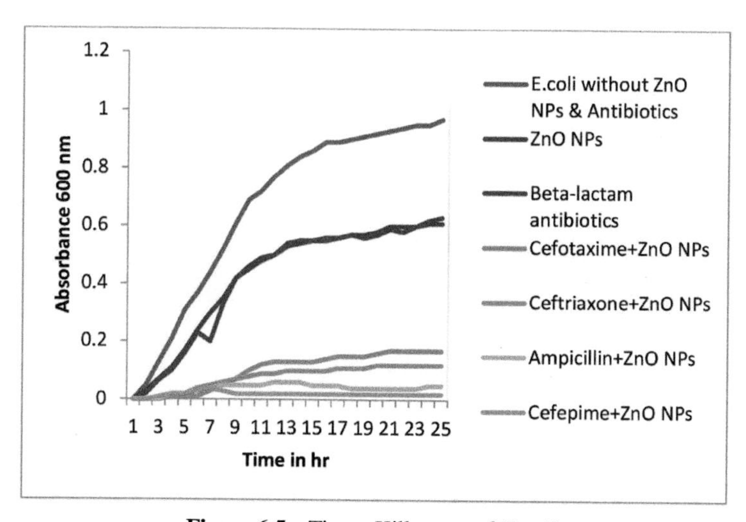

Figure 6.5 Time –Kill curve of *E.coli.*

A probable cause of this synergy may be the action of ZnO NPs as a drug carrier into *E.coli* cell membrane with standard β-lactam antibiotics. In case of *K.pneumoniae*, the synergy of Cefotaxime + ZnO NPs and Ceftriaxone + ZnO NPs suddenly decreased the duration of exponential as well as stationary phase and growth transition took place to a very low linear stationary phase (Figure 6.6).

The synergy of ampicillin + ZnO NPs and cefepime + ZnO NPs delayed the exponential phase. This increased antibacterial activity of ampicillin, cefepime, cefotoxime, and ceftriaxone was due to bonding reaction between antibiotic and ZnO NPs. The dynamics of time–kill assay revealed near about similar synergy pattern in both the bacteria, i.e., *P.aeruginosa* and *S.paucimobilis* (Figure 6.7, Figure 6.8).

In *P.aeruginosa*, cefetriaxone + ZnO NPs and in *S.paucimobilis*, ceftriaxone + ZnO NPs and cefepime + ZnONPs combinations exerted good synergistic inhibition and proved to be potent one for the drug therapy related to these bacterial infections. In this study, the synergy of ceftriaxone + ZnO NPs potentially inhibited almost all ESBL producers except *E.coli*. It is satisfactory that being a 3G cephalosporin, ceftriaxone therapy formulated with ZnO NPs can bring a new avenue for UTI treatment. Time–kill dynamics confirm that combination of 3G cephalosporins with ZnO NPs resulted in greater bactericidal effect on all ESBL producers than either of antibacterial agents applied alone.

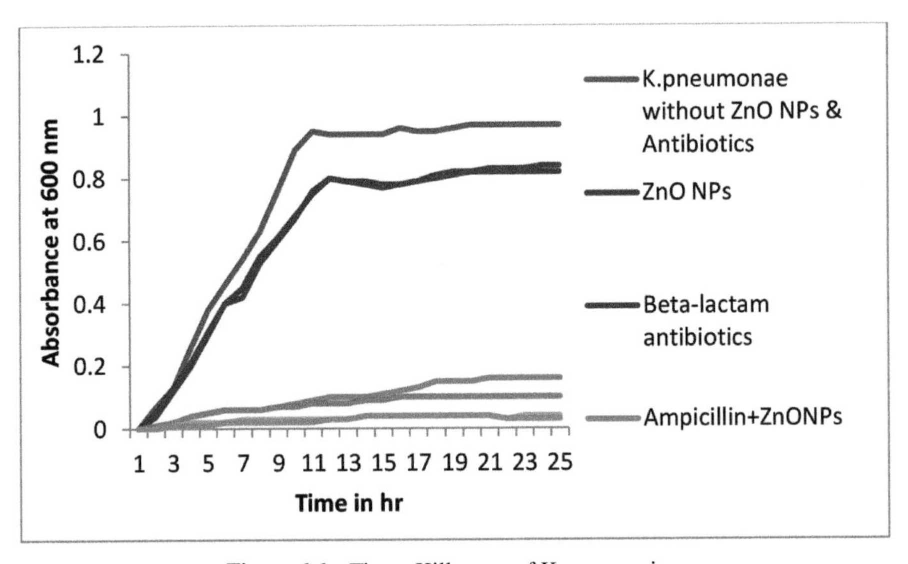

Figure 6.6 Time –Kill curve of K.pneumoniae.

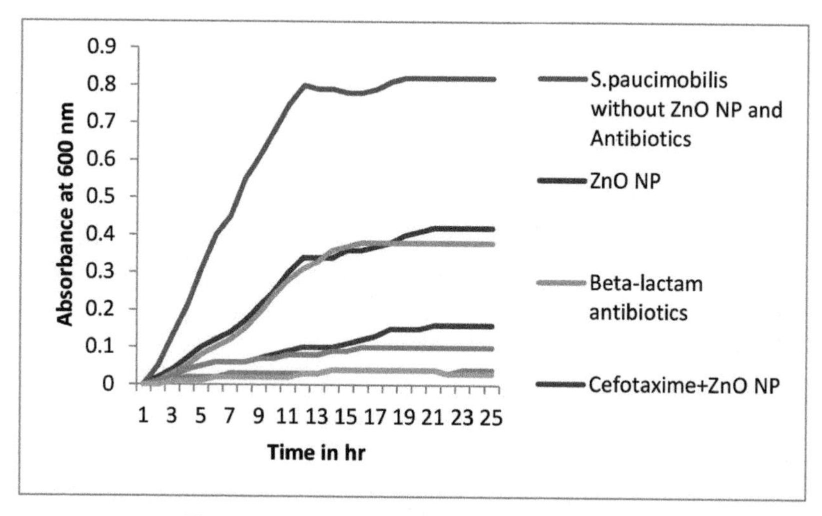

Figure 6.7 Time –Kill curve of *S.paucimobilis.*

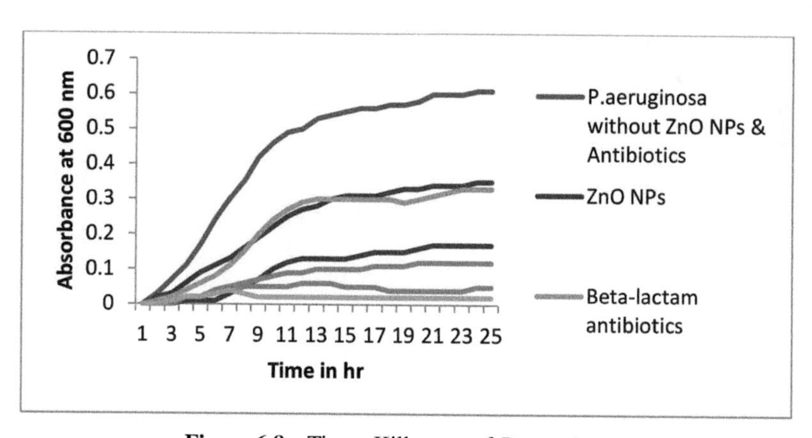

Figure 6.8 Time –Kill curve of *P.aeruginosa.*

6.5 Appications of Nanoparticles

Being an antimicrobial, antifungal agent nanoparticles possess wide applications.

- The metal based nanoparticles constitute an effective antimicrobial agent against common pathogenic microorganisms. Therefore, some of the nanoparticles such as silver, titanium dioxide and zinc oxide are receiving considerable attention as antimicrobials and additives in consumer, health-related and industrial products.

- Silver nanoparticles are used as additives in health related products such as bandages, catheters, and other materials to prevent infection, particularly during the healing of wounds and burns. An antibacterial Ag/Na carboxymethyl cotton in burn dressing by the partial cation exchange of sodium with silver has been developed and these can find applications in surgical dressings. They are currently being added to many common household products such as bedding, washers, water purification systems, tooth paste, shampoo, fabrics, deodorants, filters, paints, kitchen utensils, toys, and humidifiers to impart antimicrobial properties.
- Titanium dioxide NPs are used in cosmetics, filters that exhibit strong germicidal properties and remove odors in conjunction with silver as an antimicrobial agent. Moreover, due to the photo catalytic activity, it has been used in waste water treatment. Nowadays titanium dioxide nanoparticles are finding wide application as a self-cleaning and self-disinfecting material for surface coatings in many applications and in food industries for disinfecting equipments.
- Zinc oxide (ZnO) and copper oxide nanomaterials due to their antimicrobial property are being incorporated into a variety of medical and skin coatings. ZnO nanoparticles are used in the wallpapers in hospitals as antimicrobials. ZnO powder is an active ingredient for dermatological applications in creams, lotions and ointments on account of its antibacterial properties.
- Vaccine Development:The effectiveness of a vaccine is measured by its ability to interact and stimulate the immune system. The nano-engineering of vaccines allows the creation of better adjuvant and vaccine delivery systems. Currently, nanoparticles are being used in the design of nasal and transcutaneous vaccines. A nasal vaccine attempts to generate an immune response by exposing the nasal mucosa to antigens. Similarly, Transcutaneous vaccines target the immature dendritic cells (professional antigen-presenting cells) found in high density in the epidermis and dermis of the skin. Some of the promising nasal vaccines under development include Parainfluenza, Hepatitis B, Measles, *Yersinia pestis*, and HIV.
- Other applications:Drug delivery systems designed to interact with tissue in specific locations and times are currently being used in engineering. These systems should allow for more accurate targeting of therapeutic agents – allowing greater therapeutic effects through increased activity, and decreased adverse effects.

6.6 Conclusions

To overcome antibiotic resistance by developing more powerful drugs can lead to an only limited and temporary success and eventually contribute to developing greater resistance. Because of high surface area to volume ratio and unique physicochemical properties, nonmaterials are promising antimicrobial agents of a new class. Nanotechnology allows for the creation of unique carrier systems that enhance molecular interactions, thus allowing nanoparticles to facilitate the body's response to foreign pathogens. This includes surpassing microbial resistance mechanisms and improving the innate and adaptive immune response. Incorporating conventional antimicrobials or other materials into nanoparticle platforms also allows for targeted drug delivery and minimizes drug resistance. In the case of antimicrobial agents, the large surface-to-volume ratio of nanoparticles increases drug penetration by disrupting the microbial cell wall or cytoplasmic membrane. While most of the nanoparticle based antimicrobial drug delivery systems are currently in preclinical development, several have been approved for clinical use. With the ongoing efforts in this field, there is no doubt that nanoparticle-based drug delivery systems will continue to improve treatment to bacterial infections, especially in life-threatening diseases.

References

[1] Cohen, M.L.,"Changing patterns of infectious disease", Nature,Vol. 406, 2000, pp. 762–767.
[2] Gold, H.S., Moellering, R.C., "Antimicrobial-drug resistance", N. Engl. J. Med,Vol. 335, 1996, pp. 1445–1453.
[3] Walsh, C., "Molecular mechanisms that confer antibacterial drug resistance", Nature, Vol.406, 2000, pp. 775–781.
[4] Boucher, H.W., Talbot, G.H., Bradley, J.S., Edwards Jr., J.E., Gilbert, D., Rice, L.B., Scheld, M., Spellberg, B., Bartlett, J., "Bad bugs, no drugs: no ESKAPE! An update from the Infectious Diseases Society of America", Clin. Infect. Dis. Vol.48, 2009, pp. 1–12.
[5] Rice, L.B., "The clinical consequences of antimicrobial resistance", Curr. Opin. Microbiol. Vol.12, 2009, pp. 476–481.
[6] Talyor, P.W., Stapleton, P.D., Luzio, J.P. "New ways to treat bacterial infections, Drug Discov". Today, Vol7, No.21, 2002, pp. 1086–1091.
[7] Rai, M., Yadav, A., Gade, A., "Silver nanoparticles as a new generation of antimicrobials", Biotechnol. Adv. Vol.27, 2009, pp. 76–83.

[8] Goodman, C.M., McCusker, C.D., Yilmaz, Rotello, V.M., "Toxicity of gold nanoparticles functionalized with cationic and anionic side chains", Bioconjug. Chem. Vol.15, 2004, pp. 897–900.

[9] Schaller, M., Laude, J., Bodewaldt, H., Hamm, H.C., Korting, "Toxicity and antimicrobial activity of a hydrocolloid dressing containing silver particles in an ex vivo model of cutaneous infection", Skin Pharmacol. Physiol. Vol.17, 2004, pp. 31–36.

[10] Pal, S., Tak, Y.K., Song, J.M. "Dose the antibacterial activity of silver nanoparticles depend on the shape of the nanoparticle? A study of the gram-negative bacterium Escherichia coli", Appl. Environ. Microbiol. Vol.27, No. 6, 2007, pp. 1712–1720.

[11] Weir, E., Lawlor, A., Whelan, A., Regan, F., "The use of nanoparticles in anti-microbial materials and their characterization", Analyst, Vol.133, 2008, pp. 835–845.

[12] Allaker, R.P., Ren, G., "Potential impact of nanotechnology on the control of infectious disease", Trans. R. Soc. Trop. Med. Hyg. Vol.102, No. 1, 2008, pp. 1–2.

[13] Kolar, M., Urbanek, K., Latal, T., "Antibiotic selective pressure and development of bacterial resistance". Int J Antimicrob Ag. Vol. 17, 2001, pp. 357–363.

[14] Gajjar, P., Pettee, B., Britt, D.W., Huang, W., Johnson, W.P., Anderson, J., "Antimicrobial activities of commercial nanoparticles against an environmental soil microbe, Pseudomonas putida KT2440". Journal of Biological Engineering. Vol.3, 2009, pp. 9–22.

[15] Feynman, R., "plenty of room at the bottom". Science. Vol.254, 1991, pp. 1300–1301.

[16] Parak, W.J., Gerion, D., Pellegrino, T., Zanchet, D., Micheel, C., Williams, C.S., Boudreau,R., Le Gros, M.A., Larabell, C.A., Alivisatos, A.P. "Biological applications of colloidal nanocrystals". Nanotechnology.Vol.14, 2003, pp. 15–27.

[17] Whitesides, G.M., "The 'right' size in Nanobiotechnology". Nature Biotechnology. Vol.21, 2003, pp. 1161–1165.

[18] Gutierrez, F.M., Olive, P.L., Banuelos, A., Orrantia, E., Nino, N., Sanchez, E.M., Ruiz, F., Bach, H., Gay, Y.A., "Synthesis, characterization, and evaluation of antimicrobial and cytotoxic effect of silver and titanium nanoparticles". Nanomedicine. Vol.6, 2006, pp. 681–688.

[19] Coates, A., Hu, Y., Bax, R., Page, C., "The future challenges facing the development of new antimicrobial drugs". Nat Rev Drug Discov,Vol.1,2002, pp. 895–910.
[20] Walker, C.B., "Selected antimicrobial agents: mechanisms of action, side effects and drug interactions". Periodontol 2000, Vol.10,1996, pp. 12–28.
[21] Zhang, L., Gu, F.X., Chan, J.M., Wang, A.Z. Langer, R.S. Farokhzad, O.C., "Nanoparticles in medicine: therapeutic applications and developments". Clin. Pharmacol. Ther. Vol.83, 2008, pp. 761–769.
[22] Davis, M.E. Chen, Z.G. Shin, D.M., "Nanoparticle therapeutics: an emerging treatment modality for cancer". Nat Rev Drug Discov, Vol.7, 2008, pp. 771–782.
[23] Peer, D. Karp, J.M. Hong, S., Farokhzad, O.C., Margalit, R., Langer, R., "Nanocarriers as an emerging platform for cancer therapy". Nat Nanotechnol Vol.2, 2007, pp. 751–760.
[24] Wagner, V., Dullaart, A., Bock, A.K., Zweck, A., "The emerging nanomedicine landscape". Nat. Biotechnol.Vol.24,2006, pp. 1211–1217. [25] French, G.L., "The continuing crisis in antibiotic resistance". Int J Antimicrob Agents. Vol.36, No.3, 2010, pp. S3–S7.
[25] Zhang, L., Jiang, Y., Ding, Y., Povey, M., York, D., "Investigation into the antibacterial behaviour of suspensions of ZnO nanoparticles (ZnOnanofluids)". J Nanoparticle Res. Vol.9, No.3, 2007, pp. 479–489.
[26] Arbab, A.S., Wilson, L.B., Ashari, P., Jordan, E.K., Lewis, B.K., Frank, J.A., "A model of lysosomal metabolism of dextran coated super paramagnetic iron oxide (SPIO) nanoparticles: implications for cellular magnetic resonance imaging". NMR Biomed. Vol.18, No.6, 2005, pp. 383–389.
[27] Baker, C., Pradhan, A., Pakstis, L., Pochan, D.J., Shah, S.I., "Synthesis and antibacterial properties of silver nanoparticles". J NanosciNanotechnol. Vol.5, No.2, 2005, pp. 244–249.
[28] Suci, P.A., Berglund, D.L., Liepold, L., "High-density targeting of a viral multifunctional nanoplatform to a pathogenic, biofilm-forming bacterium". Chem Biol. Vol.14, No.4, 2007, pp. 387–398.
[29] Kim, B.Y.S., Rutka, J.T., Chan, W.C., "Nanomedicine". N En J Med, Vol.363, 2010; pp. 2434–2443.
[30] Zhang,L.,Huang,C.M.,"Development of Nanoparticles for antimicrobial drug delivery" Vol.17,2010, pp. 585–594.

[31] Pinto-Alphandary, H., Andremont, A., Couvreur, P., "Targeted delivery of antibiotics using liposomes and nanoparticles: research and applications". Int J Antimicrob Agents, Vol.13, 2000, pp. 155–168.
[32] Murray, J., Brown, L., Langer, R. "Controlled release of micro quantities of macromolecules", Cancer Drug Deliv. Vol.1, No.2, 1984, pp. 119–123.
[33] Bosman, A. W.; Janssen, H. M.; Meijer, E. W. About Dendrimers: Structure, Physical Properties, and Applications. Chem. Rev. Vol. 99, 1999, pp. 665–88.
[34] Mihu, M.R., Sandkovsky, U., Han, G., Friedman, J.M., Nosanchuk, J.D., Martinez, L.R., "Nitric oxide releasing nanoparticles are therapeutic for Acinetobacterbaumanni wound infections". Virulence, Vol.1, 2010, pp. 62–67.
[35] Vieira, D.B., Carmona-Ribeiro, A.M., "Cationic nanoparticles for delivery of amphotericin B: preparation, characterization and activity in vitro". J Nanobiotechnology Vol.6, 2008, pp. 6.
[36] Abeylath, S.C., Turos, E., Dickey, S., Lim, D.V., "Glyconanobiotics: Novel carbohydrated nanoparticle antibiotics for MRSA and Bacillus anthracis". Bioorg Med Chem, Vol.16, 2008, pp. 2412–2418.
[37] Fattal, E., Rojas, J., Youssef, M., Couvreur, P., Andremont, A., "Liposome-entrapped ampicillin in the treatment of experimental murine listeriosis and salmonellosis". Antimicrob Agents Chemother,Vol. 35, 1991, pp. 770–772.
[38] Sanpui, P., Murugadoss, A., Prasad, P.V., Ghosh, S.S., Chattopadhyay, A., "The antibacterial properties of a novel chitosan-Ag-nanoparticle composite". Int J Food Microbiol, Vol.124, 2008, pp. 142–146.
[39] Pal, S.T.Y., Song, J.M., "Does the antibacterial activity of silver nanoparticles depend on the shape of the nanoparticles? A study of the gram-negative bacterium Escherechia coli". Appl Environ Microbiol, Vol.27, 2007, pp. 1712–1720.
[40] Ruparelia, J.P., Chatterjee, A.K., Duttagupta, S.P., Mukherji, S., "Strain specificity in antimicrobial activity of silver and copper nanoparticles". ActaBiomater, Vol.4, 2008, pp. 707–716.
[41] Martinez-Gutierrez, F., Olive, P.L., Banuelos, A., Orrantia, E., Nino, N., Sanchez, E.M., "Synthesis, characterization and evaluation of antimicrobial and cytotoxic effect of silver and titanium nanoparticles". Nanomedicine, Vol.6, 2010, pp. 681–688.

[42] Lellouche, J., Kahana, E., Elias, S., Gedanken, A., Banin, E., "Antibiofilm activity of nanosized magnesium fluoride". Biomaterials, Vol.30, 2009, pp. 5969–5978.

[43] Liu, Y., He, L., Mustapha, A., Li, H., Hu, Z.Q., Lin, M., "Antibacterial activities of zinc oxide nanoparticles against Escherichia coli O157:H7". J ApplMicrobiol, Vol.107, 2009, pp. 1193–1201.

[44] Han, G., Martinez, L.R., Mihu, M.R., Friedman, A.J., Friedman, J.M., Nosanchuk, J.D., "Nitric oxide releasing nanoparticles are therapeutic for Staphylococcus aureus abscesses in a murine model of infection". PLoS One, Vol.4, 2009, pp. 7804.

[45] Plummer, E.M., Manchester, M., "Viral nanoparticles and virus-like particles: platforms for contemporary vaccine design". Wiley Interdiscip Rev Nanomed Nanobiotechnol, Vol. 3, 2010, pp. 174–196.

[46] Huang, M., "Emulsified nanoparticles containing inactivated influenza virus and cpgoligodeoxynucleotides critically influence the host immune response in mice". PLoS One, Vol. 5, 2010, pp. 12270.

[47] Nasir, A., "Nanotechnology in vaccine development: a step forward". J Invest Dermatol Vol.129, 2009, pp. 1055–1059.

[48] Liu, L., Zhong, Q., Tian, T., Dubin, K., Athale, S.K., Kupper, T.S., "Epidermal injury and infection during poxvirus immunization is crucial for the generation of highly protective T cell-mediated immunity". Nat Med, Vol.16, 2010, pp. 224–227.

[49] Combadiere, B., Vogt, A., Mahe, B., Costagliola, D., Hadam, S., Bonduelle, O., "Preferential amplification of CD8 effector-T cells after transcutaneous application of an inactivated influenza vaccine: a randomized phase I trial". PLoS One, Vol.5,2010, pp. 10818.

[50] Mahe, B., Vogt, A., Liard, C., Duffy, D., Abadie, V., Bonduelle, O., "Nanoparticle-based targeting of vaccine compounds to skin antigen-presenting cells by hair follicles and their transport in mice". J Invest Dermatol, Vol.129, 2009, 1156–1164.

[51] Csaba, N., Sanchez, A., Alonso, M.J., "PLGA:poloxamer and PLGA:poloxamine blend nanostructures as carriers for nasal gene delivery". J Control Release, Vol.113, 2006, 164–172.

[52] Rajananthanan, P., Attard, G.S., Sheikh, N.A., Morrow, W.J., "Evaluation of novel aggregate structures as adjuvants: composition, toxicity studies and humoral responses." Vaccine, Vol.17, 1999, pp. 715–730.

[53] Tolentino, M., "Systemic and ocular safety of intravitreal anti-VEGF therapies for ocular neovascular disease". SurvOphthalmol, Vol.56, 2011, pp. 95–113.

[54] Makidon, P.E., Bielinska, A.U., Nigavekar, S.S., Janczak, K.W., Knowlton, J., Scott,A.J., "Pre-clinical evaluation of a novel nanoemulsion-based hepatitis B mucosal vaccine". PLoS One, Vol.3, 2008, pp. 2954.
[55] Li, Q., Mahendra, S., Lyon, D.Y., Brunet, L., Liga, M.V., Li, D., Alvarez, P.J., "Antimicrobial nanomaterials for water disinfection and microbial control: potential applications and implications", Water Res,Vol. 42, 2008.pp. 4591–4602.
[56] Tsao, N., Luh, T.Y., Chou, C.K., Chang, T.Y., Wu, J.J., Liu, C.C., Lei, H.Y. "In vitro action of carboxyfullerene", J. Antimicrob. Chemother,Vol. 49, 2002, pp. 641–649.
[57] Brant, J.A., Labille, J., Bottero, J.Y., Wiesner, M.R., "Characterizing the impact of preparation method on fullerene cluster structure and chemistry", Langmuir,Vol. 22, 2006, pp. 3878–3885.
[58] Lyon, D.Y., Brunet, L., Hinkal, G.W., Wiesner, M.R., Alvarez, P.J., "Antibacterial activity of fullerene water suspensions (nC60) is not due to ROS-medicated damage", NanoLett. Vol. 8, No. 5, 2008, pp. 1539–1543.
[59] Hyung, H., Fortner, J.D., Hughes, J.B., Kim, J.H. "Natural organic matter stabilizes carbon nanotubes in the aqueous phase", Environ. Sci. Technol.Vol. 41, No.1, 2007, pp. 179–184.
[60] Arias, L.R., Yang, L., "Inactivation of bacterial pathogens by carbon nanotubes in suspensions", Langmuir,Vol. 25, 2009, pp. 3003–3012.
[61] Warheit, D.B., Laurence, B.R., Reed, K.L., Roach, D.H., Reynolds, G.A., Webb, T.R., "Comparative pulmonary toxicity assessment of single-wall carbon nanotubes in rats", Toxicol. Sci.Vol. 77, No.1, 2004, pp. 117–125.
[62] Makidon, P.E., Bielinska, A.U., Nigaverkar, S.S., Janczak, K.W., Knowlton, J., Scott, A.J., Mank, N., Cao, Z., Rathinavelu, S., Beer, M.R., Wilkinson, J.E., Blanco, L.P. Landers, J.J. Baker Jr., J.R., "Pre-clinical evaluation of a novel nanoemulsion-based hepatitis B mucosal vaccine", PLoS One Vol.3, No. 8, 2008, pp. e2954.
[63] Zhang, L., Pornpattananangkul, D., Hu, C.M., Huang, C.M., "Development of nanoparticles for antimicrobial drug delivery", Curr. Med. Chem. Vol.17, 2010, pp. 585–594.
[64] Torchilin, V.P., "Recent advances with Liposomes as Pharmaceutical Carriers". Nature Rev., Vol.4, 2005, pp. 145–160.
[65] Torchilin, V.P., "MicellarNanocarriers: Pharmaceuticals perspective". PharmaceutRes,Vol. 24,2007, pp. 1–16.
[66] Mundargi, R.E.C., Babu, V.R., Raghavendra, C.M., Pate, P., Aminabhavi, T.M., "Nano/micro technologies for delivering macromolecular

therapeutics using poly (D/L-Lactic-coglycolide) and its derivatives".J Control Release,Vol.125,2008, pp. 193–209.

[67] Beracchia, M.T., "Stealth nanoparticles for intravenous administration". S.T.P. PharmScience, Vol.13, 2003, pp. 155–161.

[68] Agrawal, A., Tripp, R.A., Anderson, L.J., Nie, S., "Real-time detection of virus particles and viral protein expression with two-color nanoparticle probes", J. Virol. Vol.79, 2005, pp. 8625–8628.

[69] Bentzen, E.L., House, F., Utley, T.J., Crowe, J.E., Wright, D.W., "Progression of respiratory syncytial virus infection monitored by fluorescent quantum dot probes", NanoLett. Vol.5 2005, pp. 591–595.

[70] Rashmi M Bhande, C N Khobragade, R S Mane, S Bhande (2012) Enhanced synergism of antibiotics with Zinc oxide nanoparticles against Extended spectrum β-lactamase producers implicated in Urinary Tract Infections. Journal of nanoparticle research, pp. 5–15.

7

Nanomedicine for the Treatment of Oxidative Stress Injuries

Toru Yoshitomi[1] Long Binh Vong[1] and Yukio Nagasaki[1,2,3]

[1]Department of Materials Science, Graduate School of Pure and Applied Sciences, University of Tsukuba, Tsukuba, Ibaraki 305-8573, Japan
[2]Master's School of Medical Sciences, Graduate School of Comprehensive Human Sciences, University of Tsukuba, Tsukuba, Ibaraki 305-8573, Japan
[3]Satellite Laboratory, International Center for Materials Nanoarchitectonics (WPI-MANA), National Institute for Materials Science, University of Tsukuba, Tsukuba, Ibaraki 305-8573, Japan

7.1 Introduction

While reactive oxygen species (ROS) are produced as a consequence of cellular respiration, excessively generated ROS causes several pathological conditions in many cellular systems [1]. ROS can directly damage cellular components such as lipids, DNA, and proteins via oxidation. Excessive levels of ROS continuously aggravate inflammation, thereby increasing the risk of potentially life-threatening disorders such as stroke, myocardial infarction, acute kidney failure, colitis, and so on. Oxidative damage in living organisms is governed by the balance between production of ROS and the effectiveness of antioxidant-protective systems including vitamin C, vitamin E, and glutathione, as well as antioxidant enzymes such as superoxide dismutase, catalase, and glutathione peroxidase. Under normal conditions, ROS are eliminated by these antioxidant protective systems. However, under condition of oxidative stress, free radical production increases remarkably and even endogenous antioxidants fail to scavenge all of the ROS, resulting

Post-genomic Approaches in Cancer and Nano Medicine, 199–218.

in severe injuries. One of the plausible ways to prevent oxidative stress injuries is the administration of exogenous ROS scavengers; for example, edaravone, a low-molecular-weight (LMW) ROS scavenger used for the treatment of stroke, which has been shown to inhibit lipid peroxidation and vascular endothelial cell injury, and to ameliorate brain edema, tissue injury, delayed neuronal death, and neurological deficits [2]. Edaravone is the first neuroprotective agent approved for clinical use in Japan [3]. However, such LMW ROS scavengers have limitations for clinical applications because of their severe adverse effects such as renal toxicity and rapid clearance by the kidney, resulting in lowered accumulation efficiency of the ROS scavenger in the target organ. To overcome the problems associated with conventional LMW ROS scavengers, development of new technology for the delivery of ROS scavengers is required. Recently, we have been focusing on ROS scavenging by a redox polymeric nanoparticle, designed in our laboratory which acts as a nanosized ROS scavenger. In this chapter, the development of redox nanoparticle therapeutics is discussed in detail. The topics include the synthesis of redox polymers, the preparation of redox nanoparticles, and the treatment of oxidative stress injuries by using redox polymer therapeutics.

7.2 Preparation and Characterization of Redox Polymers and Nanoparticles

For the preparation of redox nanoparticles as effective nanosized ROS scavengers, the molecular design of the block copolymer as a basal material is very important in the control of biodistribution and the on-off regulation of drug efficacy. Here we describe the preparation and characterization of redox polymers and nanoparticles possessing nitroxide radicals.

7.2.1 Design and Preparation of Redox Polymers and Redox Nanoparticles

Nitroxide radicals are chemically synthesized organic compounds possessing an unpaired electron, resulting in paramagnetic properties and redox reactions. They are known to have many applications such as ROS scavengers [4, 5], electron spin resonance (ESR) spectroscopic probes such as spin label/oxymetry [6], ESR and magnetic resonance (MR) imaging agents [7], and oxidizing agents [8]. 4-Hydroxy-2,2,6,6-tetramethylpiperidine-1-oxyl

(TEMPOL) is the most extensively utilized nitroxide radical in biological fields. TEMPOL is a cell membrane-permeable compound that reacts with superoxide, hydroxyl radicals and alkyl peroxyl radical, and so on. It has been shown to have a broad effect on detoxifying ROS in cell and animal studies of gastric mucosal injury [9], colitis [10], cerebral ischemia-reperfusion (IR) [11], and myocardial IR [12]. However, application of LMW nitroxide radicals *in vivo* is difficult because of the preferential renal clearance and adverse effects such as antihypertensive activity. To solve these issues, we have developed a redox nanoparticle (RNP) prepared using a self-assembling amphiphilic block copolymer possessing nitroxide radicals in the hydrophobic segment for the treatment of oxidative stress injuries and bioimaging (Figure 7.1) [13–24]. Thus far, we have designed two types of RNPs: one is RNP^N, which consists of poly(ethylene glycol)-b-poly[4-(2,2,6,6-tetramethylpiperidine-1-oxyl)aminomethylstyrene] (PEG-b-PMNT) and has a pH-sensitive disintegration character, and the other is RNP^O, which

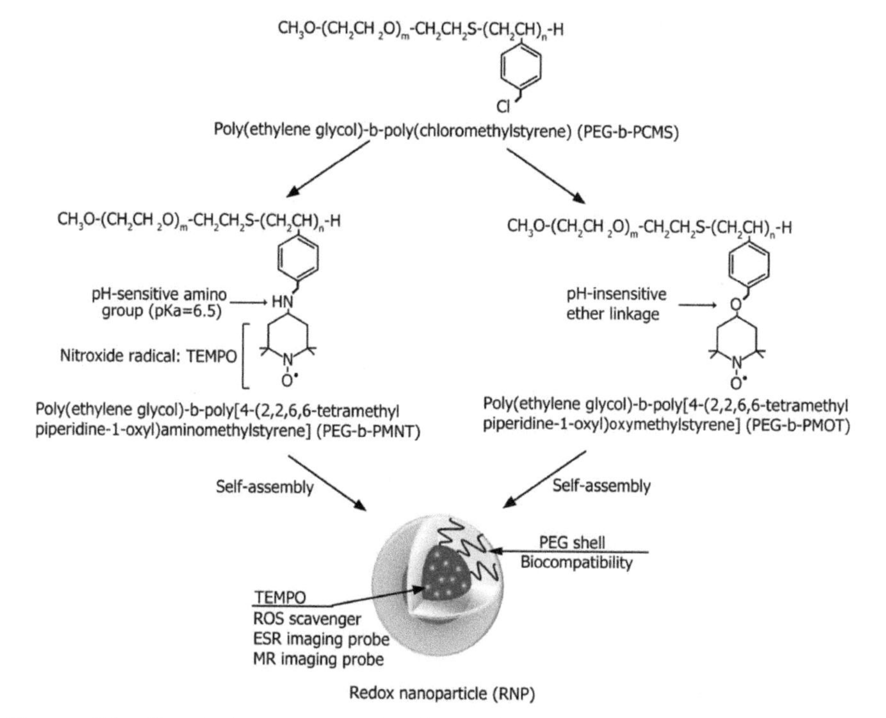

Figure 7.1 Chemical structures of redox polymers possessing nitroxide radicals, PEG-b-PMNT and PEG-b-PMOT, and a redox nanoparticle (RNP).

consists of poly(ethylene glycol)-b-poly[4-(2,2,6,6-tetramethylpiperidine-1-oxyl)oxymethylstyrene] (PEG-b-PMOT) and has a pH-insensitive character. The important points of design for these polymers are described. PEG, which is the hydrophilic segment of the amphiphilic diblock copolymer, is one of the most widely used polymers as remarkable biocompatible materials due to its strong hydration ability and conformational flexibility [25]. To introduce nitroxide radicals in the polymer, poly(chloromethylstyrene) (PCMS) is employed as the hydrophobic segment of the amphiphilic diblock copolymer because nitroxide radicals are easily introduced as side chains of the hydrophobic segment via two types of nucleophilic reactions using 4-amino-TEMPO and TEMPOL. PEG-b-PCMS is synthesized by the free-radical telomerization of chloromethylstyrene by using PEG derivatives possessing a sulfanyl group at one end as a telogen [13]. When PEG-b-PCMS is mixed with TEMPOL in the presence of sodium hydride in dry *N,N*-dimethylformamide, PEG-b-PMOT is obtained almost quantitatively. When 4-amino-TEMPO is used instead of TEMPOL in dimethylsulfoxide, PEG-b-PMNT is obtained.

Both PEG-b-PMNT and PEG-b-PMOT form core-shell-type polymeric micelles, RNP^N and RNP^O, respectively, in aqueous media by the dialysis method. Both RNPs are 40 nm in diameter, as determined by dynamic light scattering, and show unique ESR spectra. In contrast with the free-TEMPO signal, which is a clear triplet ESR signal corresponding to an interaction between ^{14}N nuclei and the unpaired electron in the dilute solution, the ESR signals of both RNPs broaden after dialysis, which is attributable to the restricted mobility and exchange interaction of the nitroxide radicals in the hydrophobic core of RNPs. This is one of the indications of the core-shell-type structure of RNPs.

7.2.2 Safety of RNPs

The entrapment of nitroxide radicals in the hydrophobic core of polymeric micelles via covalent conjugation is one of the important strategies for suppressing the toxicity and adverse effects of nitroxide radicals. Many types of physically drug-loaded nanoparticles have been reported and studied in the field of drug delivery systems because drug delivery by nanoparticles such as polymeric micelles and liposomes can alter the pharmacokinetics of drugs. However, as the leakage of drug from the nanoparticle occurs, it leads to adverse effects. As RNPs conjugate with nitroxide radicals in the hydrophobic core via covalent linkage, there is no leakage of nitroxide radicals *in vivo* and the toxicity of the nitroxide radical is suppressed. The toxicity of TEMPO derivatives and RNP^N has been reported in a cell study [15]. With an increase in the concentrations of LMW TEMPO and 4-amino-TEMPO, cell viabilities are gradually reduced at 24 h after drugs are added to cultures. The median inhibitory concentrations (IC_{50}) of TEMPO and 4-amino-TEMPO are 8.3 and 4.8 mM, respectively. In contrast, RNP^N shows almost no cytotoxicity up to a concentration of 8 mM, which is in sharp contrast to the finding obtained for LMW TEMPO and 4-amino-TEMPO. The median lethal dose (LD_{50}) of RNP^N in ICR mice has also been reported [15]. No mice die after the intravenous administration of RNP^N at a concentration of up to 300 mg/kg for 2 weeks. Even at 600 mg/kg (concentration of 4-amino-TEMPO moieties: 960 μmol N [N means the number of 4-amino-TEMPO moieties]/kg), 60% of the mice remain alive. The LD_{50} of poly($_L$-lysine) with a molecular weight of 28,000–42,000 (polymerization degree = 135 – 203) has been reported to be between 15 and 30 mg/kg (concentration of amino groups: 72.5 – 145 μmol N/kg) [26]. The extremely low toxicity of RNP^N, *viz.*, IC_{50} of $>$8 mmol N/L and LD_{50} of $>$600 mg/kg ($>$960 μmol N/kg), is considered to be attributable to the confinement of the 4-amino-TEMPO moieties in the hydrophobic core of RNP^N. Moreover, the antioxidant character of the RNPs might also contribute to the reduced toxicity.

LMW nitroxide radical compounds are also known to show a dose-related antihypertensive action accompanied by reflex tachycardia, increased skin temperature, and seizures [27]. Hypotensive action would lead to serious complications during treatment and surgery. Interestingly, both RNPs inhibit any decrease in arterial blood pressure, although LMW TEMPOL and 4-amino-TEMPO reduce arterial blood pressure dramatically. These results indicate the importance of the encapsulation of nitroxide radicals in the hydrophobic core of the nanoparticles via covalent linkage, because of the lack of leakage of nitroxide radicals from RNPs. Moreover, LMW TEMPO derivatives have been reported to induce mitochondrial dysfunction because they pass through the cell membrane and cause inessential redox reactions. Given that the nanoparticles can suppress uptake into normal cells unlike LMW compounds, RNPs must suppress LMW nitroxide radical-induced mitochondrial dysfunction. This may be one of the reasons for the extremely low toxicity of RNPs both *in vivo* and *in vitro*.

7.2.3 pH-Sensitive Disintegration of RNP^N

Polyamine with the appropriate hydrophobicity is known to show phase transition as a function of pH [28], for example, the poly[2-(*N*,*N*-diethylamino)ethyl methacrylate] (PEAMA) homopolymer exhibits precipitation above pH 7.5 owing to the deprotonation of the amino groups of PEAMA [29]. As the PMNT segment possesses both a hydrophobic phenyl group and an amino group in each repeating unit, RNP^N shows a similar phase-transition phenomenon in response to pH, owing to the protonation of the amino groups. The pK_a value of PMNT is approximately 6.5, indicating that it is precipitated above pH 6.5, whereas it is soluble below pH 6.5. As shown in Figure 7.2a, the scattering intensity of RNP^N drastically decreases at pH below 7.0, indicating that RNP^N disintegrates in an acidic environment. In contrast, the scattering intensity of RNP^O does not change in response to pH because of the lack of amino groups in its hydrophobic segment. Interestingly, ESR spectra of RNPs give information about the morphological change of nanoparticles. Figures 7.2b and 7.2c show the ESR spectra of both RNPs from pH 5.6 to 8.2. The ESR spectra of RNP^O do not change in response to pH. In contrast, although the ESR spectra of RNP^N broaden at pH values above 7.4, the ESR signals change to typical triplet signals under acidic pH ($pH < 6.5$) conditions, which is consistent with the disintegration region of RNP^N. Thus, RNP^N disintegrates in acidic regions such as tumor and inflammation sites. Given this property, it can expose nitroxide radicals from the RNP^N core and effectively scavenge ROS in response to low pH environments due to the disintegration of RNP^N. Furthermore, the ESR spectra of RNPs enable the evaluation of the biodistribution and morphological changes of RNPs without the need for specific labelling with fluorescence or radioisotope for example.

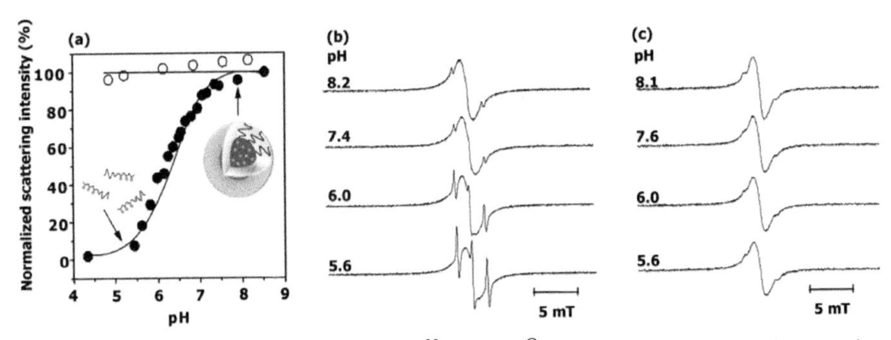

Figure 7.2 *In vitro* characterization of RNP^N and RNP^O **(a)** Effect of pH on the light scattering intensities of RNP^N (closed circle) and RNP^O (open circle). The normalized scattering intensity (%) is expressed as the value relative to that at pH 8.2 **(b,c)** X-band ESR spectra of (b) RNP^N and (c) RNP^O from pH 5.6 to 8.2. Reprinted with permission from [15] and [19] and modification. © American Chemical Society (2009) and Elsevier B.V. (2011), respectively.

7.3 Treatment of Ischemia-Reperfusion Injuries with pH-Sensitive Redox Nanoparticles

Nanoparticles with highly dispersible and biocompatible properties are known to accumulate in specific regions such as tumors and inflamed areas along with long circulation in blood, referred to as the enhanced permeability and retention effect, because of the changes in the specific vascular microenvironment, such as leaking neovascular vessels and immature lymphatic systems [30]. At most, they accumulate at levels of up to 10% of the injected dose. However, more than 90% of the nanoparticles are distributed non-specifically *in vivo* after systemic administration [31]. Active targeting is one of the challenges to improve the accumulation of nanoparticles in specific disease areas; however, thus far, no remarkable effect on their biodistribution has been reported. One promising strategy to improve the efficiency of nanotherapy is "on-off regulation," whereby the nanoparticle is dormant in non-target tissue and is activated in the target area, thus improving treatment outcomes and decreasing adverse effects [32]. To date, drug-loaded stimuli-sensitive nanoparticles, which are sensitive to pH, temperature, or ionic strength, have been proposed [33]. However, the leakage of drug from the nanoparticle into the blood stream would decrease its therapeutic efficiency and cause severe adverse effects because the signal-switching characteristics often decrease the entrapping efficiency and/or loading stability of the drug in the nanoparticle [34]. To solve these issues, we have paid attention to the covalent conjugation of the drug to the polymer backbone to prevent drug leaking. In particular, RNP^N with their on-off regulation by pH are suitable design for the treatment of oxidative stress injuries after intravenous administration [14, 18, 19, 23]. We have considered the disintegration of the nanoparticle and the exposure of nitroxide radicals to effectively scavenge ROS in response to low pH environments such as tumors, ischemia, and inflammation regions, without leakage of drugs due to the covalent conjugation of active species to polymer backbone. Thus, this strategy improves the therapeutic efficiency and suppresses severe adverse effects, referred to as environmental-signal-enhanced polymer drug therapy (Figure 7.3).

7.3.1 Ischemia-Reperfusion Injury

Ischemia occurs when the blood supply to a tissue area is interrupted. Ischemic injury is closely related to various diseases such as myocardial infarction, stroke, acute kidney failure, and other thrombotic events. Ischemic injury can

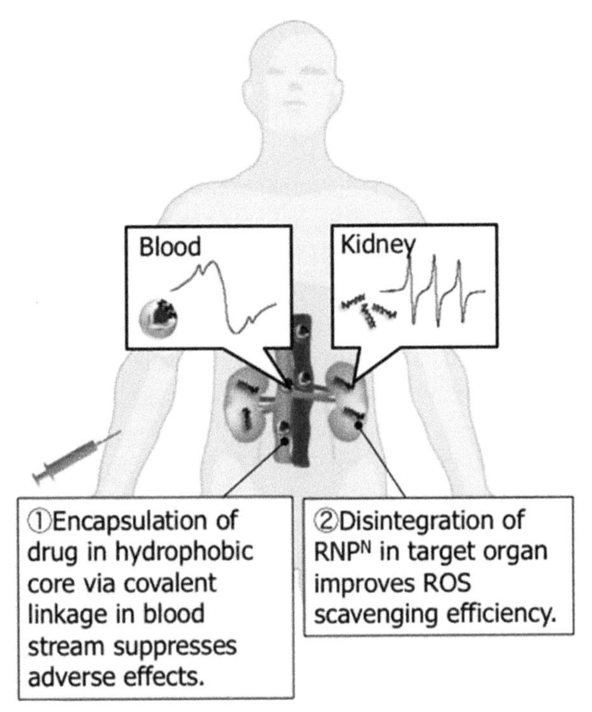

Figure 7.3 Environmental-signal-enhanced polymer drug therapy using RNP^N for the treatment of oxidative stress injuries.

also occur during surgery including organ transplantation when blood vessels are cross-clamped. During ischemia, oxygen deprivation occurs in tissue, and eventually, all ischemic tissue becomes necrotic. Restoration of the blood supply should minimize the damage, but the injured area is often increased when the blood supply is restored. It is widely accepted that this additional damage, reperfusion injury, is due to the excessive generation of ROS. To prevent reperfusion injury, RNP^N is used in the treatment of IR injury because the ischemic region has a low pH due to an increase in the glycolytic system.

7.3.2 Biodistribution and Morphological Change of RNP^N

The therapeutic effect of RNPs has been reported using a renal IR-acute kidney failure model in mice [19]. Utilizing the *ex vivo* ESR assessment, the biodistribution and morphological changes of RNPs have been evaluated after their intravenous administration to mice with kidney IR injury (Figures 7.4a and 7.4b). The ESR signals of LMW TEMPOL disappear rapidly after

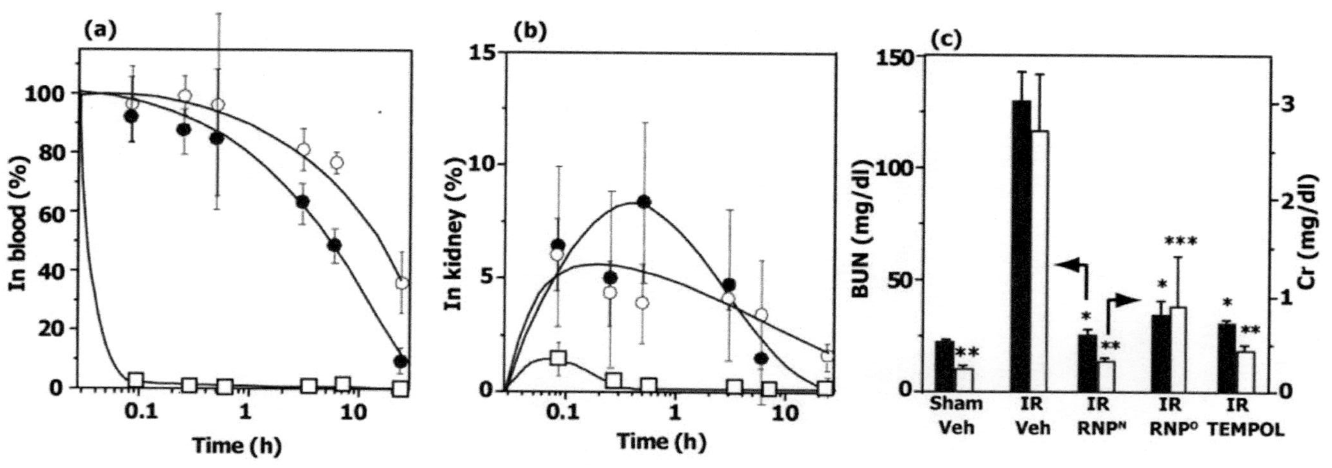

Figure 7.4 Time profile of drug concentration in **(a)** blood and **(b)** injured kidney. (white circle, RNP^O; black circle, RNP^N; white square, TEMPOL) **(c)** Therapeutic effect of RNP on renal IR. BUN and Cr levels in the plasma of mice are measured at 24 h after reperfusion following 50 min of ischemia. Drugs are administered at 5 min after reperfusion. Sham veh, sham-operated and vehicle-treated groups; IR veh, vehicle treated group; IR RNP^N, RNP^N-treated group; IR RNP^O, RNP^O-treated group; IR TEMPOL, TEMPOL-treated group. (Values are expressed as mean ± SE. *$P < 0.0001$ as compared to IR veh. **$P < 0.005$ as compared to IR veh. ***$P < 0.05$ as compared to IR veh. n = 7, ANOVA) Reprinted with permission from [19] and modification. © Elsevier B.V. (2011).

intravenous administration in blood stream. The half-life of TEMPOL in the blood has been reported to be approximately 15 s [35]. In contrast, when RNP^N prepared by PEG-b-PMNT is administered intravenously, ESR signals in the blood are observed even at 10 h. The half-life of RNP^N is 1440-fold longer (ca. 6 h) than that of LMW TEMPOL. RNP^O shows a much longer circulation. The half-life of RNP^O is approximately 18 h, which is 4320-fold longer than that of LMW TEMPOL. The rapid clearance of LMW TEMPOL from the blood stream is probably a result of its preferential renal clearance in the blood. The ESR signals of both RNPs in blood are broad singlets, indicating that the nitroxide radical moieties are still located in the solid core of the RNPs. Thus, the self-assembling structure of the amphiphilic block copolymer, which confines the nitroxide radicals in the solid core, is one of the important design features for improving the retention of nitroxide radicals in the blood stream. In the injured kidney, a similar tendency is observed. LMW TEMPOL disappears within 1 h in the injured kidney, whereas the ESR signals of RNP^N are detectable for more than 10 h after intravenous administration. RNP^O shows the longest retention in kidney, as well as in blood. The ESR signals of RNP^O are broad even in injured kidneys. In contrast, it is interesting to note that the administration of RNP^N shows a triplet ESR signal in the IR kidney. This result strongly indicates the disintegration of RNP^N at the injured renal lesion in response to the lowered pH.

7.3.3 Therapeutic Effect of RNPs on Renal Ischemia-Reperfusion Injury in Mice

The therapeutic effect of RNPs on acute kidney failure has been evaluated by considering parameters that reflect renal function, such as the levels of blood urea nitrogen (BUN) and creatinine (Cr) (Figure 7.4c). The RNP^N-treated mice show extremely lowered BUN and Cr levels, compared to those shown by both non-treated mice and mice treated with LMW TEMPO compounds at 24 h after the reperfusion. It is interesting to note that the therapeutic effect of RNP^N is greater than that of RNP^O, although the amount of RNP^O found in the kidney is much higher than that of RNP^N. This finding indicates that the site-specific disintegration of RNP^N in the targeted organs leads to an improved therapeutic effect on the IR-acute kidney injury, regardless of the longer retention time of RNP^O in model of ischemic renal injury.

A similar therapeutic effect of RNP^N has been confirmed for cerebral IR injury [18]. The intravenous administration of RNP^N in a rat model of transient middle-cerebral-artery occlusion resulted in a significantly smaller infarction area, compared to that after treatments using polymeric micelles without TEMPO or LMW TEMPOL. Further, comparison of neurological symptom scores shows that the RNP^{N}-treated group has a significantly higher therapeutic efficiency than those of the groups treated with polymeric micelle without TEMPO or TEMPOL-treated groups. From these data, RNP^{N} appears to have therapeutic effects on both renal and cerebral IR injuries.

7.4 Oral Nanotherapy with pH-Insensitive Redox Nanoparticle for the Treatment of Inflammatory Bowel Disease

7.4.1 Inflammatory Bowel Disease (IBD)

IBD, including Crohn's disease and ulcerative colitis (UC), affects millions of patients worldwide [36, 37]. Since the etiology and pathogenesis of IBD are not well understood, it is considered as an intractable disease. For many years, there have been only two treatment options for IBD, corticosteroids and mesalamine [38]. Although these drugs are effective in treating IBD to some extent, their severe adverse effects have raised significant concerns among both physicians and patients, and have also limited their use. Furthermore, an anti-tumor necrosis factor α antibody has been employed to effectively suppress the inflammation in UC, although it is a cost-oriented therapy with multiple adverse effects [39].

The intestinal mucosa of patients with IBD has been reported to be characterized by ROS overproduction and an imbalance of important antioxidants, leading to oxidative damage. Self-sustaining cycles of oxidant production amplify inflammation and mucosal injury [40, 41]. In several experimental models, ROS scavengers have improved colitis [42]. The oral route has several advantages such as inexpensive, non-invasive, and convenient for patients, making it the preferred route of drug administration. However, LMW compounds are not completely effective when administered orally due to non-specific drug distribution, lower retention in the colon, and adverse effects. If ROS scavengers are specifically targeted to the diseased sites and effectively scavenge excess generated ROS, they represent a safe and effective treatment for IBD.

Although pH-sensitive RNP^N works effectively in IR injuries via intravenous injection, its pH-disintegrative character is not suitable for the treatment of UC via oral administration because it disintegrates in the stomach by the protonation of its amino groups. Additionally, the gastrointestinal tract (GIT), which is approximately 9 m in humans, is a harsh environment that involves exposure to gastric juices with strong acid, digestive enzymes, and bile acids. To adapt the RNP system for oral administration, the pH-insensitive redox nanoparticle RNP^O is utilized for the treatment of UC (Figure 7.5) [20].

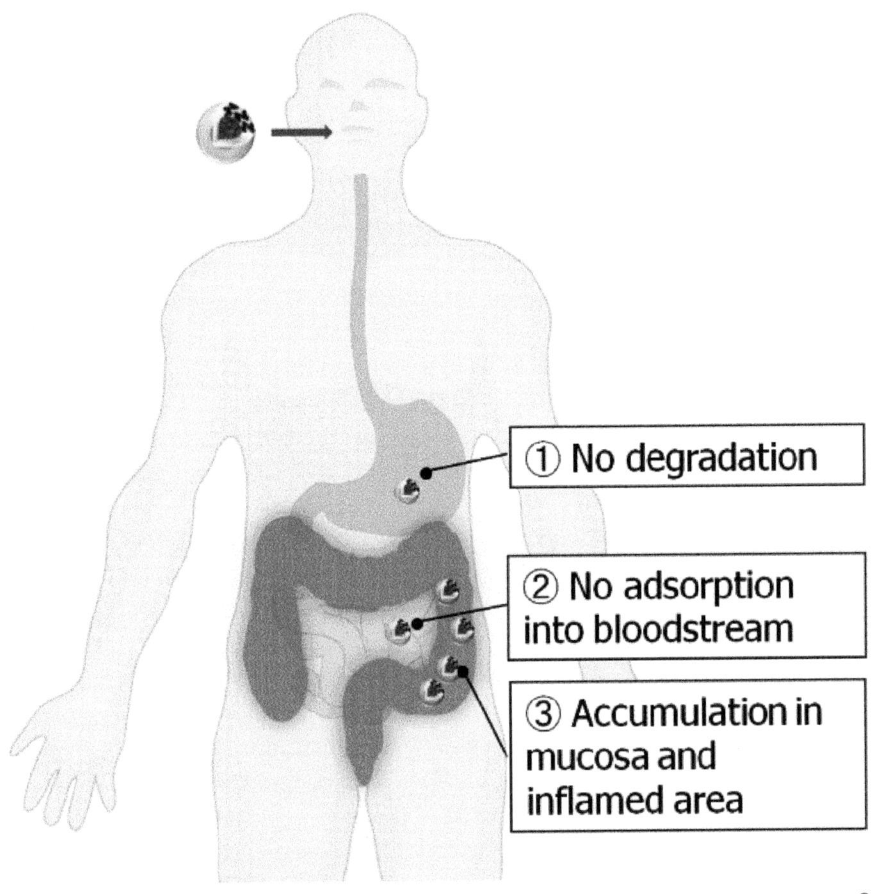

Figure 7.5 Oral nanotherapy with RNP^O reduces the inflammation in UC patients. RNP^O is stable, withstands the harsh conditions of the GIT, and reaches the colon to scavenge ROS, especially at sites of inflammation.

7.4.2 Specific Accumulation of Orally Administered RNP^O without Uptake in the Blood Stream

The stability of orally administered RNP^O in the GIT has been evaluated using the ESR spectra of RNP^O in the colon of dextran sodium sulfate (DSS)-induced colitis in mice. The ESR signals of RNP^O in the colon are consistently broad, indicating that RNP^O remains as a core-shell-type micelle even in the colon. The RNP^O with pH-insensitive character is sufficiently stable under the harsh environments of the GIT after oral administration. Therefore, the nitroxide radical is in the core of RNP^O when it is delivered to the colon area. Interestingly, orally administered RNP^O is not absorbed into the bloodstream because of the nanosize and non-degradable characteristics of RNP^O in the GIT; this means that orally administered RNP^O does not cause the adverse effects associated with nitroxide radicals in the entire body.

The accumulation of nanoparticles in the colon area is one of the most important feature for effective nanomedicine against UC. The accumulation of orally administered RNP^O in the colon of normal mice has been evaluated using ESR measurement (Figure 7.6a). LMW TEMPOL shows lowered accumulation and is excreted from the colon area almost completely 12 h after oral administration, because of metabolism in the GIT and absorption from the mesentery in the blood stream. In contrast, commercially available polystyrene latex particles accumulated in a size-dependent manner. The smallest polystyrene nanoparticle (40 nm) shows the most effective accumulation as compared to other polystyrene nanoparticles (100 nm, 0.5 μm, and 1 μm), indicating that the optimal size for the accumulation of a particle in the colon is several tens of nanometers, which allows easier diffusion of the particles in the mucosa compared to that of larger-sized particles. Interestingly, RNP^O accumulates remarkably in the colon, as compared to polystyrene latex particles with the same size (40 nm). The colloidal stability of RNP^O is much higher than that of polystyrene latex particles because PEGylation on the surface of RNP^O inhibits the aggregation of particles in the GIT. Therefore, PEGylated RNP^O shows higher accumulation and longer retention in the colonic mucosa, as compared to commercially available polystyrene latex particles. In fact, orally administered rhodamine-labeled RNP^O shows a strong fluorescent signal at the surface of the colonic mucosa, indicating that RNP^O accumulates in the mucosa area specifically.

In addition, the amount of accumulated RNP^O in the colon of mice with DSS-induced colitis is 1.5-fold higher than that in the colon of normal mice under the same administration conditions. In the colon of patients with UC, the

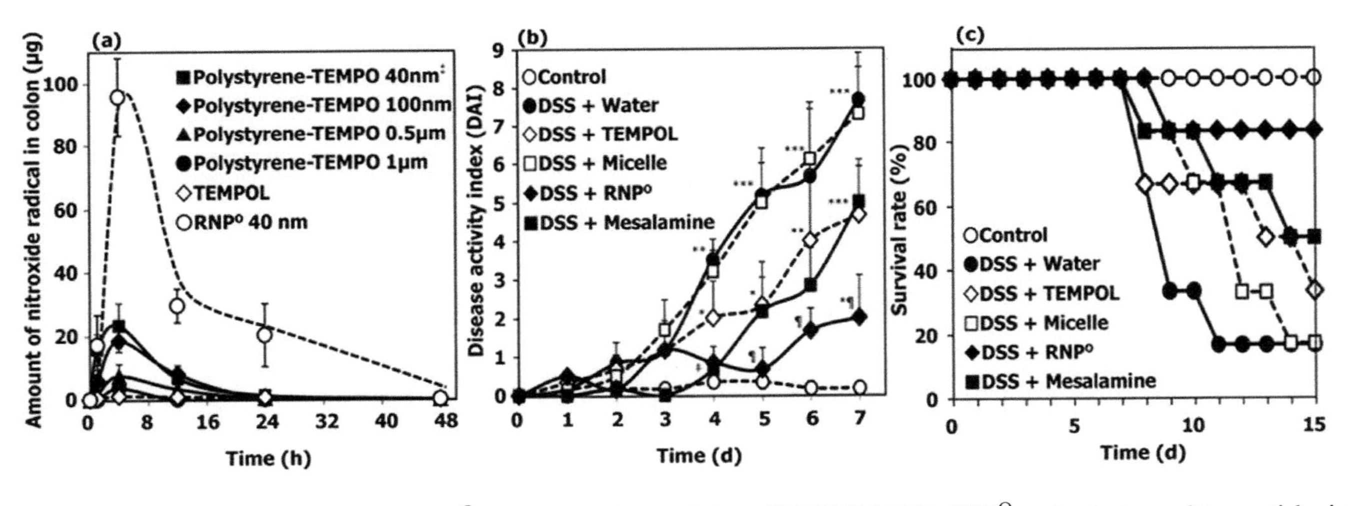

Figure 7.6 **(a)** Specific accumulation of RNP^O in the colon. Accumulation of LMW TEMPOL, RNP^O, and polystyrene latex particles in the colon. The data are expressed as mean ± SE, $n = 3$. **(b,c)** Therapeutic effect of RNP^O on DSS-induced colitis in mice. (b) Changes in DAI. DAI is the summation of the stool consistency index (0–3), the fecal bleeding index (0–3), and the weight loss index (0–4). The data are expressed as mean ± SE, $^*P < 0.05$, $^{**}P < 0.01$, and $^{***}P < 0.001$ vs. control group; $^{\ddagger}P < 0.05$ and $P < 0.001$ vs. DSS groups, $n = 6$–7, two-way ANOVA, followed by the Bonferroni post-hoc test. **(c)** The survival rate of mice with 3% (wt/vol) DSS treatment for 15 d. From 5 d, drugs are orally administered daily until 15 d. The number of surviving mice is counted until 15 d, $n = 6$. Reprinted with permission from [20]. © Elsevier B.V. (2012).

overall thickness of the adherent mucus layer is reduced due to the reduction of goblet cells and the exposed open tight junctions of epithelium cells, resulting in the facile penetration of RNP^{O} in inflammatory tissues.

7.4.3 Therapeutic Effect of RNP^{O} in a Mouse Model of Colitis

Given that orally administered RNP^{O} is shown to accumulate greatly in the colonic mucosa of DSS-injured mice and avoid uptake into the bloodstream, it is anticipated to be an ideal nanomedicine for the treatment of UC. The suppressive effect of RNP^{O} in mice model of colitis has been evaluated, as compared with LMW TEMPOL and mesalamine, a commercially available drug for the treatment of UC. While RNP^{O} is orally administered daily during the 7 d of DSS treatment, the severity of colitis is assessed based on the disease active index (DAI), which is the summation of the stool consistency index, the fecal bleeding index, and the body weight loss index (Figure 7.6b). Mice treated with DSS show a remarkably higher DAI level. Treatment with LMW TEMPOL or mesalamine decreases in the DAI as compared to DSS-treatment, although their efficiency is not significant. On the contrary, RNP^{O}-treated mice show a much lower DAI as compared to DSS-treated mice and mice treated with other LMW drugs. Compared to LMW drugs, RNP^{O} effectively suppresses pro-inflammatory mediators such as myeloperoxidase, interleukin-1β, and superoxide in the colon of mice with DSS-induced colitis. These pro-inflammatory mediators are well-known markers of inflammation and play an important role in UC. In addition, the therapeutic effect of orally administered RNP^{O} on the survival rate of mice with colitis induced by the administration of DSS for 5 d has been reported (Figure 7.6c). Orally administered LMW TEMPOL and mesalamine slightly enhance the survival rate (33.3% and 50%, respectively) compared with treatment of DSS and micelle (16.7%). In contrast, RNP^{O} treatment significantly enhances the survival rate of DSS-treated mice to 83.3%. This result indicates that RNP^{O} has not only suppressive but also therapeutic effects on mice with DSS-induced colitis, indicating that the therapeutic efficiency of nitroxide radicals could be successfully enhanced by using redox nanoparticle RNP^{O}.

7.5 Treatment of Other Oxidative Stress Injuries

RNPs have been shown to have therapeutic effects in other disease models. Both RNPs have been reported to suppress excessive inflammation after cerebral hemorrhage induced by focusing ultrasound beam radiation coupled with fluorocarbon microbubbles after intravenous administration [21]. Given both the scavenging of ROS in penumbra areas and the suppression of intracerebral hemorrhage, the RNP system is the ideal therapy for cerebral infarction. This redox nanotherapeutic strategy can be further extended to Alzheimer's disease models *in vitro* [22, 23]. RNPs also improve the gene transfection efficiency associated with a polyplex system [24].

7.6 Conclusion

RNP therapeutics, based on amphiphilic block copolymers possessing nitroxide radicals, has been developed. RNPs overcome the issues associated with LMW nitroxide radicals such as toxicities, adverse effects, and rapid elimination from the kidneys, due to the confinement of nitroxide radicals in the hydrophobic core of polymeric micelles via covalent conjugation. In addition, as a result of its pH-sensitivity of RNP^N due to the protonation of amino groups in the hydrophobic segment, RNP^N shows remarkable therapeutic effects on IR injury after its intravenous administration. On the other hand, the pH-insensitive character of RNP^O enables it to accumulate in the mucosa and inflamed areas of the GIT after oral administration without uptake in blood, resulting in remarkable therapeutic effects on colitis. Thus, this RNP-based therapeutics is anticipated to be high-performance nanomedicine for the treatment of oxidative stress-related diseases.

References

[1] Harman, D., "Aging: a theory based on free radical and radiation chemistry," Journal of Gerontology, Vol. 11, No. 3, 1956, pp. 298–300.
[2] Watanabe, T., Yuki, S., Egawa, M., and Nishi, H., "Protective effects of MCI-186 on cerebral ischemia: Possible involvement of free radical scavenging and antioxidant actions," J Pharmacol Exp Ther, Vol. 268, No. 3, 1994, pp. 1597–1604.
[3] The Edaravone Acute Brain Infarction Study Group (Chair: Eiichi Otomo, MD), "Effect of a novel free radical scavenger, Edaravone (MCI-186), on acute brain infarction. Randomized, placebo-controlled, double-blind study at multicenters," Cerebrovasc Dis, Vol. 15, No. 3, 2003, pp. 222–229.
[4] Krishna, M.C., Russo, A., Mitchell, J.B., Goldstein, S., Dafni, H., and Samuni, A., "Do Nitroxide Antioxidants Act as Scavengers of O2-. or as SOD Mimics?," J. Biol. Chem, Vol. 271, No. 42, 1996, pp. 26026–26031.

[5] Miura, Y., Utsumi, H., and Hamada, A., "Antioxidant activity of nitroxide radicals in lipid peroxidation of rat liver microsomes," Arch. Biochem. Biophys, Vol. 300, No. 1, 1993, pp. 148–156.

[6] Morse, P.D., and Swartz, H.M., "Measurement of intracellular oxygen concentration using the spin label TEMPOL," Magn Reson Med. Vol. 2, No. 2, 1985, pp.114–127.

[7] Soule, B.P., Hyodo, F., Matsumoto, K., Simone, N.L., Cook, J.A., Krishna, M.C., and Mitchell, J.B.," The chemistry and biology of nitroxide compounds," Free Radic. Biol. Med. Vol. 42, No. 11, 2007, pp. 1632–1650.

[8] Adam, W., SahaMoller, C.R., and Ganeshpure, P.A., "Synthetic applications of nonmetal catalysts for homogeneous oxidations," Chem Rev, Vol. 101, No. 11, 2001, pp. 3499–3548.

[9] Rachmilewitz, D., Karmeli, F., Okon, E., and Samuni, A., "A novel antiulcerogenic stable radical prevents gastric mucosal lesions in rats," Gut, Vol. 35, No. 9, 1994, pp. 1181–1188.

[10] Karmeli, F., Eliakim, R., Okon, E., Samuni, A., and Rachmilewitz, D., "A stable nitroxide radical effectively decreases mucosal damage in experimental colitis," Gut, Vol. 37, No. 3, 1995, pp. 386–393.

[11] Rak, R., Chao, D. L., Pluta, R. M., Mitchell, J.B., Oldfield, E.H., and Watson, J. C., "Neuroprotection by the stable nitroxide Tempol during reperfusion in a rat model of transient focal ischemia," J Neurosurg, Vol. 92, No. 4, 2000, pp. 646–651.

[12] McDonald, M. C., Zacharowski, K., Bowes, J., and Cuzzocrea, S., Thiemermabnn, C., "Tempol reduces infarct size in rodent models of regional myocardial ischemia and reperfusion," Free Radic. Biol. Med, Vol. 27, No. 5–6, 1999, pp. 493–503.

[13] Yoshitomi, T., Miyamoto, D., and Nagasaki, Y., "Design of Core-shell-type Nanoparticles Carrying Stable Radicals in the Core," Biomacromolecules, Vol.10, No. 3, 2009, pp. 596–601.

[14] Yoshitomi, T., and Nagasaki, Y., "Nitroxyl radical-containing nanoparticles for novel nanomedicine against oxidative stress injury," Nanomedicine, Vol. 6, No. 3, 2011, pp. 509–518.

[15] Yoshitomi, T., Suzuki, R., Mamiya, T., Matsui, H., Hirayama, A., and Nagasaki, Y., "pH-Sensitive Radical-Containing-Nanoparticle (RNP) for the L-Band-EPR Imaging of Low pH Circumstances," Bioconjugate Chem, Vol. 20 No. 9, 2009, pp.1792–1798.

[16] Yoshitomi, T., and Nagasaki, Y., "Design and preparation of a nanoprobe for imaging inflammation sites," Biointerphase, Vol. 7, No. 1–4, 2012, 7.

[17] Nagasaki, Y., "Nitroxide radicals and nanoparticles: A partnership for nanomedicine radical delivery," Therapeutic Delivery, Vol. 3, No. 2, 2012, pp. 1–15.

[18] Marushima, A., Suzuki, K., Nagasaki, Y., Yoshitomi, T., Toh, K., Tsurushima, H., Hirayama, A., and Matsumura, A., "Newly Synthesized Radical-Containing Nanoparticles (RNP) Enhance Neuroprotection After Cerebral Ischemia-Reperfusion Injury," Neurosurgery, Vol. 68, No. 5, 2011, pp. 1418–1428.

[19] Yoshitomi, T., Hirayama, A., and Nagasaki, Y., "The ROS scavenging and renal protective effects of pH-responsive nitroxide radical-containing nanoparticles," Biomaterials, Vol. 32, No. 31, 2011, pp. 8021–8028.

[20] Vong, L.B., Tomita, T., Yoshitomi, T., Matsui, H., and Nagasaki, Y., "An Orally Administered Redox Nanoparticle that Accumlates in the Colonic Mucosa and Reduces Colitis in Mice," Gastroenterology, Vol. 143, No. 4, 2012, pp. 1027–1036.

[21] Chonpathompikunlert, P., Fan, C.H., Ozaki, Y., Yoshitomi, T., Yeh, C.K., and Nagasaki, Y., "Redox Nanoparticle Treatment Protects Against Neurological Deficit in Focused Ultrasound-Induced Intracerebral Hemorrhage," Nanomedicine, Vol. 7, No. 7, 2012, pp. 1029–1043.

[22] Chonpathompikunlert, P., Yoshitomi, T., Han, J., Toh, K., Isoda, H., and Nagasaki, Y., "Chemical nanotherapy: Nitroxyl radical-containing nanoparticle (RNP) protects neuroblastoma SH-SY5Y cells from Aβ-induced oxidative stress," Therapeutic Delivery, Vol. 2, No. 5, 2011, pp. 585–597.

[23] Chonpathompikunlert, P., Yoshitomi, T., Han, J., Isoda, H. and Nagasaki, Y., "The use of nitroxide radical-containing nanoparticles coupled with piperine to protect neuroblastoma SH-SY5Y cells from Aβ-induced oxidative stress," Biomaterials, Vol. 32, No. 33, 2011, pp. 8605–8612.

[24] Toh, K., Yoshitomi, T., Ikeda, Y., and Nagasaki, Y., "Novel redox nanomedicine improves gene expression of polyion complex vector," Science and Technology of Advanced Materials, Vol. 12, No. 6, 2011, 065001.

[25] Harris, J.M., Eds., Poly(ethylene glycol) chemistry, Plenum Press, New York and London, 1992.

[26] Stefano, G.D., Busi, C., Mattioli, A., and Fiume, L., "Selective delivery to the liver of antiviral nucleoside analogs coupled to a high molecular mass lactosaminated poly-L-lysine and administered to mice by intramuscular route," Biochemical Pharmacology, Vol. 49, No. 12, 1995, pp. 1769–1775.

[27] Ankel, E.G., Lai, C.S., Hopwood, L.E., and Zivkovic, Z., "Cytotoxicity of commonly used nitroxide radical spin probes," Life Sci, Vol. 40, No. 5, 1987, pp. 495–498.

[28] Lee, A.S., Gast, A.P., Butun, V., and Armes, S.P., "Characterizing the Structure of pH Dependent Polyelectrolyte Block Copolymer Micelles," Macromolecules, Vol. 32, No. 13, 1999, pp. 4302–4310.

[29] Asayama, S., Maruyama, A., Cho, C., and Akaike, T., "Design of Comb-Type Polyamine Copolymers for a Novel pH-Sensitive DNA Carrier," Bioconjugate Chem, Vol. 8, No. 6, 1997, pp. 833–838.

[30] Matsumura, Y., and Maeda, H., "A new concept for macromolecular therapeutics in cancer chemotherapy: mechanism of tumoritropic accumulation of proteins and the antitumor agent smancs," Cancer Res, Vol. 46, No. 12, 1986, pp. 6387–6392.

[31] Yamamoto, Y., Nagasaki, Y., Kato, Y., Sugiyama, Y., and Kataoka, K., "Long-circulating poly(ethylene glycol)-poly(D,L-lactide) block copolymer micelles with modulated surface charge," J Controlled Release, Vol. 77, No. 1–2, 2001, pp. 27–38.

[32] Bae, Y., Nishiyama, N., Fukushima, S., Koyama, H., Yasuhiro, M., and Kataoka, K., "Preparation and biological characterization of polymeric micelle drug carriers with intracellular pH-triggered drug release property: tumor permeability, controlled subcellular drug distribution, and enhanced in vivo antitumor efficacy," Bioconjugate Chem, Vol. 16, No. 1, 2005, pp. 122–130.

[33] Lee, E., Na, K., and Bae, Y., "Doxorubicin loaded pH-sensitive polymeric micelles for reversal of resistant MCF-7 tumor," J Controlled Release, Vol. 103, No. 2, 2005, pp. 405–418.

[34] Liu, J., Li, H., Jiang, X., Zhang, C., and Ping, Q., "Novel pH-sensitive chitosan-derived micelles loaded with paclitaxel," Carbohydr Polym, Vol. 82, No. 2, 2010, pp. 432–439.

[35] Takechi, K., Tamura, H., Yamaoka, K., and Sakurat, H., "Pharmacokinetic analysis of free radicals by in vivo BCM (Blood Circulation Monitoring)-ESR method," Free Rad. Res. Vol. 26, No. 6, 1997, 483–496.

[36] Khor, B., Gardet, A., and Xavier, R.J., "Genetics and pathogenesis of inflammatory bowel disease," Nature, Vol. 474, No. 7351, 2011, pp. 307–317.

[37] Abraham, C., and Cho, H.J., "Mechanism of disease inflammatory bowel disease," N Engl J Med, Vol. 361, No. 21, 2009, pp. 2066–2078.

[38] Stephen, B.H., "Medical therapy for ulcerative colitis 2004," Gastroenterology, Vol. 126, No. 6, 2004, pp. 1582–1592.

[39] Singh, K., Chaturvedi, R., Barry, D.P., Coburn, L.A., Asim, M., Lewis, N.D., Piazuelo, M.B., Washington, M.K., Vitek, M.P., and Wilson, K.T., "The apolipoprotein E-mimetic peptide COG112 inhibits NF-kappaB signaling, proinflammatory cytokine expression, and disease activity in murine models of colitis," J. Biol. Chem, Vol. 286, No. 5, 2011, pp. 3839–3850.
[40] Babbs, C.F., "Oxygen radicals in ulcerative colitis," Free Radic. Biol. Med., Vol. 13, No. 2, 1992, pp. 169–182.
[41] McCord, J.M., "The evolution of free radicals and oxidative stress," Am J Med, Vol. 108, No. 8, 2000, pp. 652–659.
[42] Aggarwal, B.B., and Harikumar, K.B., "Potential therapeutic effects of curcumin, the anti-inflammatory agent, against neurodegenerative, cardiovascular, pulmonary, metabolic, autoimmune and neoplastic diseases," Int J Biochem Cell Biol, Vol. 41, No. 1, 2009, pp. 40–59.

8

Rational Design of Multifunctional Nanoparticles for Targeted Cancer Imaging and Therapy

Arun K. Iyer

Department of Pharmaceutical Sciences
Eugene Applebaum College of Pharmacy and Health Sciences
Wayne State University
259 Mack Ave, Detroit, MI 48201

8.1 Introduction

The treatment of several forms of cancer has remained elusive even after persistent efforts by researchers and clinicians to combat this deadly disease. In an insightful review, Hanahan and Weinberg have described the hallmark features of cancers and the multitude of deregulation and signaling pathways that are involved in the development and progression of tumors [1]. Although very different in nature, all forms of cancer are derived from normal cells that have acquired abnormal phenotypes [2]. Either by enhanced proliferation or reduced apoptosis, these neoplastic cells begin to neo-vascularize forming new blood vessels, establishing subpopulation of virulent cells, with the capability to invade and infiltrate into adjacent normal cells and tissues. Tumor cells also possess the ability to infiltrate into lymphatic system and/or blood vessels enabling them to circulate (called circulating tumor cells or CSCs) and migrate to multiple distant sites or organs (a phenomenon called tumor metastasis), forming secondary tumors [1, 3]. At this stage of the disease, surgical resection or other forms of treatment are rarely successful. Moreover, at an advanced stage, tumor resection presents a higher risk of relapse of the disease possessing unfavorable multidrug resistant (MDR) phenotypes and/or tumor "stem cell like" features that are implicitly involved in self-renewal and formation of new phenotypes of tumors [4–7]. Also, the genomic instability in cancers

Post-genomic Approaches in Cancer and Nano Medicine, 219–266.

causes several types of mutations that lead to intra-tumoral heterogeneity in successive generations of tumors [8, 9]. In addition, tumor cells are smart in reprogramming themselves to cater to demanding conditions, such as the ability to survive and often thrive in hypoxic conditions and/or highly acidic low pH microenvironment of the tumors [10–15].

While the microenvironmental selection pressures leads to development of multidrug resistance (MDR) [16], the derivation of tumors from normal tissue poses another major barrier in the safe and effective eradication of tumors and cancerous cells. Both molecular and physical boundaries between normal and cancerous tissue are often diffused, making surgical resection incomplete and rendering many of the current treatments highly unspecific. Also, conventional forms of chemotherapy using small molecule drugs and anticancer agents do not differentiate tumor and normal tissues and cells causing acute systemic side effects, necessitating the need for the development of novel targeted strategies and therapeutics for more effective management of cancers.

In response to the failures of conventional chemotherapies in the clinics, there has been a growing interest in devising advanced targeted delivery systems based on nanomedical technologies for early detection, diagnosis and treatment of cancers [17–22]. Since the microenvironment and physiological properties of solid tumors are uniquely different from the (healthy) normal tissues, it is possible to design nanosized delivery platforms that specifically transport drugs and genes to the sites of tumors [23]. Furthermore, nanoparticle systems have the ability to achieve temporal and spatial site-specific drug delivery. These unique properties of nanoparticle systems not only enhance the efficacy of chemotherapy but also significantly reduce the adverse side effects associated with the free drug, and allow for fewer doses of the drug administration to cancer patients. In this review, we will discuss the design and development of some such nanoparticles-based delivery systems such as polymeric micelles and nanoparticles, liposomes, and multifunctional organic/inorganic hybrid nanoparticle systems that have demonstrated tremendous potentials for targeted cancer therapy. Also, the design of "theranostic systems" are discussed that can combine multiple agents such as anticancer drugs, genes and imaging agents using a single nanoconstruct, that have immense potentials for simultaneous diagnosis, imaging and therapy of cancers.

8.1.1 Mechanism of Tumor Selective Delivery: Passive and Active Targeting

The development of tumors involves several anatomical and pathophysiological changes that could be utilized for selective tumor targeting [25]. As the tumor cells multiply and grow to a size of ~1 mm in diameter, they start to form neovasculatures [26]. In order to sustain their growth, the so formed tumor nodules develop a more complex network of blood vessels around them, a process called angiogenesis [27]. The endothelial cells lining the tumor blood vessels are highly disorganized and defective in architecture. Moreover, the tumor blood vessels have wide gap junctions leading to "leaky" vascular architectures [28, 29]. Apart from the unique anatomical features described above, tumors cells secrete elevated levels of permeability mediators such as vascular endothelial growth factor (VEGF) (also known as vascular permeability factor (VPF)), bradykinin (BK), prostaglandins (PGs), matrix metalloproteinases (MMPs), nitric oxide (NO) and peroxynitrite) [30–33]. The increase in permeability-mediators coupled with the anatomical and pathophysiological abnormalities leads to extensive accumulation of blood plasma components, macromolecules and nanoparticles into the tumor interstitium [30, 31, 34, 35]. This phenomenon was coined the *"enhanced permeability and retention"* (EPR) effect, first discovered by Matsumura and Maeda more than three decades ago [36] (**Figure 8.1**). In this regard, it is also important to note that polymeric drugs and macromolecules with molecular weight >40 kDa (which are above the renal excretion threshold) are able to circulate longer in the blood and show prolonged accumulation in the solid tumors

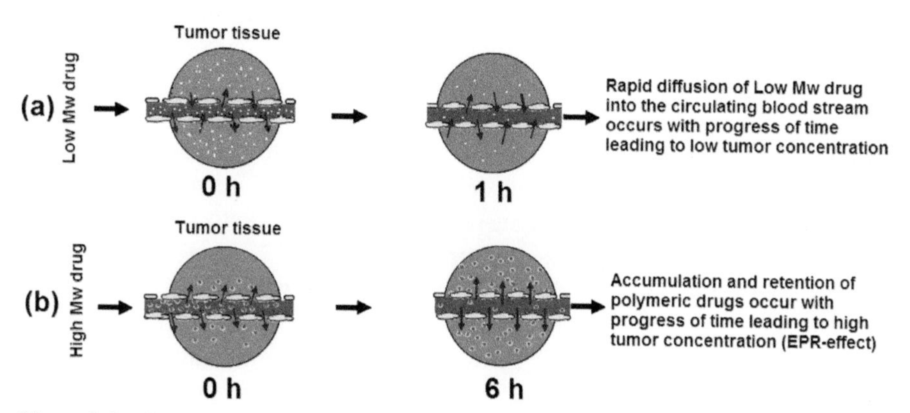

Figure 8.1 Concept of EPR effect for tumor targeted drug delivery. (Adapted from Ref. [25] with permissions Elsevier B.V.).

[37–41]. The EPR phenomenon was later observed for many types of polymer-drug conjugates, micelles, nanoparticles and liposomal delivery systems [25, 42–45]. Also, the tumors tissues were found to have impaired lymphatic clearance due to which the accumulation and retention of nanoparticles continued to occur in the tumor as long as they could circulate in the blood. Furthermore, the EPR effect was found to be more effective if the nanoparticles could escape mononuclear phagocytic systems (MPS) and show prolonged circulation half-life in the blood. In this regard, incorporation of amphiphatic molecules such as poly(ethylene glycol) (PEG) on the surface of nanosystems was found very useful in facilitating MPS escape and rendering long plasma residence time, thus enhancing tumor accumulation [46, 47]. For instance, PEG-modified "stealth" liposomes encapsulated with doxorubicin could circulate for prolonged periods in the blood and exhibited improved tumor accumulation in addition to lowering the toxicities associated with free form of doxorubicin [48]. In general, nanosystems in the size range of 20–200 nm have been found effective in permeating and accumulating in the solid tumor tissue [23], although there has been some indication that the hyperpermeability of tumor vasculature with wide gap junctions can facilitate particles as large as 700 nm to accumulate effectively in solid tumors [49–51].

Although the EPR effect provides a "first pass" for selective accumulation of nanoparticles, micelles, and liposomal formulations into the tumor interstitium, their intracellular delivery still remains challenging [52]. It is critical for nanoparticle system to enter into the cancer cells and more importantly, release the drug/gene cargo in the right location, for effective cell killing. Intracellular delivery to specific location within the cells and organelles is essential for almost all anticancer drugs and genes. For instance, intracellular delivery of siRNAs and its release in the cytoplasm is key for the success of RNA interference (RNAi)-based gene silencing strategies [53]. In this regard, the specificity and targeting ability of nanosystems can be remarkably improved when tumor-targeting ligands are used as part of the nano-delivery systems (**Figure 8.2**) [52, 54]. Such targeted delivery systems can selectively bind and internalize into tumor cells that overexpress specific receptors or antigens [55], thereby promoting more specific tumor targeting [52, 53, 56, 57]. Use of such mechanism is called "active" tumor targeting (**Figure 8.2**) [23]. Thus, active targeting in effect aids in more specific "secondary" targeting after "primary" targeting based on the EPR-effect and in combination, passive and active targeting provide for increased accumulation and penetration of nanoparticles at the tumor site thereby facilitating improved maintenance of high intracellular drug concentrations. Such systems provide several fold

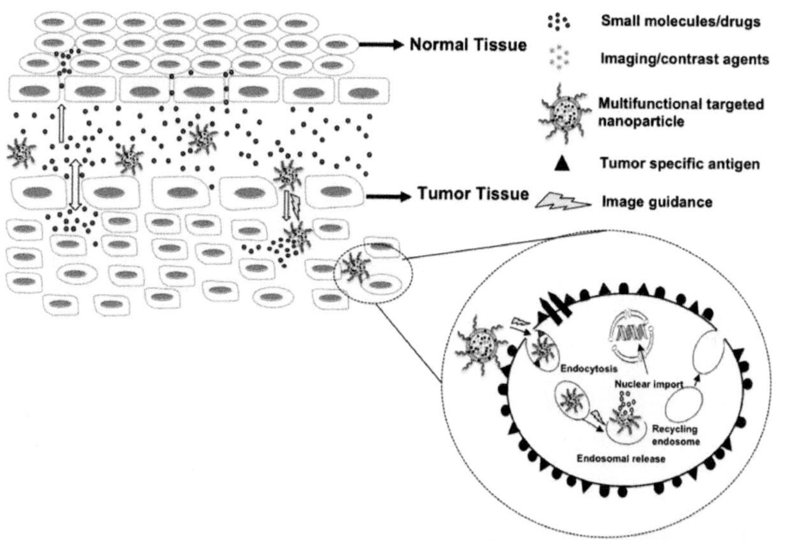

Figure 8.2 The schematic shows the passive and active targeting mechanisms of multifunctional image guided nanoparticles and the difference in the vasculature of normal and tumor tissues; drugs and small molecules diffuse freely in and out of the normal and tumor blood vessels due to their small size and thus the effective drug concentration in the tumor drops rapidly with time. However, macromolecular drugs and nanoparticles can passively target tumors due to the leaky vasculature or the EPR effect, however they cannot diffuse back into blood stream due to their large size and impaired lymphatic clearance, leading to enhanced tumor accumulation and retention. Targeting molecules such as antibodies or peptides decorated on the surface of nanoparticles can selectively bind to cell surface receptors/antigens overexpressed by tumor cells and can be taken up by receptor-mediated endocytosis (active targeting). The image guiding molecules and contrast agents conjugated/encapsulated in the nanoparticles can be useful for targeted imaging and (non-invasive) visualization of nanoparticle accumulation/localization, as well as for mechanistic understanding of events and efficacy of drug treatment simultaneously (Reprinted with permissions from Ref. [58] Bentham Science.

increased effectiveness as compared to free drug administration [23]. Indeed currently, active targeting has become a widely recognized potential route for increasing therapeutic indexes in cancer treatment.

Several ligands or targeting agents can be used to surface decorate nanoparticle systems for active targeting to tumors [59–61]. For example, RGD peptide coupled nanoparticles can target integrin receptors ($\alpha_v\beta_5$ or $\alpha_v\beta_3$) overexpressed on vascular endothelial cells of angiogenic blood vessels and tumor cells [62]. In a recent study, we functionalized poly(epsilon caprolactone) (PCL) microparticles containing colloidal gold with RGD peptide that

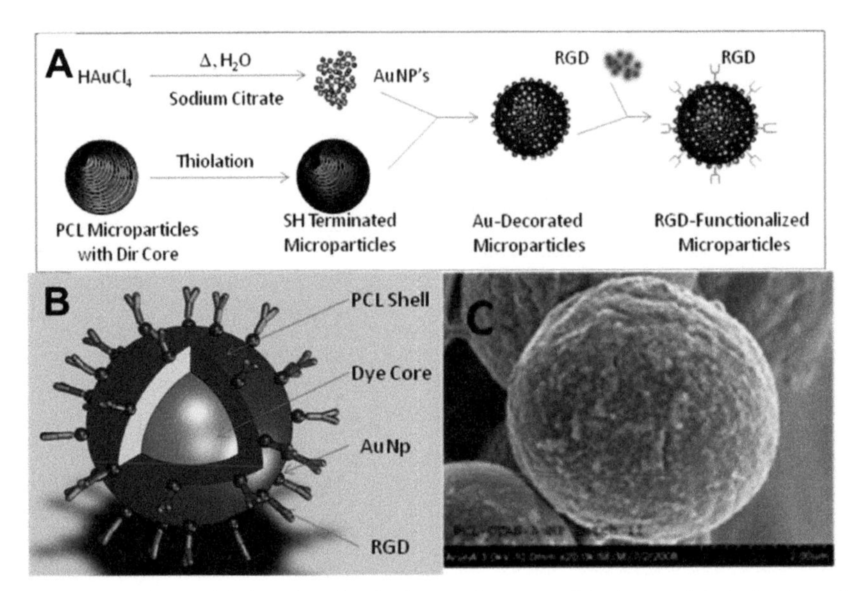

Figure 8.3 RGD peptide functionalized polycaprolactone-gold microparticle design for colon cancer screening. A. Fabrication steps; **B**. 3D design and; **C**. Scanning Electron Micrograph (SEM) revealing a size of ~1.5 μm for the microparticle. Please see Ref. [59] and [63] for details.

specifically homes to colon cancers overexpressing integrin receptors **(Figure 8.3)** [59, 63]. These multifunctional microparticles could increase the localization of fluorescent probes loaded in them for the diagnostic imaging and screening of colon cancers wherein they could assist in differentiating colon adenocarcinomas from normal colon mucosa. There are several such examples in the literature where a combination of active and passive targeting has been utilized for tumor targeted drug delivery [23, 52, 53, 57,64–67].

8.2 Rational Design of Nanoparticles

The design of nanoparticles for applications in drug delivery systems especially for clinical use requires a rational and calculated approach. There are many variables affecting safety, delivery efficiency and targeting that needs to be addressed for systemic cancer therapy. For instance, the biological safety and inertness of the materials used for fabrication of nanoparticles have to be thoroughly tested. In this regard, naturally occurring polymers or polymers derived from biological sources such as dextran, hyaluronic acid, chitosan and gelatin that are either water soluble, biological inert or nontoxic in nature can be ideal for systemic drug delivery proposes [53, 68–71]. Also, utilization of synthetic polymers such as polyethylene glycol (PEG) and poly(lactic-co-glycolic acid) (PLGA) that are FDA approved for human use can fast-track the regulatory process for scale-up manufacturing of nanoparticle constructed using such polymers. Apart for the safety aspect, another important criteria for designing nanoparticles requires consideration of the compatibility of the drug of choice to the polymeric system in question. Such analysis requires thorough investigation on the chemical structure and hydrophilicity and/or hydrophobicity (or log P value) of the drug molecule as well as the polymer properties in order to establish if the molecular interaction would yield favorable outcome [53, 72]. It is also imperative to test and optimize the drug loading, size, charge and other nanoparticle parameters for the intended application [73]. In this regard, in many cases it becomes essential to perform chemical modification of the backbone polymer used in constructing nanoparticles to obtain desired properties such as stable drug encapsulation and optimal drug loading [53, 73, 74].

Despite the improvement in conceptual understanding and evolution of nanomaterial for *in vitro* and *in vivo* drug delivery, majority of the studies do not assess their long-term effects. For example, cationic lipids are attractive materials for gene delivery due to their electrostatic interaction with negatively charged nucleic acids. However, some lipids have been shown to induce immune response [75] and some such liposomes can alter gene expression [76]. In this regard, the use of neutral lipids such as 1,2-dioleoyl-snglycero-3-phosphatidylcholine in the liposomes might overcome such toxicities issues [77]. Similarly, polymeric micelles constructed using some types of monomer blocks produce immunological reactions and cytotoxicity [78, 79]. As discussed earlier, MPS uptake of nanosystems (that are not coated with PEG) can also present liver and spleen toxicity [80]. Such conceptual understanding on the materials of choice could allow designing nanosystems with desired surface properties that can exhibit predictable systemic clearance and localization. Regulatory guidance from the US food and drug administration (FDA) or other agencies is not yet fully defined regarding the use of nanosystems in the clinical setting. However, some consensus has been reached regarding the testing of nanoparticle systems for their safety in preclinical settings. In such cases, *in vitro* cell viability assays are used to assess toxicity profiles of the nanosystems. However, these results cannot be directly translated for pre-clinical and clinical setting because of lack of true dynamic conditions and active immunizing components. For such purposes, *in vivo* preclinical evaluation of nanoparticles using small animal models has become the norm. Although tumor models of mice and rats are relatively useful for testing several aspects of nanoparticles properties such as tumor targeting, pharmacokinetics and biodistribution profiles, efficacy as well as safety, precautions must be taken when translating the obtained *in vivo* preclinical data to human trials, because there is always a risk of intra-and inter-species differences. Irrespective of such challenges, the research and development of nanoparticle-based delivery systems are intensely being pursued due to their promising potentials especially for diseases such as cancer. We discuss below some of the properties of nanoparticle systems such as size, shape and surface charge that are most critical in dictating their localization to tumors [81, 82]

8.2.1 Shape of Nanoparticles

There has been compelling evidence that shape plays an important role in dictating the preferential uptake of nanoparticles by specific cell types. For these studies, researchers have resorted to the use of model systems consisting of organic, inorganic or hybrid nanoconstructs. It has been reported that particle shape could be used to inhibit phagocytosis by macrophages that in turn can enable its enhanced circulation in the blood. The tangential angle of the initial point of contact between the cell and the nanoparticle has also shown a positive correlation with inhibition of phagocytosis. Elliptical disk-shaped particles were shown to avoid phagocytosis when oriented correctly, and with a greater surface area, allowing for more targeting molecules to be attached to the surface [82]. Chitrani *et al.*, have investigated the intracellular uptake of varying size and shape of colloidal gold nanoparticles in mammalian cells [83]. In this study, it was observed that the kinetics and saturation concentrations were highly dependent upon the physical dimensions of the nanoparticles. These findings indicate the importance of design parameters for fabrication of nanoparticles for applications in drug delivery, including cancer imaging and therapy.

Mitragotri's group has performed extensive work on the effect of nanoparticle shape on their circulation in the blood and preferential uptake by cells [84–86]. In a recent study, Barua *et al.*, have reported that changing the shape of nanoparticles (coated with tumor targeting antibodies) from spherical to rod-like enhanced their specificity and targeting to breast cancer cells up to 10,000 times [87]. These data present unique opportunities for designing antibody-conjugated nanoparticles for cancer therapeutics.

In another recent study, Jain's group has developed quantum-dot-based nanospheres and nanorods with different aspect ratios and identical hydrodynamic size and surface properties for deciphering shape dependent tumor penetration of such nanoparticles [88]. The nanospheres and nanorods were grown using CdSe/CdS with a silica coating, decorated with PEG. The diameter of the nanospheres was approximately 35 nm, with the rods having a diameter of 15 nm and a length of 54 nm. The nanorods showed an improvement in tumor penetration over spherical nanoparticles by a factor of 1.7, which may be attributed to the size of their smallest dimension [88]. Recently, there has been a boom in the use of nanorods in cancer therapeutics that were produced using inert materials such as silica or gold for imaging and targeted delivery to cancer cells [89, 90]. Taken together, these studies clearly indicate that nanoparticle shape needs to be explored in great detail for designing nanomedicines for application in cancer imaging and therapy.

8.2.2 Size and Surface Charge of Nanoparticles

As discussed earlier, particle size is one of the most important factors taken into account in designing nanosystems, especially for tumor targeting. By 2004, many studies had shown that majority of particles over 500 nm are internalized into cells through phagocytosis, while particles under 500 nm are taken up preferentially via non-degenerative endocytosis [81]. With regard to tumor delivery, many studies indicate that nanoparticles smaller than 250 nm are more adept at extravasation into tumor tissues. However, particles in excess of 200 nm more readily undergo splenic elimination [91]. At the same time, it was found that nanoparticles less than 10 nm were too small and could be efficiently cleared via the kidneys. It was thus found that particle size in a range of 10–250 nm was most desirable for application in drug delivery and tumor targeting [92, 93]. Perrault *et al.*, conducted a detailed study on designing nanoparticles ranging in size from 10–100 nm and found that the permeation of nanoparticles within the tumor is highly dependent on the overall size of the nanoparticle [94]. It was concluded from the study that smaller nanoparticles rapidly diffuse throughout the tumor matrix whereas larger ones appeared to stay near the vasculature.

In addition to shape and size, the pharmacokinetics and biodistribution of nanoparticles are also affected by surface charge. In general, it is observed that the clearance of positively charged molecules occurs very quickly from the body. This phenomenon has been observed with nanoparticles with positively charged surfaces as well [95, 96]. Researchers from He's group have shown that positively charged nanoparticles were taken up by macrophages much more rapidly than negatively charged nanoparticles of the same composition

[87]. On the other hand, the same study also showed that positively charged particles show higher accumulation in certain organs, such as lung tissues, than their negatively charged counterparts. It can be inferred from this study that pharmacokinetics and biodistribution depends on several parameters and one must use caution in selecting the nanoparticles for each and every application. Nevertheless, over the years, there has been quite a bit of data and a widespread understanding on the effect of size and charge for the design of nanosystems. As a follow up, recent efforts on the shape effects of nanoparticles have also shed light on the tumor penetration of such nanoparticles [88]. Readers are directed to a more comprehensive review by Albanese *et al.*, on the effect of nanoparticle parameters and surface chemistry for their application in biological systems [97].

8.2.3 Surface Functionalization of Nanoparticles

As discussed earlier, apart from designing nanosystems with optimal size, shape and surface charge for passive tumor targeting, it is further possible to achieve improved specificity and cell penetration based on "active targeting" (Figure 8.2). Target-specific nanosystems rely on targeting moieties such as antibodies [98], peptides [99] or aptamers, that are often attached to the surface of nanoparticle. However, they could as well be conjugated onto the backbone of polymers (such as block copolymers that self assemble to form nanosized micelles [100]. In one study, an anti-nucleosome monoclonal antibody, mAb2C5, was used to target the antibody conjugated micelles to ovarian cancer cell spheroids [101]. The system selected for this experiment incorporated a hydrophilic PEG shell with a hydrophobic phosphatidylethanolamine (PE) core. These micelles were loaded with doxorubicin in an attempt to improve the therapeutic index for ovarian cancer treatment, which is often performed using both doxorubicin and paclitaxel. The targeted micelles in this study showed a two-fold improvement in accumulation in spheroids as compared to free drug or untargeted micelles, while also showing improved cytotoxicity for the cancer model under study [101]. Peptides have also shown promising potentials as targeting moieties. For example, Kohno *et al.*, successfully showed improved tumor cell killing by using a "hybrid" lytic peptide D-K_6-L_9 conjugated to the epidermal growth factor receptor (EGFR) peptide. The EGFR peptide has been shown to bind to small unilamellar vesicles that are overexpressed in many cancer cells. The conjugation of D-K_6-L_9lytic peptide improved the binding of EGFR peptide to these tumor cell vesicles, whereas its cytotoxicity for normal cells was also found to be decreased [102]. In the case of nanoparticles-based targeting, the peptide is either conjugated or impregnated onto the surface of nanoparticles and liposomes. In one such example, Milane *et al.*, showed improved therapeutic efficacy in treating multidrug resistant tumors using paclitaxel/lonidamine loaded in EGFR peptide surface decorated blend nanoparticles [103]. It is however, not always necessary to conjugate a ligand in order to achieve tumor targeting. For such systems, the backbone material used to construct the nanoparticle is chosen to inherently target tumor cells, based on its chemical structure. For example, hyaluronic acid (HA), a naturally occurring mucopolysaccharide composed of alternating disaccharide units of D-glucuronic acid and N-acetyl D glucosamine (NAG) with β (1–4) interglycosidic linkage binds specifically to CD44 receptors, which are often overexpressed on several types of tumor cells, including tumor initiating stem cells [53, 104–107]. Yoon *et al.*, utilized this methodology to create a multifunctional nanoparticle system for tumor-targeted imaging and therapy [108]. Ligand conjugated nanosystems are thus proving to be one of the most effective ways to achieve tumor cell specific targeting and imaging that are being intensely pursued for the better management of diseases such as cancer. Some such multifunctional nanoparticle systems are covered in the latter sections.

8.3 Types of Nanoparticle Systems for Cancer Therapy

8.3.1 Polymeric Nanoparticles

Polymers are highly versatile materials with unique properties and wide range of applications in diverse fields. Polymeric nanoparticles represent a distinct class of delivery system that can be designed specifically for the disease target and serve to carry a range of payloads. For example, polymeric nanoparticles can be designed to encapsulate diverse molecules such as hydrophilic as well as hydrophobic drugs, genes, oligonucleotides, and small interfering RNAs [109]. In addition, use of biologically approved polymers affords construction of nanoparticles relatively easily, using methods that allow scale-up and manufacturing under regulatory guidelines. Indeed, one of the most interesting aspects of polymeric nanoparticles is the ability to achieve spatial and temporal control on release of the encapsulated drug or gene payload, in order to optimize cancer therapy. In addition, there have been significant recent developments, on parallel synthesis of degradable and biocompatible polymeric material libraries and high-throughput screening methodologies specifically intended for biomedical applications [110]. Out of several application of nanotechnology in medicine, target-specific drug/gene delivery and early diagnosis in cancer treatment using nanoparticles has been identified as one of the priority research areas [111].

Polymer conjugates and nanoparticles have become a popular choice for cancer research community due to their history of safety and utility in preclinical and clinical setting (Table 8.1). In this regard, natural polymers are often preferred for fabricating nanosystems due to their solubility, biodegradability or biologically inert nature. More importantly, several natural polymers fall in the classification of "generally regarded as safe" (GRAS) category for human application which makes the translation of such materials relatively easier for clinical use. Natural polymers, such as dextran [112], gelatin [113], hyaluronic acid [114], albumin, heparin, and chitosan [67] have been utilized for construction of nanoparticles. Synthetic polymers have also been studied extensively, and both natural and synthetic polymers offer unique opportunities for construction of nanoparticles, that have favorable properties such as stability, biodegradability, relative ease of chemical modification, and with the opportunity for high drug loading capacities in the nanoparticles [114]. Examples of such polymers include poly(L-lactic acid) (PLA), poly(glycolic acid) (PGA) or their copolymers [i.e.,poly(D,L-lactic-co-glycolic acid) (PLGA)]. These classes of polyesters are the most widely studied and use polymers for drug delivery applications especially for cancer, including some in the clinical trials [111]. Another widely used polymer from the polyester family, is poly(ϵ-caprolactone) (PCL) [115]. PCL is considered non-toxic and tissue compatible material with greater degree of hydrophobicity than PLA. As such, PCL degrades at a much lower rate than PLA and is useful for developing long-term sustained drug release formulations or implantable delivery systems. PCL nanoparticles and their copolymers are widely used for delivery of anticancer drugs because of their hydrophobic nature, which can be utilized to efficiently encapsulate hydrophobic anticancer drugs [116]. Poly(ortho esters) and polyanhydrides are other classes of synthetic biodegradable polymers developed and investigated for applications in drug delivery.

Langer and colleagues were the first to use the hydrolytic instability of aliphatic polyanhydrides for sustained drug release. Owing to the hydrophobic nature, polyanhydrides degrade by surface erosion makes them highly attractive for controlled-release applications [117]. Several modified polymers, such as polyethyleneoxide(PEO)-modified poly (b-amino ester) (PbAE) nanoparticles that in addition to being biodegradable are also pH-responsive which allow for pH-triggered drug release [118]. Among the polymers used for gene delivery, cationic polymers such as polyethyleneimine (PEI),

Table 8.1 Polymer–Drug Conjugates in Clinical Trials

Conjugates	Indication	Status
HPMA copolymer–doxorubicin (PK1; FCE28068)	Cancer, in particular lung, breast cancers	Phase II
HPMA copolymer–doxorubicin galactosamine (PK2;FCE28069)	Hepatocellular carcinoma	Phase I/II
HPMA copolymer–camptothecin (MAG–CPT; PNU166148)	Clinical evaluation on several solid cancers	Phase I
HPMA copolymer–paclitaxel (PNU166945)	Clinical evaluation on several solid cancers	Phase I
HPMA copolymer–platinate (AP5280)	Clinical evaluation on several solid cancers	Phase II
HPMA copolymer–platinate (AP5346)	Clinical evaluation on several solid cancers	Phase I/II
Polyglutamate–paclitaxel (XYOTAX; CT- 2103)	Cancer, in particular lung, ovarian and esophageal cancers	Phase II/III
Polyglutamate–camptothecin (CT-2106)	Clinical evaluation on colorectal, lung, ovarian cancer	Phase I/II
PEG–camptothecin (PROTHECAN)	Clinical evaluation on several solid cancers	Phase II
PEG-paclitaxel	Clinical evaluation of several solid cancers	Phase I
Dextran–doxorubicin (AD-70, DOX-OXD)	Clinical evaluation of several solid cancers	Phase I
Modified dextran–camptothecin (DE-310)	Clinical evaluation of several solid cancers	Phase I

chitosan, Poly(L-lysine) (PLL) and their various copolymers are commonly used
[53, 119–122]. In one study, corona-stabilized inter-polyelectrolyte complexes of siRNA was utilized for the enhanced stability and delivery of siRNAs [123]. Amiji's group utilized PEO modified PbAE and PEO-PCL nanoparticles for the efficient encapsulation of MDR-1 gene silencing siRNAs and the anticancer drug, paclitaxel, respectively [124].

8.3.2 Polymeric Micelles and Dendrimers

Polymeric micelles are one of the most attractive delivery vehicles for cancer targeting due the optimal size and the relative ease of drug encapsulation with high loading. In addition, polymeric micelles are attractive to encapsulate hydrophobic drugs and poorly soluble anticancer compounds. Many of the polymeric micelles are fabricated using synthetic block copolymers that can efficiently self-assemble to form stable nanosystems. Some examples include di-block, tri-block or graft copolymers [125]. Some studies have indicated that micelles from lipophilc-cored amphiphilic block copolymers can be more effective for drug encapsulation than hydrophilic shell and hydrophobic core structures. Akter *et al.*, made micelles using polyethylene glycol modified with polyaspartate or polyaspartate hydrazide to deliver the glycolytic enzyme inhibitor 3PO (3-(3- pyridinyl)-1-(4-pyridinyl)-2-propen-1-one). In this study, the lipophilic modification by polyaspartate hydrazide allowed higher drug loading as well as increased stability after encapsulation [126].

We found that amphiphatic styrene-maleic acid (SMA) copolymer could efficiently form nanosized micelles with several types of anthracycline anticancer agents such as doxorubicin [127], aclarubicin and pirarubicin [128], as well as with heme oxygenase inhibitor-zinc protoporphyrin (ZnPP) [129]. The micelles were constructed by subtle pH adjustments to form non-covalent interaction between the hydrophobic drug and amphiphilic SMA. The micelles thus formed were nanoparticles with narrow size distribution in water, having tunable loading with remarkable aqueous solubility. More importantly, we found significant antitumor response with the SMA micelles. In one such study, we found synergistic effects of light induced photosensitizing and HO-1 inhibitory potentials of the SMA-ZnPP micelles that resulted in remarkable tumor cell killing *in vitro* and *in vivo* [130]. These results demonstrate the potentials of SMA micellar platform technology as a delivery vehicle for diverse forms of anticancer agents.

Koo *et al.* have used pH responsive polymeric micelles for simultaneous *in vivo* imaging and photodynamic therapy of tumors in mice models [131]. In this study a Michael type addition reaction was utilized to conjugate a pH-sensitive polymer, poly(β-amino ester) with mPEG. The block copolymer could self assemble with hydrophobic radiosensitizer protoporphyrin IX (PpIX), forming stable nanosized particles. More importantly, the nanoparticles showed pH-responsive demicellization and triggered release of the photosensitizer in the tumors due to the acidic pH conditions resulting in targeted tumor sensitization. Furthermore, the nanosystem showed clear tumor accumulation as assessed by fluorescence imaging and complete tumor ablation when irradiated with laser light in tumor bearing mice models demonstrating their potentials for photodynamic theranostics [131].

Dendrimers are a class of hyperbranched polymers that have immense potentials for drug and gene delivery and for the development of multifunctional nano-theranostics [132–139]. Dendrimers are highly monodisperse polymers that are produced using very controlled branching chemistry, resulting in homogenous products that can form discrete and uniform nanostructures. Furthermore, while dendrimers were initially produced by a divergent method, with branches stemming from one central point, a higher level of modification and branching has been made possible using convergent methods of chemical synthesis [140]. In one example, a new class poly(L-lysine) dendrimer of generation 3 called "nanoglobules" was developed by Kaneshiro and Lu for the co-delivery of siRNA and doxorubicin to glioblastoma cells [141]. Furthermore, in order to target these dendrimers to glioblastoma tumors, they were surface decorated with RGD-peptide that homes to $\alpha_v\beta_3$-integins receptors overexpressed on the tested U87 glioblastoma cells. The targeted dendrimers showed marked gene silencing activity and tumor growth inhibition due to their efficient internalization in the tested brain tumor cells demonstrating their utility in cancer therapy [141].

One major advantage of dendrimers and hyperbranched polymer architectures is the abundance of functional groups available for chemical conjugation of drugs, genes and imaging agents [139]. Furthermore, the method of payload delivery can also involve encapsulation of drug/gene within the core of the dendrimers. For example, Wang *et al.*, encapsulated methoxyestradiol within the pockets of polyamidoamine (PAMAM) dendrimers decorated with folic acid. These nanoconstructs were able to target human endothelial carcinoma cell line (show KB-HFAR overexpressing folate receptors) [142]. Readers are directed to reviews on the utility of dendrimers for biomedical and nanomedical applications reported elsewhere [143, 144].

8.3.3 Lipid-Based Nanoparticles

Liposomes are probably one of the most widely investigated lipid-based nanosystems, especially for cancer therapy. Liposomes are very attractive because of their biological inertness and safety. Furthermore, the inner core of the liposomes can be used to load water-soluble drugs while the hydrophobic phospholipid bilayer can be used to load poorly water soluble or hydrophobic drugs. In fact, liposomes were one of the few early candidate formulations to enter clinical trials and subsequently into the clinics [145, 146]. For example, liposomal anthracycline Doxil® and DaunoXome® were clinically approved for treating patients with AIDS related Kaposi sarcoma [147].

Some of the recent development have been focused on developing pH-sensitive liposomes that are stable in blood but undergo phase-transition in the acidic environment of the endosomes. In one such case, phosphatidylethanolamine (PE) has been incorporated into liposomes for the delivery of oligonucleotide and drugs [148]. Similarly, poly(organophosphazenes) and cholesteryl hemisuccinate have also been used as a pH-sensitive lipid for the design of liposomal delivery systems [149, 150]. Researchers have also utilized the temperature-responsive behavior of lipids to construct temperature sensitive liposomes. For such purposes, phase transition lipid such as dipalmitoylphosphatidylcholine (DPPC) have been incorporated into liposomes for achieving drug release at the site of hyperthermia induction [151]. In other instances, thio-cholesterol-based liposome have been formulated with redox-responsive TAT-peptides for targeted *in vitro* gene delivery [152]. These lipoplexes could release its payload in the presence of small concentration of reducing agents.

In an effort to overcome multi-drug resistance in breast cancer, Zhao *et al.*, used liposomal systems for delivering epirubicin to breast cancers [153]. For this purpose, epirubicin was loaded into the interior spaces of

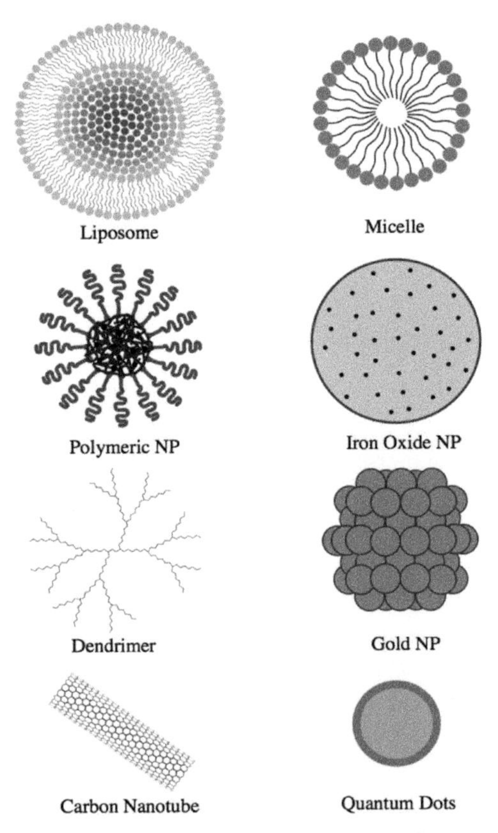

Figure 8.4 Schematic of Type of Nanoparticles (NP) Used for Cancer Imaging and Therapy.

propylene glycol liposomes. It was reported that the cells took up the propylene glycol liposomes, presumably via endocytosis, whereas the free drug could not enter the target cells. Additionally, *in vivo* experiments showed a marked improvement in the therapeutic index by use of the propylene glycol-based liposomal delivery vehicles [153]. Overall, liposomal and lipid-based delivery systems have shown tremendous potentials for disease diagnosis and therapy of cancers. Readers are directed to a comprehensive review on such systems and several other types of lipid-based delivery systems including emulsions and solid lipid nanoparticles discussed elsewhere [154–159]. Figure 8.4 shows some of the nanoconstructs that are commonly utilized for cancer therapeutics.

8.4 Multifunctional Theranostic Nanosystems

As discussed in earlier sections, the development of polymeric and hybrid nano-delivery systems can perform several individual tasks such as cancer therapy or diagnostic imaging separately. However, the commonalities in devising such system has made it possible to bring diagnostic and therapeutic interventions into one realm [160]. The so-called field of "theranostics" encompasses the current efforts to develop more specific, individualized therapies for various diseases where the diagnostic and therapeutic potentials are combined into a single component delivery system. The ultimate goal of these efforts is to achieve simultaneous disease diagnosis and treatment in a more efficient and safe way [161–164]. Figure 8.5 shows the recently increasing interest among research community towards the utilization of multifunctional and theranostic systems for disease diagnosis and therapy.

In general, nanosystems used in theranostics have three major functions: disease diagnosis, delivery of drug/genes and monitoring the therapeutic response [165, 166]. The multifunctional properties of theranostic nanoparticles confer distinctive advantages over conventional form of drug delivery such as cancer-specific delivery of imaging and therapeutic agents and the ability to study the release of the entrapped drug, observe its biodistribution pattern and understand and foresee the efficiency of the drug treatment [167]. Such functions can thus be performed non-invasively and recorded simultaneously in real-time, using a single nanoconstruct.

In addition, the aim of developing a multi-functional nanoparticulate delivery system can aid in fulfilling critical interrelated goals and functions using a single *"nanoconstruct"*. For instance, a nanoparticle delivery system can be designed with low, moderate, or high level of complexity depending on the requirements of the target disease or intended function in the body

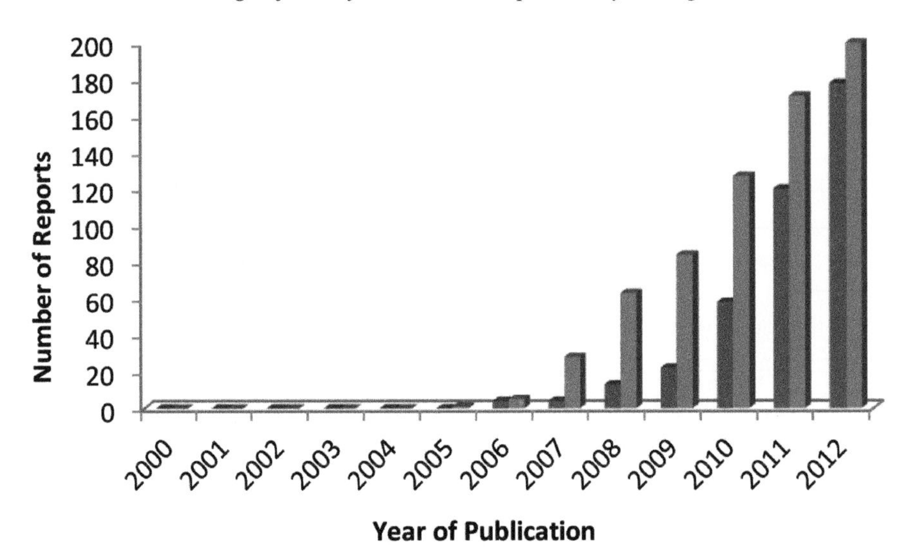

Figure 8.5 Publications on Multifunctional Theranostic Nanosystems. Number of annual publications on theranostic (red bars) and multifunctional (blue bars) nanosystems in medical applications, for the years 2000 to 2012. (Search results obtained from Google scholar and Highwire press).

such as: (a) protecting a labile payload (such as the drug, DNA and siRNA) from degradation on systemic delivery and elimination of off-target effects associated with therapeuticals; [168] (b) enabling long circulation half-life *in vivo* by surface functionalization of nanoparticles with amphiphilic molecules such as PEG that can help evade the immune system, and reduce clearance by the MPS; [169, 170] (c) take advantage of passive tumor targeting or the EPR effect; [171] (d) load/encapsulate nanoparticles with single and often large doses of anticancer drugs or co-deliver a cocktail of drug combinations or co-agents such as siRNA and drugs that works synergistically to have a better therapeutic outcome [172, 173]. In this regard, polymeric nanoparticles also offer the flexibility to devise the system to have spatial and temporal control or regulate the kinetics of payload release (based on the nature and choice of the biodegradable polymer used for constructing the nano-systems) [174–176]. For example, the delivery vehicle can be engineered to release one drug or siRNA concomitantly with another agent or if desired, after a time delay, or in a controlled manner, based on the requirement of the disease target [174]. Along these lines, we are developing a combinatorial-designed library of novel biodegradable polymers that can self assemble in the presence of oligonucleotides and drugs to form stable nanoparticles

[177, 178]. The advantage of such systems is the diverse choice of polymers with varying functionality that could be synthesized and screened simultaneously based on a "mix and match" type screening [177, 178] (e) The nanoparticles could also be surface decorated with targeting ligands such as antibodies or peptides for active targeting as discussed (Figure 8.2), or can be incorporated with agents that can trigger the release of the payload (such as pH, enzymatic or redox-agents) in specific location or facilitate the drug escape from organelles (such as lysosomes) within the cells [179–184]. More importantly, intracellular trafficking of drugs and genes is more challenging

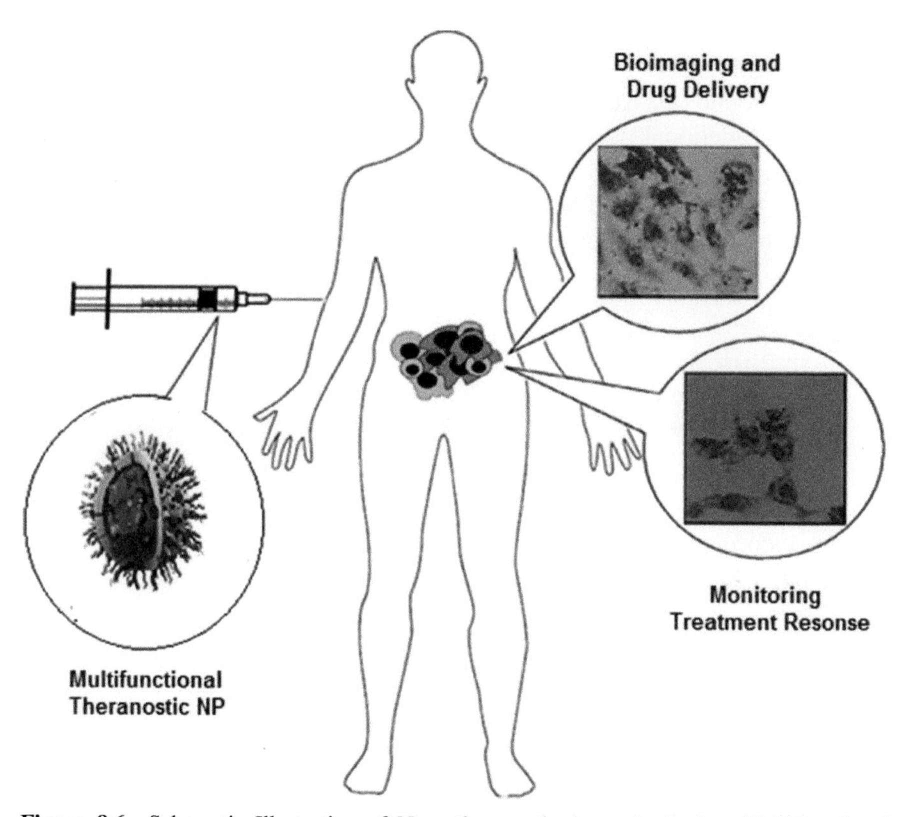

Figure 8.6 Schematic Illustration of Nano-theranostic Agent in Action. Multifunctional theranostic nanoparticle (NP) endowed with tumor homing ligands encapsulated with drugs, genes and imaging agents when inject intravenously into the blood stream, can selectively target tumor tissues and cells. Furthermore, the single nano-construct can be used for simultaneous cancer detection/imaging, diagnosis and measurement of treatment response, all performed non-invasively, in real-time.

for refractory diseases such as cancers that are equipped with active drug efflux pumps such as P-glycoprotein and multi-drug resistant proteins, which are responsible for developing multi-drug resistance (MDR) [185]. In such cases, nanoparticles-based delivery systems have shown to evade the drug *efflux* pathway, thus increasing the intracellular delivery efficiency of drugs and genes [186, 187]. Lastly, (f) nanoparticles can be conjugated or labeled with imaging agents such as radioactive/optical ligands and contrast agents for simultaneous disease detection, imaging and therapy [163, 164].

Polymer nanoparticles, liposomes, micelles, dendrimers and organic-inorganic hybrid materials represent the major multifunctional nanocarriers being explored in cancer therapy (Figure 8.4). Nuclear imaging (PET/SPECT), optical imaging, magnetic resonance imaging (MRI), computed tomography (CT), and ultrasound (US) are the major imaging modalities used in conjunction with multifunctional nanosystems [188]. A combination of nanoprobes and imaging intervention results in multi-modal theranostics being developed today (Figure 8.6). Discussed below are some examples of multifunctional organic-inorganic hybrid nanoparticles systems used for cancer imaging and therapy.

8.5 Illustrative Examples of Multifunctional Nanosystems

8.5.1 Iron Oxide Nanoparticles-Based Theranostic Systems

The development of inorganic nanoparticles has come a long way since its identification as a potential candidate for application in disease diagnosis and therapy. Its incorporation in polymeric nanosystems was intended to supplement or complement the existing favourable properties of the nanoparticle system. For instance, incorporation of iron oxide nanoparticles in polymeric particles containing therapeutic drug could be used for magnetic resonance imaging (MRI) as well as hyperthermia-based therapeutic interventions. In one such study, iron oxide nanoparticles modified with poly (TMSMA-*r*-PEGMA) could evade immune response and accumulate quickly (within an 1 hour) in the tumor after intravenous injection [189].

In another example, Das *et al.*, used phosphonic acid chemistry in creating stealth multifunctional NPs to selectively distinguish and kill cancer cells that overexpress folate receptors [190]. These nanoparticles simultaneously enabled real-time monitoring of the effects of drug treatment on the tumor by means of dual-modal fluorescence and magnetic resonance imaging (MRI). This multifunctional nanosystem was designed using ultrasmall supermagnetic iron oxide crux altered by reaction with N-phosphonomethyl iminodiacetic acid (PMIDA) to form a hydrophilic, biocompatible, and biodegradable coating. Employing suitable conjugation techniques, functional molecules such as rhodamine-B isothiocyanate, folic acid and methotrexate, were attached to the amine-derivatized USPIO–PMIDA. An *in vitro* drug release study indicated a pH dependent release, with the highest release in the pH range of 2–5 and almost no release in the physiological pH, suggesting that the release of the drug would be greater in the acidic tumor environment as compared to normal tissues [190]. In another report, Heidari Majd *et al.*, developed theranostic nanoparticles by grafting mitoxantrone to Fe_3O_4 magnetic nanoparticles (MNPs) modified by a dopamine-polyethylene glycol-folic acid (DPA-PEG-FA) moiety [191]. The superparamagnetic property of the Fe_3O_4 nanoparticle allowed magnetic resonance imaging (MRI), and the folate component helped to target cells expressing folate receptors. The so-developed Fe_3O_4-DPA-PEGFA-MTX nanoparticles had an average particle size of 35 nm and exhibited selective binding and inhibition of folate receptor (FR)-positive MCF-7 cells, but not the FR-negative A549 cells. These nanoparticles also demonstrated enzyme-dependent drug release of the mitoxantrone, due to overexpression of esterases in the cancer cells [191]. In yet another novel study by Santra *et al.*, gadolinium-diethylenetriaminepentaacetic acid (Gd-DTPA) was encapsulated within a poly (acrylic acid) (PAA) polymer coating of a superparamagnetic iron oxide nanoparticle (IO-PAA). The iron oxide core of the IO-PAA-Gd-DTPA nanoparticle quenched the longitudinal spin-lattice magnetic relaxation (T1) of Gd-DTPA when encapsulated within the nanoparticles, however, in an acidic environment (such as tumor microenvironment), the Gd-DTPA could be released from the iron oxide's (IO) polymer coating, thereby resorting its relaxivity, resulting in a corresponding increase in the T1-weighted MRI signal [192]. Readers are directed to a more detailed account of iron oxide based nanoparticles and its biomedical application reported elsewhere [193].

8.5.2 Quantum-Dots-Based Nano-Theranostic Agents

Semiconductor quantum dots (QDs) are yet another class of nanomaterial that has drawn tremendous interest due to their versatility such as tuneable emission (ranging from the visible to infrared region), large absorption coefficient, high photo-stability and discrete emission bands. QDs have therefore been incorporated as hybrid nanosystems for various diagnostic-imaging applications. In one study Gao *et al.*, designed a multifunctional nanosystems containing CdSe-ZnS QDs encapsulated using amphiphilic triblock copolymer. The prostrate-specific membrane antigen (PSMA) tagged core-shell nanoparticles could target prostate tumors utilizing both active and passive tumor targeting mechanisms (Figure 8.2) [194]. In another report, Dubertret *et al.*, encapsulated CdSe-ZnS in the hydrophobic core of a mixture of PEG-PE and phosphotidylcholine polymer-lipid micelles for *in vivo* imaging [195]. Huang *et al.*, developed a novel light-triggered theranostic carbon-dot-based system conjugated with a chlorin e6 photosensitizer (C-dots-Ce6) for multimodality imaging employing photosensitizer fluorescence detection (PFD) and photodynamic therapy (PDT) by FRET mechanism [196]. For this study, the C-dots surfaces were first passivated with PEG diamine (PEG2000N). The PEG-coated C-dots (C-dots-NH_2) were covalently linked to Ce6 using a modified EDC–NHS reaction. C-dots-Ce6 particles were characterized by transmission electronic microscopy (TEM), and the size of C-dots-Ce6 was in the range of 2.5 to 10 nm while showing excellent stability in various media such as deionized water, PBS (pH 7.4), saline and serum. These particles showed excellent tumor homing and imaging ability suitable for NIR fluorescence imaging directed PDT treatment [196]. Chu *et al.*, recently reported their findings regarding the therapeutic efficiency of red, brown or black CdTe and CdSe fluorescent quantum dots in the treatment of cancer by means of photothermal therapy. When irradiated with 671 nm laser source the CdTe and CdSe QDs convert the light energy into heat efficiently both *in vitro* and *in vivo* and also resulted in generation of reactive oxygen species (ROS). Interestingly, the silica shell coating did not influence the photothermal effect as well as the intracellular ROS generation by the QDs. More importantly, the mouse treated with 50 mL (5 mg/mL) of deep brown colored CdTe(710) QDs followed by laser irradiation showed significant inhibition of melanoma tumor [197].

In another approach, quantum dot (QD)-aptamer conjugates were engineered for concurrent cancer imaging and therapy of drug delivery based on bi-fluorescence resonance energy transfer (Bi-FRET) technique [198]. For this purpose, the surface of QDs was functionalized with a RNA aptamer that could recognize the extracellular domains of PSMA. The anticancer drug, doxorubicin, was loaded by intercalation in the double-stranded stem of aptamer forming the QD-Apt-Doxorubicn conjugate with reversible self-quenching properties based on a Bi-FRET mechanism [198]. The engineered smart multifunctional nanosystems were able to deliver doxorubicin to the targeted prostate cancer cells as well as image the delivery of the drug to the cancer cells [198]. Such advanced delivery systems demonstrate the versatility of QD-based nanotheranostics for highly specific, sensitive and therapeutically effective strategies for cancer therapy.

A recent study by Yong *et al.*, demonstrated the use of PEGylated phospholipid micelles to encapsulate PbS QDs, thereby decreasing the toxic effects of PbS QDs [199]. These theranostic nanosystems served as NIR optical probes for *in vitro* and *in vivo* imaging. The folic acid conjugated micelles containing PbS QDs could efficiently target and image pancreatic cancer cells, indicating their potentials as a powerful nano-platform for theranostics research [199]. Yong *et al.*, in another study, illustrated the effectiveness of surface modified and functionalized bioconjugated QDs as theranostic agents. In their study, synthesis of a carboxylated pluronic F127 triblock polymer micelle encapsulating CdTe/ZnS core/shell QDs was reported, wherein the encapsulated nanoprobe was functionalized with folic acid for effective tumor targeting [200]. The nanoprobe showed excellent optical and colloidal stability under physiological conditions with negligible *in vitro* toxicity. Furthermore, an *in vivo* imaging studies showed that the system effectively targeted the tumor site after intravenous injection in a pancreatic tumor mouse model demonstrating their utility for targeted tumor delivery and imaging. There are several such reports of QD-based delivery systems reviewed elsewhere for application in imaging and therapy [201].

8.5.3 Gold Nanoparticles-Based Theranostic Agents

Metallic nanoparticles, especially the ones based on colloidal gold have been extensively researched due to their biological inertness [202], favourable surface chemistry and precise control on their shape and size for imaging. Recent advances in the manipulation of materials at the nanometer length scale have made it possible to combine metallic gold and polymeric systems into one, bringing together the advantages that colloidal gold presents with the chemical versatility of polymeric systems [203]. For instance, the synthesis of gold nanoparticles as small as 2 nm allows for the efficient attachment of large number of gold particles on the surface of polymeric nanoparticles for tumor imaging/targeting [204]. In other examples, PEG modified gold nanoparticles was used for scattering based imaging [205]. Anisotropic nanoparticles such as gold nanoshells [206], nanorods [207] and nanocubes [207] have also been researched for their imaging as well as hyperthermia-based cancer ablation therapy. Cha *et al.*, developed a pH-sensitive calcium phosphate-coated gold nanoparticle with theranostic potential for CT imaging and cancer treatment [208]. The gold nanoparticle (AuNP) was stabilized with

PEG-Asp-Cys copolymers and a coating of calcium phosphate (CaP) was deposited by mineralization. Doxorubicin (Dox) was encapsulated between the gold nanocore and the calcium phosphate outer coating. PEGylated Dox-AuNP@CaP was stable at physiological condition for long durations and exhibited up to 81% HeLa tumor cell killing when incubated for 24 hrs. The Dox free PEGylated AuNP@CaP nanocarrier showed negligible cytotoxicity in a parallel study. The drug release studies showed that the prepared nanoparticles released the drug at higher rate in endosomal fluid (pH 4.5) when compared to extracellular fluid (pH 7.4) due to the dissolution of CaP in acidic endosomal conditions [208]. C. Kojima *et al.*, used polyethylene glycol (PEG)-modified dendrimers (PEGylated dendrimers) as carriers of AuNP [209]. Gold nanoparticle-loaded PEGylated dendrimers can be used for photothermal therapy and CT imaging. The seeding conditions and reducing agents involved during the AuNP preparation determined both the size and surface plasmonic properties, and use of formaldehyde during preparation of AuNP resulted in gold nanoparticle-loaded PEGylated dendrimers capable of absorbing at NIR region [209]. Heo *et al.*, investigated the use of gold nanoparticles (AuNP) surface-functionalized with PEG forming a solvated anti-fouling shell, with biotin serving as a cancer-specific targeting ligand, and beta-cyclodextrin (β-CD) as a drug pocket [210]. The β-CD formed inclusion complexes with paclitaxel (PTX). Furthermore β-CD linked to rhodamine facilitated imaging and tracking of the nanoparticles. The AuNP, surface modified through thiol chemistry, released PTX under intracellular conditions when glutathione (GSH) level were at 10 mM concentration. Confocal laser scanning microscopy (CLSM) and fluorescence-activated cell-sorting (FACS) analysis indicated that the surface functionalized gold nanoparticles have higher affinity to cancer cells such as HeLa, A549, and MG63 in comparison to normal NIH3T3 fibroblast cells. This can be attributed to the overexpression of biotin receptors on their tumor cell surface relative to normal cells [210]. Day *et al.*, presented data which confirmed that NIR-absorbing resonant gold-gold sulfide nanoparticles (GGS-NPs) functionalized with anti-HER2 antibodies could be used as a dual contrast and therapeutic agent in cancer treatment [211]. The study showed that GGS-NPs decorated with anti-HER2 selectively bound to SK-BR-3 breast cancer cells, while the GGS-NPs functionalized with IgG or mPEG-SH did not bind to the SK-BR-3 breast cancer cells. These GGS-NPs, when excited with a low-intensity pulsed laser, emit luminescence that could be used for imaging. When excited with a higher intensity laser, the particles generate heat in cancerous cells and cause photo-abalation, making these nanoparticles suitable for hyperthermia based nanotheranostics [211]. A recent review summarizes the utility of hybrid gold nanoparticles-based systems in nanomedicine [212].

8.5.4 Silica Nanoparticles and Carbon Nanotubes-Based Theranostic Agents

Cheng *et al.*, reported the synthesis and evaluation of the theranostic potential of tri-functionalized mesoporous silica nanoparticles (MSNs) with optical tracking, photodynamic therapeutic ability (PDT) and cell specific targeting ability [213]. The optical-tracking property of the MSNs was achieved using NIR fluorescent contrast agent (ATTO647N) integrated directly into the silica framework of MSNs. An oxygen-sensing, palladium-porphyrin-based photosensitizer (Pd-porphyrin; PdTPP) conferred the ability to cause photoablation of cancerous cells and cRGDyK peptides were tethered onto the outermost surface of the MSNs which aided in selective targeting of cancerous cells overexpressing $\alpha_v\beta_3$ integrins. Evaluation of this multifunctional theranostic platform *in vitro* cell studies suggested excellent targeting specificity with very little non-specific binding to healthy cells [213]. Koole *et al.* demonstrated a novel approach to coat silica nanoparticles with a thick monolayer of paramagnetic PEGylated lipid. The silica nanoparticles in the core harbored quantum dots and the particles decorated with $\alpha_v\beta_3$-integrin-specific RGD-peptides conferred tumor specificity to the nanoparticles [214]. The paramagnetic lipid-coated silica nanoparticles with fluorescent QD core exhibited multi-modal imaging properties capable of being visualized using both quantitative fluorescence and magnetic resonance imaging [214].

Multifunctional theranostic systems employing carbon nanotubes (CNTs) have also gained recent attention for cancer imaging and therapy due to their unique properties. In one study, Shuba *et al.*, presented their findings that single-walled carbon nanotubes (SWCNT) in suspension can be used for medical imaging and thermal killing of cancer cells in the radiofrequency and microwave ranges based on their effective dielectric permittivity and relative absorptivity properties [215]. Delogu *et al.*, showed the potential of functionalized multiwalled carbon nanotubes (MWCNTs) as ultrasound contrast agents for their future development as theranostic nanoparticles [216]. In yet another study Kosuge *et al.*, showed the promising properties of single-walled carbon nanotubes for NIR imaging and photothermal ablation of macrophages [217]. This finding, though not studied on a tumor model, backs several other studies demonstrating the potential of carbon nanotubes as theranostic agents in cancer management.

8.6 Conclusions and Future Directions

Cancer has remained a major killer of mankind and despite advancements in understanding of the disease at the cellular and molecular level, clinicians still continue to face an uphill task of managing several forms of recurrent and metastatic cancers with multi-drug resistant phenotypes. Researchers over the past few decades have been actively developing alternative strategies for better management of cancers using drug cocktails and drug-gene combination therapies employing organic and inorganic nanoparticle systems. Among them, polymeric nanoparticles, micelles and liposomal delivery systems have been the most successful, including some clearing preclinical and clinical stages of development. These first generation nanosystems predominantly utilize the aberrant tumor vasculature to selectively target solid tumor tissues. In an effort to maximize the clinical utility and capabilities of nanoparticles based delivery systems, researchers have been engineering more sophisticated delivery systems by incorporating tumor homing ligands and integrating stimuli responsive features to multifunctional (targeted) nanosystems. This second generation of nanosystems are more complex and utilize a combination of passive and active tumor targeting principles for achieving better therapeutic outcomes. More recently, a highly sophisticated third generation of multifunctional nanoparticles are intensely being pursued and researched upon that incorporate several diagnostic and therapeutic components including drugs/genes, imaging/contrast agents, tumor targeting peptides/antibodies all built into the same nano delivery system using organic-inorganic hybrid nanomaterials. These multifunctional "theranostic" nanosystems" have thus far demonstrated tremendous utility both *in vitro* and *in vivo* for imaging and therapy of cancers. These precision-guided theranostic nanosystems portent to have promising potentials for early detection, diagnosis and treatment of cancers in the clinics.
Abbreviations: DACH, diaminocyclohexane; HPMA, N-(2-hydroxypropyl)-methacrylamide; PEG, poly(ethyleneglycol).

References

[1] Hanahan, D., and R. A. Weinberg, "Hallmarks of cancer: the next generation," Cell, vol. 144, No. 5, March 2011, pp. 646–674.

[2] Varmus, H. E., and R. A. Weinberg, Genes and the Biology of Cancer, New york, NY: Scientific American Library, 1993.

[3] Hanahan, D., and R. A. Weinberg, "The hallmarks of cancer," Cell, vol. 100, No. 1, January 2000, pp. 57–70.

[4] Al-Hajj, M., and M. F. Clarke, "Self-renewal and solid tumor stem cells," Oncogene, vol. 23, No. 43, September 2004, pp. 7274–7282.

[5] Dutour, A., D. Leclers, J. Monteil, F. Paraf, J. L. Charissoux, R. Rousseau and M. Rigaud, "Non-invasive imaging correlates with histological and molecular characteristics of an osteosarcoma model: application for early detection and follow-up of MDR phenotype," Anticancer Res, vol. 27, No. 6B, November/December 2007, pp. 4171–4178.

[6] Biedler, J. L., "Genetic aspects of multidrug resistance," Cancer, vol. 70, no. 6 Suppl, September 15, 1992, pp. 1799–1809.

[7] Jabr-Milane, L. S., L. E. van Vlerken, S. Yadav and M. M. Amiji, "Multi-functional nanocarriers to overcome tumor drug resistance," Cancer Treat Rev, vol. 34, No. 7, November 2008, pp. 592–602.

[8] Cahill, D. P., K. W. Kinzler, B. Vogelstein and C. Lengauer, "Genetic instability and darwinian selection in tumours," Trends Cell Biol, vol. 9, No. 12, December 1999, pp. M57–60.

[9] Loeb, L. A., K. R. Loeb, and J. P. Anderson, "Multiple mutations and cancer," Proc Natl Acad Sci U S A, vol. 100, No. 3, February 2003, pp. 776–81.

[10] Milane, L., Z. Duan, and M. Amiji, "Role of hypoxia and glycolysis in the development of multi-drug resistance in human tumor cells and the establishment of an orthotopic multi-drug resistant tumor model in nude mice using hypoxic pre-conditioning," Cancer Cell Int, vol. 11, No. 3, February 2011, pp. 1–16.

[11] Vaupel, P., and A. Mayer, "Hypoxia in cancer: significance and impact on clinical outcome," Cancer Metastasis Rev, vol. 26, No. 2, June, 2007, pp. 225–239.

[12] Dang, C. V., B. C. Lewis, C. Dolde, G. Dang and H. Shim, "Oncogenes in tumor metabolism, tumorigenesis, and apoptosis," J Bioenerg Biomembr, vol. 29, No. 4, August 1997, pp. 345–354.

[13] Asosingh, K., H. De Raeve, M. de Ridder, G. A. Storme, A. Willems, I. Van Riet, B. Van Camp and K. Vanderkerken, "Role of the hypoxic bone marrow microenvironment in 5T2MM murine myeloma tumor progression," Haematologica, vol. 90, No. 6, June 2005, pp. 810–817.

[14] Martinez-Zaguilan, R., E. A. Seftor, R. E. Seftor, Y. W. Chu, R. J. Gillies and M. J. Hendrix,"Acidic pH enhances the invasive behavior of human melanoma cells," Clin Exp Metastasis, vol. 14, No. 2, March 1996, pp. 176–86.
[15] Gatenby, R. A., and E. T. Gawlinski, "The Glycolytic Phenotype in Carcinogenesis and Tumor Invasion Insights through Mathematical Models," Cancer research, vol. 63, No. 14, July 2003, pp. 3847–3854.
[16] Donnenberg, V. S., and A. D. Donnenberg, "Multiple drug resistance in cancer revisited: the cancer stem cell hypothesis," J Clin Pharmacol, vol. 45, No. 8, August 2005, pp. 872–877.
[17] Printz, C., "Keeping it small: nanotechnology enables scientists to target cancer in new ways," Cancer, vol. 118, No. 16, August2012, pp. 3879–3880.
[18] Bakht, M. K., M. Sadeghi, M. Pourbaghi-Masouleh and C. Tenreiro, "Scope of nanotechnology-based radiation therapy and thermotherapy methods in cancer treatment," Current cancer drug targets, vol. 12, No. 8, October, 2012, pp. 998–1015.
[19] Parhi, P., C. Mohanty, and S. K. Sahoo, "Nanotechnology-based combinational drug delivery: an emerging approach for cancer therapy," Drug Discov Today, vol. 17, No. 17–18, pp. September 2012, 1044–1052.
[20] Kolhe, S., and K. Parikh, "Application of nanotechnology in cancer: a review," Int J Bioinform Res Appl, vol. 8, No. 1–2, March 2012, pp. 112–125.
[21] Schroeder, A., D. A. Heller, M. M. Winslow, J. E. Dahlman, G. W. Pratt, R. Langer, T. Jacks and D. G. Anderson, "Treating metastatic cancer with nanotechnology," Nat Rev Cancer, vol. 12, No. 1, January 2012 pp. 39–50.
[22] Wang, X., Y. Wang, Z. G. Chen and D. M. Shin, "Advances of cancer therapy by nanotechnology," Cancer Res Treat, vol. 41, No.1, March 2009, pp. 1–11.
[23] Danhier, F., O. Feron, and V. Preat, "To exploit the tumor microenvironment: Passive and active tumor targeting of nanocarriers for anti-cancer drug delivery," J Control Release, vol. 148, No. 2, December 2010, pp. 135–146.
[24] Couvreur, P., and C. Vauthier, "Nanotechnology: intelligent design to treat complex disease," Pharm Res, vol. 23, No. 7, July 2006, pp. 1417–1450.

[25] Iyer, A. K., G. Khaled, J. Fang and H. Maeda, "Exploiting the enhanced permeability and retention effect for tumor targeting," Drug discovery today, vol. 11, No. 17, August 2006, pp. 812–818.
[26] Voest, E. E., "Neovascularization: the Achilles' heel of tumors?," Drug resistance updates: Reviews and Commentaries on Antimicrobial and Anticancer Chemotherapy, vol. 1, No. 2, February 1998, pp. 86–87.
[27] Folkman, J., "What is the evidence that tumors are angiogenesis dependent?," J Natl Cancer Inst, vol. 82, No. 1, January 1990, pp. 4–6.
[28] Dvorak, H. F., "Leaky tumor vessels: consequences for tumor stroma generation and for solid tumor therapy," Prog Clin Biol Res, vol. 354A, January 1990, pp. 317–330.
[29] Nagy, J. A., L. F. Brown, D. R. Senger, N. Lanir, L. Van de Water, A. M. Dvorak and H. F. Dvorak, "Pathogenesis of tumor stroma generation: a critical role for leaky blood vessels and fibrin deposition," Biochim Biophys Acta, vol. 948, No. 3, February, 1989, pp. 305–326.
[30] Greish, K., J. Fang, T. Inutsuka, A. Nagamitsu and H. Maeda, "Macromolecular therapeutics: advantages and prospects with special emphasis on solid tumour targeting," Clin. Pharmacokinet., vol. 42, No. 13, December 2003, pp. 1089–1105.
[31] Iyer, A. K., G. Khaled, J. Fang, H. Maeda, "Exploiting the enhanced permeability and retention effect for tumor targeting," Drug Discov. Today, vol. 11, No. 17–18, September 2006, pp. 812–818.
[32] Maeda, H., T. Sawa, and T. Konno, "Mechanism of tumor-targeted delivery of macromolecular drugs, including the EPR effect in solid tumor and clinical overview of the prototype polymeric drug SMANCS," J. Control. Release, vol. 74, No. 1–3, July 2001, pp. 47–61.
[33] Maeda, H., J. Wu, T. Sawa, Y. Matsumura and K. Hori, "Tumor vascular permeability and the EPR effect in macromolecular therapeutics: a review," J. Control. Release, vol. 65, No. 1–2, March 2000, pp. 271–284.
[34] Yuan, F., M. Dellian, D. Fukumura, M. Leunig, D. A. Berk, V. P. Torchilin and R. K. Jain, "Vascular permeability in a human tumor xenograft: molecular size dependence and cutoff size," Cancer Res., vol. 55, No. 17, September 1995, pp. 3752–3756.
[35] Maeda, H., "The enhanced permeability and retention (EPR) effect in tumor vasculature: the key role of tumor-selective macromolecular drug targeting," Adv. Enzyme Regul., vol. 41, pp. 189–207, 2001.

[36] Matsumura, Y., and H. Maeda, "A new concept for macromolecular therapeutics in cancer chemotherapy: mechanism of tumoritropic accumulation of proteins and the antitumor agent smancs," Cancer research, vol. 46, No. 12 Part 1, December 1986, pp. 6387–6392.
[37] Maeda, H., "Macromolecular therapeutics in cancer treatment: the EPR effect and beyond," J Control Release, vol. 164, No. 2, December 2012, pp. 138–144.
[38] Torchilin, V., "Tumor delivery of macromolecular drugs based on the EPR effect," Adv Drug Deliv Rev, vol. 63, No. 3, March 2011, pp. 131–135.
[39] Greish, K., "Enhanced permeability and retention (EPR) effect for anticancer nanomedicine drug targeting," Methods Mol Biol, vol. 624, 2010, pp. 25–37.
[40] Maeda, H., T. Sawa, and T. Konno, "Mechanism of tumor-targeted delivery of macromolecular drugs, including the EPR effect in solid tumor and clinical overview of the prototype polymeric drug SMANCS," J Control Release, vol. 74, No. 1–3, July 2001 pp. 47–61.
[41] Maeda, H., "The enhanced permeability and retention (EPR) effect in tumor vasculature: the key role of tumor-selective macromolecular drug targeting," Adv Enzyme Regul, vol. 41, 2001, pp. 189–207.
[42] Duncan, R., "The dawning era of polymer therapeutics," Nat Rev Drug Discov, vol. 2, No. 5, May 2003, pp. 347–360.
[43] Jain, R. K., "Transport of molecules in the tumor interstitium: a review," Cancer Res, vol. 47, No. 12, June 1987, pp. 3039–3051.
[44] Maeda, H., T. Sawa, and T. Konno, "Mechanism of tumor-targeted delivery of macromolecular drugs, including the EPR effect in solid tumor and clinical overview of the prototype polymeric drug SMANCS," Journal of Controlled Release, vol. 74, No. 1, 2001, pp. 47–61.
[45] Noguchi, Y., J. Wu, R. Duncan, J. Strohalm, K. Ulbrich, T. Akaike and H. Maeda, "Early phase tumor accumulation of macromolecules: a great difference in clearance rate between tumor and normal tissues," Japanese journal of cancer research : Gann, vol. 89, No. 3, 1998, pp. 307–314.
[46] Pasut, G., and F. M. Veronese, "PEGylation for improving the effectiveness of therapeutic biomolecules," Drugs Today (Barc), vol. 45, No. 9, September 2009, pp. 687–695.
[47] Veronese, F. M., and G. Pasut, "PEGylation, successful approach to drug delivery," Drug Discov Today, vol. 10, No. 21, November 2005, pp. 1451–1458.

[48] Gabizon, A. A., "Selective tumor localization and improved therapeutic index of anthracyclines encapsulated in long-circulating liposomes," Cancer Res, vol. 52, No. 4, February 1992, pp. 891–896.

[49] Gao, G. H., Y. Li, and D. S. Lee, "Environmental pH-sensitive polymeric micelles for cancer diagnosis and targeted therapy," J Control Release, November 2012.

[50] Hobbs, S. K., W. L. Monsky, F. Yuan, W. G. Roberts, L. Griffith, V. P. Torchilin and R. K. Jain, "Regulation of transport pathways in tumor vessels: role of tumor type andmicroenvironment," ProcNatlAcad Sci U S A, vol. 95, No. 8, April 1998, pp. 4607–4612.

[51] Yuan, F., H. A. Salehi, Y. Boucher, U. S. Vasthare, R. F. Tuma, and R. K. Jain, "Vascular permeability and microcirculation of gliomas and mammary carcinomas transplanted in rat and mouse cranial windows," Cancer Res, vol. 54, No. 17, September 1994, pp. 4564–8.

[52] Torchilin, V. P., R. Rammohan, V. Weissig and T. S. Levchenko, "TAT peptide on the surface of liposomes affords their efficient intracellular delivery even at low temperature and in the presence of metabolic inhibitors," Proc Natl Acad Sci U S A, vol. 98, No. 15, 2001, pp. 8786–8791.

[53] Ganesh, S., A. K. Iyer, D. V. Morrissey and M. M. Amiji, "Hyaluronic acid based self-assembling nanosystems for CD44 target mediated siRNA delivery to solid tumors," Biomaterials, vol. 34, No. 13, 2013, pp. 3489–3502.

[54] Pasqualini, R., E. Koivunen, and E. Ruoslahti, "Alpha v integrins as receptors for tumor targeting by circulating ligands," Nature biotechnology, vol. 15, No. 6, 1997, pp. 542–546.

[55] Marcucci, F., and F. Lefoulon, "Active targeting with particulate drug carriers in tumor therapy: fundamentals and recent progress," Drug Discov Today, vol. 9, No. 5, March 2004, pp. 219–228.

[56] Milane, L., Z. Duan, and M. Amiji, "Development of EGFR-targeted polymer blend nanocarriers for combination paclitaxel/lonidamine delivery to treat multi-drug resistance in human breast and ovarian tumor cells," Molecular Pharmaceutics, vol. 8, No. 1, 2011, pp. 185–203.

[57] Magadala, P., and M. Amiji, "Epidermal growth factor receptor-targeted gelatin-based engineered nanocarriers for DNA delivery and transfection in human pancreatic cancer cells," The AAPS journal, vol. 10, No. 4, 2008, pp. 565–576.

[58] Iyer, A. K., J. He, and M. M. Amiji, "Image-guided nanosystems for targeted delivery in cancer therapy," Current medicinal chemistry, vol. 19, No. 19, 2012, pp. 3230–3240.

[59] Lue, N., S. Ganta, D. X. Hammer, M. Mujat, A. E. Stevens, L. Harrison, R. D. Ferguson, D. Rosen, M. Amiji and N. Iftimia, "Preliminary evaluation of a nanotechnology-based approach for the more effective diagnosis of colon cancers," Nanomedicine (Lond), vol. 5, No. 9, November 2010, pp. 1467–1479.

[60] Rihova, B., "Receptor-mediated targeted drug or toxin delivery," Adv Drug Deliv Rev, vol. 29, No. 3, February 1998, pp. 273–289.

[61] Farokhzad, O. C., J. M. Karp, and R. Langer, "Nanoparticle-aptamer bioconjugates for cancer targeting," Expert Opin Drug Deliv, vol. 3, No. 3, May 2006, pp. 311–324.

[62] Ruoslahti, E., and M. D. Pierschbacher, "New perspectives in cell adhesion: RGD and integrins," Science, vol. 238, No. 4826, October 1987, pp. 491–497.

[63] Iftimia, N., A. K. Iyer, D. X. Hammer, N. Lue, M. Mujat, M. Pitman, R. D. Ferguson and M. Amiji, "Fluorescence-guided optical coherence tomography imaging for colon cancer screening: a preliminary mouse study," Biomedical optics express, vol. 3, No. 1, 2012, pp. 178–191.

[64] Torchilin, V. P., "Passive and active drug targeting: drug delivery to tumors as an example," Handbook of experimental pharmacology, No. 197, 2010, pp. 3–53.

[65] Keereweer, S., I. M. Mol, J. D. Kerrebijn, P. B. Van Driel, B. Xie, "Targeting integrins and enhanced permeability and retention (EPR) effect for optical imaging of oral cancer," J Surg Oncol, vol. 105, No. 7, June2012, pp. 714–718.

[66] Kim, K., J. H. Kim, H. Park, Y. S. Kim, K. Park, H. Nam, S. Lee, J. H. Park, R. W. Park, I. S. Kim, K. Choi, S. Y. Kim, K. Park and I. C. Kwon, "Tumor-homing multifunctional nanoparticles for cancer theragnosis: Simultaneous diagnosis, drug delivery, and therapeutic monitoring," J Control Release, vol. 146, No. 2, September 2010, pp. 219–227.

[67] Cho, K., X. Wang, S. Nie, Z. G. Chen and D. M. Shin, "Therapeutic nanoparticles for drug delivery in cancer," Clin Cancer Res, vol. 14, No. 5, March 2008, pp. 1310–1316.

[68] Kommareddy, S., and M. M. Amiji, "Intracellular trafficking studies using gold-encapsulated gelatin nanoparticles," CSH protocols, vol. 2008, 2008, pp. pdb.prot4886.

[69] Kommareddy, S., and M. Amiji, "Poly(ethylene glycol)-modified thiolated gelatin nanoparticles for glutathione-responsive intracellular DNA delivery," Nanomedicine : nanotechnology, biology and medicine, vol. 3, No. 1, 2007, pp. 32–42.

[70] Kommareddy, S., and M. Amiji, "Preparation and evaluation of thiol-modified gelatin nanoparticles for intracellular DNA delivery in response to glutathione," Bioconjugate Chemistry, vol. 16, No. 6, 2005, pp. 1423–1432.

[71] Susa, M., A. K. Iyer, K. Ryu, F. J. Hornicek, H. Mankin, M. M. Amiji and Z. Duan,"Doxorubicin loaded Polymeric Nanoparticulate Delivery System to overcome drug resistance in osteosarcoma," BMC cancer, vol. 9, 2009, pp. 399.

[72] Abeylath, S. C., and M. M. Amiji, "'Click'synthesis of dextran macrostructures for combinatorial-designed self-assembled nanoparticles encapsulating diverse anticancer therapeutics," Bioorganic & medicinal chemistry, vol. 19, No. 21, 2011, pp. 6167–6173.

[73] Brigger, I., C. Dubernet, and P. Couvreur, "Nanoparticles in cancer therapy and diagnosis," Adv Drug Deliv Rev, vol. 54, No. 5, Sepember 2002, pp. 631–651.

[74] Abeylath, S. C., S. Ganta, A. K. Iyer and M. Amiji, "Combinatorial-designed multifunctional polymeric nanosystems for tumor-targeted therapeutic delivery," Acc Chem Res, vol. 44, No. 10, October 2011, pp. 1009–17.

[75] Ma, Z., J. Li, F. He, A. Wilson, B. Pitt and S. Li, "Cationic lipids enhance siRNA-mediated interferon response in mice," Biochem Biophys Res Commun, vol. 330, No. 3, May 2005, pp. 755–759.

[76] Omidi, Y., J. Barar, and S. Akhtar, "Toxicogenomics of cationic lipid-based vectors for gene therapy: impact of microarray technology," Curr Drug Deliv, vol. 2, No. 4, October 2005, pp. 429–441.

[77] Gutierrez-Puente, Y., A. M. Tari, R. J. Ford, R. Tamez-Guerra, R. Mercado-Hernandez, M. Santoyo-Stephano and G. Lopez-Berestein, "Cellular pharmacology of P-ethoxy antisense oligonucleotides targeted to Bcl-2 in a follicular lymphoma cell line," Leuk Lymphoma, vol. 44, No. 11, November 2003, pp. 1979–1985.

[78] Savic, R., L. Luo, A. Eisenberg and D. Maysinger "Micellar nanocontainers distribute to defined cytoplasmic organelles," Science, vol. 300, No. 5619, April 2003, pp. 615–618.

[79] Nishiyama, N., F. Koizumi, S. Okazaki, Y. Matsumura, K. Nishio and K. Kataoka, "Differential gene expression profile between

PC-14 cells treated with free cisplatin and cisplatin-incorporated polymeric micelles," Bioconjug Chem, vol. 14, No. 2, March/April 2003, pp. 449–457.
[80] Demoy, M., S. Gibaud, J. P. Andreux, C. Weingarten, B. Gouritin and P. Couvreur,"Splenic trapping of nanoparticles: complementary approaches for in situ studies," Pharm Res, vol. 14, No. 4, April 1997, pp. 463–468.
[81] Rejman, J., V. Oberle, I. S. Zuhorn and D. Hoekstra, "Size-dependent internalization of particles via the pathways of clathrin- and caveolae-mediated endocytosis," Biochem J, vol. 377, No. Pt 1, January 2004, pp. 159–169.
[82] Champion, J. A., and S. Mitragotri, "Shape induced inhibition of phagocytosis of polymer particles," Pharm Res, vol. 26, No. 1, 2009, pp. 244–249.
[83] Chithrani, B. D., A. A. Ghazani, and W. C. W. Chan, "Determining the size and shape dependence of gold nanoparticle uptake into mammalian cells," Nano Lett, vol. 6, No. 4, 2006, pp. 662–668.
[84] Venkataraman, S.,J. L. Hedrick, Z. Y. Ong, C. Yang, P. L. Ee, P. T. Hammond and Y. Y. Yang,"The effects of polymeric nanostructure shape on drug delivery," Adv Drug Deliv Rev, vol. 63, No. 14–15, November 2011, pp. 1228–1246.
[85] Champion, J. A., Y. K. Katare, and S. Mitragotri, "Particle shape: a new design parameter for micro- and nanoscale drug delivery carriers," J Control Release, vol. 121, No. 1–2, August 2007, pp. 3–9.
[86] Yoo, J.-W., E. Chambers, and S. Mitragotri, "Factors that control the circulation time of nanoparticles in blood: challenges, solutions and future prospects," Current pharmaceutical design, vol. 16, No. 21, 2010, pp. 2298–2307.
[87] Barua, S., J. W. Yoo, P. Kolhar, A. Wakankar, Y. R Gokarn and S. Mitragotri, "Particle shape enhances specificity of antibody-displaying nanoparticles," Proc Natl Acad Sci U S A, vol. 110, No. 9, February 2013, pp. 3270–3275.
[88] Chauhan, V. P., Z. Popovic, O. Chen, J. Cui, D. Fukumura, M. G. Bawendi and R. K. Jain, "Fluorescent nanorods and nanospheres for real-time in vivo probing of nanoparticle shape-dependent tumor penetration," Angew Chem Int Ed Engl, vol. 50, No. 48, November 2011, pp. 11417–11420.
[89] Liu, Q., Y. Cui, D. Gardner, X. Li, S. He, "Self-alignment of plasmonic gold nanorods in reconfigurable anisotropic fluids for tunable

bulk metamaterial applications," Nano Lett, vol. 10, No. 4, April 2010, pp. 1347–1353.

[90] Wang, Z., S. Zong, J. Yang, J. Li and Y. Cui, "Dual-mode probe based on mesoporous silica coated gold nanorods for targeting cancer cells," Biosens Bioelectron, vol. 26, No. 6, February 2011, pp. 2883–2889.

[91] Ferrari, M., "Cancer nanotechnology: opportunities and challenges," Nat Rev Cancer, vol. 5, No. 3, March 2005, pp. 161–171.

[92] Yoo, J. W., N. Doshi, and S. Mitragotri, "Adaptive micro and nanoparticles: temporal control over carrier properties to facilitate drug delivery," Adv Drug Deliv Rev, vol. 63, No. 14–15, November 2011, pp. 1247–56.

[93] Fang, C., B. Shi, Y. Y. Pei, M. H. Hong, J. Wu and H. Z. Chen, "In vivo tumor targeting of tumor necrosis factor-alpha-loaded stealth nanoparticles: effect of MePEG molecular weight and particle size," Eur J Pharm Sci, vol. 27, No. 1, January 2006, pp. 27–36.

[94] Perrault, S. D., C. Walkey, T. Jennings, H. C. Fischer and W. C. W. Chan, "Mediating tumor targeting efficiency of nanoparticles through design," Nano Lett, vol. 9, No. 5, 2009, pp. 1909–1915.

[95] He, C., Y. Hu, L. Yin, C. Tang and C. Yin, "Effects of particle size and surface charge on cellular uptake and biodistribution of polymeric nanoparticles," Biomaterials, vol. 31, No. 13, May 2010, pp. 3657–3666.

[96] Wunderbaldinger, P., L. Josephson, and R. Weissleder, "Tat peptide directs enhanced clearance and hepatic permeability of magnetic nanoparticles," Bioconjug Chem, vol. 13, No. 2, March/April 2002, pp. 264–268.

[97] Albanese, A., P. S. Tang, and W. C. W. Chan, "The effect of nanoparticle size, shape, and surface chemistry on biological systems," Annual review of biomedical engineering, vol. 14, 2012, pp. 1–16.

[98] Montenegro, J.-M., V. Grazu, A. Sukhanova, S. Agarwal, J. M. de la Fuente, I, Nabiev, A, Greiner and W. J. Parak, "Controlled antibody/(bio-) conjugation of inorganic nanoparticles for targeted delivery," Adv Drug Deliv Rev, 2012.

[99] Ruoslahti, E., "Peptides as targeting elements and tissue penetration devices for nanoparticles," Advanced Materials, vol. 24, No. 28, pp. 3747–3756, 2012.

[100] Wu, X. L., J. H. Kim, H. Koo, S. M. Bae, H. Shin, M. S. Kim, B. H. Lee, R. W. Park, I. S. Kim, K. Choi, I. C. Kwon, K. Kim and D. S. Lee, "Tumor-targeting peptide conjugated pH-responsive micelles as a

potential drug carrier for cancer therapy," Bioconjug Chem, vol. 21, No. 2, February 2010, pp. 208–213.
[101] Perche, F., N. R. Patel, and V. P. Torchilin, "Accumulation and toxicity of antibody-targeted doxorubicin-loaded PEG-PE micelles in ovarian cancer cell spheroid model," J Control Release, vol. 164, No. 1, November 2012, pp. 95–102.
[102] Kohno, M., T. Horibe, M. Haramoto, Y. Yano, K. Ohara, O. Nakajima, K. Matsuzaki and K. Kawakami, "A novel hybrid peptide targeting EGFR-expressing cancers," Eur J Cancer, vol. 47, No. 5, March 2011, pp. 773–783.
[103] Milane, L., Z. Duan, and M. Amiji, "Therapeutic efficacy and safety of paclitaxel/lonidamine loaded EGFR-targeted nanoparticles for the treatment of multi-drug resistant cancer," PloS one, vol. 6, No. 9, 2011, pp. e24075.
[104] Platt, V. M., and F. C. Szoka, Jr., "Anticancer therapeutics: targeting macromolecules and nanocarriers to hyaluronan or CD44, a hyaluronan receptor," Mol. Pharm., vol. 5, No. 4, July/August, 2008, pp. 474–486.
[105] Rivkin, I., K. Cohen, J. Koffler, D. Melikhov, D. Peer and R. Margalit, "Paclitaxel-clusters coated with hyaluronan as selective tumor-targeted nanovectors," Biomaterials, vol. 31, No. 27, September 2010, pp. 7106–7114.
[106] Prince, M. E., R. Sivanandan, A. Kaczorowski, G. T. Wolf, M. J. Kaplan, P. Dalerba, I. L. Weissman, M. F. Clarke and L. E. Ailles, "Identification of a subpopulation of cells with cancer stem cell properties in head and neck squamous cell carcinoma," Proc. Natl. Acad. Sci. U. S. A., vol. 104, No. 3, January 2007, pp. 973–8.
[107] Collins, A. T., P. A. Berry, C. Hyde, M. J. Stower and N. J. Maitland, "Prospective identification of tumorigenic prostate cancer stem cells," Cancer Res., vol. 65, No. 23, December 2005, pp. 10946–10951.
[108] Yoon, H. Y., H. Koo, K. Y. Choi, S. J. Lee, K. Kim, I. C. Kwon, J. F. Leary, K. Park, S. H. Yuk, J. H. Park and K. Choi, "Tumor-targeting hyaluronic acid nanoparticles for photodynamic imaging and therapy," Biomaterials, vol. 33, No. 15, May 2012, pp. 3980–3989.
[109] Akinc, A., A. Zumbuehl, M. Goldberg, E. S. Leshchiner, V. Busini, N. Hossain, S. A. Bacallado, D. N. Nguyen, J. Fuller, R. Alvarez, A. Borodovsky, T. Borland, R. Constien, A. de Fougerolles, J. R. Dorkin, K. Narayanannair Jayaprakash, M. Jayaraman, M. John, V. Koteliansky, M. Manoharan, L. Nechev, J. Qin, T. Racie, D. Raitcheva, K. G. Rajeev,

D. W. Sah, J. Soutschek, I. Toudjarska, H. P. Vornlocher, T. S. Zimmermann, R. Langer and D. G. Anderson, "A combinatorial library of lipid-like materials for delivery of RNAi therapeutics," Nature biotechnology, vol. 26, No. 5, May 2008, pp. 561–569.

[110] Green, J. J., R. Langer, and D. G. Anderson, "A combinatorial polymer library approach yields insight into nonviral gene delivery," Accounts of Chemical Research, vol. 41, No. 6, 2008, pp. 749–759.

[111] Amass, W.,A. Amass, and B. Tighe, "A review of biodegradable polymers: Uses, current developments in the synthesis and characterization of biodegradable polyesters, blends of biodegradable polymers and recent advances in biodegradation studies," Polymer International, vol. 47, No. 2, October 1998, pp. 89–144.

[112] Kobayashi, E., A. K. Iyer, F. J. Hornicek. M. M. Amiji, and Z. Duan, "Lipid-functionalized dextran nanosystems to overcome multidrug resistance in cancer: a pilot study," Clin Orthop Relat Res, vol. 471, No. 3, March 2013, pp. 915–925.

[113] Xu, J., and M. Amiji, "Therapeutic gene delivery and transfection in human pancreatic cancer cells using epidermal growth factor receptor-targeted gelatin nanoparticles," Journal of visualized experiments : JoVE, No. 59, 2012, pp. e3612.

[114] Patel, T., J. Zhou, J. M. Piepmeier, W. M. Saltzman, "Polymeric nanoparticles for drug delivery to the central nervous system," Adv Drug Deliv Rev, vol. 64, No. 7, May 2012, pp. 701–705.

[115] Ulery,V L., S. Nair, and C. T. Laurencin, "Biomedical Applications of Biodegradable Polymers," J Polym Sci B Polym Phys, vol. 49, No. 12, June 2011, pp. 832–864.

[116] Devalapally, H., Z. Duan, M. V. Seiden and M. M. Amiji, "Paclitaxel and ceramide co-administration in biodegradable polymeric nanoparticulate delivery system to overcome drug resistance in ovarian cancer," Int J Cancer, vol. 121, No. 8, October 2007, pp. 1830–1838.

[117] Kumar, N., R. S. Langer, and A. J. Domb, "Polyanhydrides: an overview," Adv Drug Deliv Rev, vol. 54, No. 7, October 2002, pp. 889–910.

[118] Shenoy, D., S. Little, R. Langer et al., "Poly(ethylene oxide)-modified poly(beta-amino ester) nanoparticles as a pH-sensitive system for tumor-targeted delivery of hydrophobic drugs. 1. In vitro evaluations," Mol Pharm, vol. 2, No. 5, September/October 2005, pp. 357–366.

[119] Taqieddin, E., and M. Amiji, "Enzyme immobilization in novel alginate-chitosan core-shell microcapsules," Biomaterials, vol. 25, No. 10, May 2004, pp. 1937–1945.

[120] Babic, M., D. Horak, M. Trchova, P. Jendelova, K. Glogarova, P. Lesny, V. Herynek, M. Hajek and E. Sykova, "Poly(L-lysine)-modified iron oxide nanoparticles for stem cell labeling," Bioconjug Chem, vol. 19, No. 3, March 2008, pp. 740–750.
[121] Zhu, S. G., J. J. Xiang, X. L. Li et al., "Poly(L-lysine)-modified silica nanoparticles for the delivery of antisense oligonucleotides," Biotechnol Appl Biochem, vol. 39, No. Pt 2, April 2004, pp. 179–187.
[122] Zhu, S. G., K. Gan, Z. Li et al., "[Biocompatibility of poly-l-lysine-modified silica nanoparticles]," Ai Zheng, vol. 22, No. 10, October 2003, pp. 1114–1117.
[123] Scales, C. W., Y. A. Vasilieva, A. J. Convertine et al., "Direct, controlled synthesis of the nonimmunogenic, hydrophilic polymer, poly (N-(2-hydroxypropyl)methacrylamide) via RAFT in aqueous media," Biomacromolecules, vol. 6, No. 4, July/August 2005, pp. 1846–1850.
[124] Yadav, S., L. E. van Vlerken, S. R. Little, S. R. Shen, H. B. Lu, J. Zhou, W. Xiong, B. C. Zhang, X. M. Nie, M. Zhou, K. Tang and G. Y. Li, "Evaluations of combination MDR-1 gene silencing and paclitaxel administration in biodegradable polymeric nanoparticle formulations to overcome multidrug resistance in cancer cells," Cancer Chemother Pharmacol, vol. 63, No. 4, March 2009, pp. 711–722.
[125] Oerlemans, C., W. Bult, M. Bos, G. Storm, J. F. Nijsen and W. E. Hennink, "Polymeric micelles in anticancer therapy: targeting, imaging and triggered release," Pharm Res, vol. 27, No. 12, December 2010, pp. 2569–2589.
[126] Akter, S., B. F. Clem, H. J. Lee, J. Chesney and Y. Bae, "Block copolymer micelles for controlled delivery of glycolytic enzyme inhibitors," Pharm Res, vol. 29, No. 3, March 2012, pp. 847–855.
[127] Greish, K., T. Sawa, J. Fang, T. Akaike and H. Maeda, "SMA-doxorubicin, a new polymeric micellar drug for effective targeting to solid tumours," J Control Release, vol. 97, No. 2, 2004, pp. 219–230.
[128] Daruwalla, J., K. Greish, C. Malcontenti-Wilson, V. Muralidharan, A. Iyer, H. Maeda and C. Christophi, "Styrene maleic acid-pirarubicin disrupts tumor microcirculation and enhances the permeability of colorectal liver metastases," Journal of vascular research, vol. 46, No. 3, 2009, pp. 218–228.
[129] Iyer, A. K., K. Greish, J. Fang, R. Murakami and H. Maeda, "High-loading nanosized micelles of copoly(styrene-maleic acid)-zinc protoporphyrin for targeted delivery of a potent heme oxygenase inhibitor," Biomaterials, vol. 28, No. 10, 2007, pp. 1871–1881.

[130] Iyer, A. K., K. Greish, T. Seki, S. Okazaki, J. Fang, K. Takeshita and H. Maeda,"Polymeric micelles of zinc protoporphyrin for tumor targeted delivery based on EPR effect and singlet oxygen generation," Journal of Drug Targeting, vol. 15, No. 7–8, 2007, pp. 496–506.
[131] Koo, H., H. Lee, S. Lee, K. H. Min, M. S. Kim, D. S. Lee, Y. Choi, I. C. Kwon, K. Kim and S. Y. Jeong,"In vivo tumor diagnosis and photodynamic therapy via tumoral pH-responsive polymeric micelles," Chem. Commun. (Camb.), vol. 46, August 2010, No. 31, pp. 5668–5670.
[132] drug delivery," Drug Discov. Today, vol. 10, No. 1, January 2005, pp. 35–43.
[133] Luo, K., C. Li, L. Li, W. She, G. Wang and Z. Gu, "Arginine functionalized peptide dendrimers as potential gene delivery vehicles," Biomaterials, vol. 33, No. 19, June 2012, pp. 4917–4927.
[134] Pandita, D., J. L. Santos, J. Rodrigues, A. P. Pego, P. L. Granja, and H. Tomas, "Gene delivery into mesenchymal stem cells: a biomimetic approach using RGD nanoclusters based on poly(amidoamine) dendrimers," Biomacromolecules, vol. 12, No. 2, February 2011, pp. 472–481.
[135] Luo, K., C. Li, G. Wang, Y. Nie, B. He, Y. Wu and Z. Gu, "Peptide dendrimers as efficient and biocompatible gene delivery vectors: Synthesis and in vitro characterization," J. Control. Release, vol. 155, No. 1, October 2011, pp. 77–87.
[136] Xu, Q., C. H. Wang, and D. W. Pack, "Polymeric carriers for gene delivery: chitosan and poly(amidoamine) dendrimers," Curr. Pharm. Des., vol. 16, No. 21, July 2010, pp. 2350–2368.
[137] Yuan, Q., W. A. Yeudall, and H. Yang, "PEGylated polyamidoamine dendrimers with bis-aryl hydrazone linkages for enhanced gene delivery," Biomacromolecules, vol. 11, No. 8, August 2010, pp. 1940–1947.
[138] Santos, J. L., D. Pandita, J. Rodrigues, A. P. Pego,P. L. Granja, G. Balian and H. Tomas, "Receptor-mediated gene delivery using PAMAM dendrimers conjugated with peptides recognized by mesenchymal stem cells," Mol. Pharm., vol. 7, No. 3, June 2010, pp. 763–774.
[139] Khan, M. K., S. S. Nigavekar, L. D. Minc, M. S. Kariapper, B. M. Nair, W. G. Lesniak, and L. P. Balogh, "In vivo biodistribution of dendrimers and dendrimer nanocomposites – implications for cancer imaging and therapy," Technol Cancer Res Treat, vol. 4, No. 6, December 2005, pp. 603–613.

[140] Wijagkanalan, W., S. Kawakami, and M. Hashida, "Designing dendrimers for drug delivery and imaging: pharmacokinetic considerations," Pharm Res, vol. 28, No. 7, July 2011, pp. 1500–1519.

[141] Kaneshiro, T. L., and Z.-R. Lu, "Targeted intracellular codelivery of chemotherapeutics and nucleic acid with a well-defined dendrimer-based nanoglobular carrier," Biomaterials, vol. 30, No. 29, 2009, pp. 5660–5666.

[142] Wang, Y.,R. Guo, X. Cao, M. Shen and X. Shi, "Encapsulation of 2-methoxyestradiol within multifunctional poly (amidoamine) dendrimers for targeted cancer therapy," Biomaterials, vol. 32, No. 12, 2011, pp. 3322–3329.

[143] Menjoge, A. R., R. M. Kannan, and D. A. Tomalia, "Dendrimer-based drug and imaging conjugates: design considerations for nanomedical applications," Drug Discov Today, vol. 15, No. 5–6, March 2010, pp. 171–185.

[144] Patri, A. K., I. J. Majoros, and J. R. Baker, "Dendritic polymer macromolecular carriers for drug delivery," Curr Opin Chem Biol, vol. 6, No. 4, August 2002, pp. 466–471.

[145] =Sutton, G., J. Blessing, P. Hanjani and P. Kramer, "Phase II evaluation of liposomal doxorubicin (Doxil) in recurrent or advanced leiomyosarcoma of the uterus: a Gynecologic Oncology Group study," Gynecol Oncol, vol. 96, No. 3, March 2005, pp. 749–752.

[146] =O'Brien, M. E., N. Wigler, M. Inbar, R. Rosso, E. Grischke, A, Santoro, R. Catane, D. G. Kieback, P. Tomczak, S. P. Ackland, F. Orlandi, L. Mellars, L. Alland and C. Tendler, "Reduced cardiotoxicity and comparable efficacy in a phase III trial of pegylated liposomal doxorubicin HCl (CAELYX/Doxil) versus conventional doxorubicin for first-line treatment of metastatic breast cancer," Ann Oncol, vol. 15, No. 3, March 2004, pp. 440–449.

[147] =Tardi, P., M. B. Bally, and T. O. Harasym, "Clearance properties of liposomes involving conjugated proteins for targeting," Adv Drug Deliv Rev, vol. 32, No. 1–2, June 1998, pp. 99–118.

[148] Litzinger, D. C., and L. Huang, "Phosphatidylethanolamine liposomes: drug delivery, gene transfer and immunodiagnostic applications," Biochimica et biophysica acta, vol. 1113, No. 2, August 1992, pp. 201–227.

[149] =Simoes, S., J. N. Moreira, C. Fonseca, N. Duzgunes and M. C. de Lima, "On the formulation of pH-sensitive liposomes with long circulation

times," Advanced Drug Delivery Reviews, vol. 56, No. 7, April 2004, pp. 947–965.

[150] Couffin-Hoarau, A. C., and J. C. Leroux, "Report on the use of poly(organophosphazenes) for the design of stimuli-responsive vesicles," Biomacromolecules, vol. 5, No. 6, November/December 2004, pp. 2082–2087.

[151] Yatvin, M. B., J. N. Weinstein, W. H. Dennis and R. Blumenthal, "Design of liposomes for enhanced local release of drugs by hyperthermia," Science, vol. 202, No. 4374, December 1978, pp. 1290–1293.

[152] Huang, Z., W. Li, J. A. MacKay and F. C., Jr. Szoka, "Thiocholesterol-based lipids for ordered assembly of bioresponsive gene carriers," Molecular therapy : the journal of the American Society of Gene Therapy, vol. 11, No. 3, March 2005, pp. 409–417.

[153] Zhao, Y.-Z., D.-D. Dai, C.-T. Lu et al., "Epirubicin loaded with propylene glycol liposomes significantly overcomes multidrug resistance in breast cancer," Cancer Letters, vol. 330, No. 1, 2013, pp. 74–83.

[154] Sofou, S., and G. Sgouros, "Antibody-targeted liposomes in cancer therapy and imaging," Expert Opin Drug Deliv, vol. 5, No. 2, February 2008, pp. 189–204.

[155] Muthu, M. S., and S. S. Feng, "Theranostic liposomes for cancer diagnosis and treatment: current development and pre-clinical success," Expert Opin Drug Deliv, vol. 10, No. 2, February 2013, pp. 151–155.

[156] Muthu, M. S., and S. S. Feng, "Nanopharmacology of liposomes developed for cancer therapy," Nanomedicine (Lond), vol. 5, No. 7, Septeber 2010, pp. 1017–1019.

[157] Torchilin, V., "Antibody-modified liposomes for cancer chemotherapy," Expert Opin Drug Deliv, vol. 5, No. 9, September 2008, pp. 1003–1025.

[158] Kang, S. N., S. S. Hong, M. K. Lee and S. J. Lim, "Dual function of tributyrin emulsion: solubilization and enhancement of anticancer effect of celecoxib," Int J Pharm, vol. 428, No. 1–2, May 2012, pp. 76–81.

[159] Sawant, R. R., O. Vaze, G. G. D'Souza, K. Rockwell and V. P. Torchilin, "Palmitoyl ascorbate-loaded polymeric micelles: cancer cell targeting and cytotoxicity," Pharm Res, vol. 28, No. 2, February 2011, pp. 301–308.

[160] Liu, Y., H. Miyoshi, and M. Nakamura, "Nanomedicine for drug delivery and imaging: A promising avenue for cancer therapy and diagnosis using targeted functional nanoparticles," International Journal of Cancer, vol. 120, No. 12, 2007, pp. 2527–2537.

[161] Del Vecchio, S., A. Zannetti, R. Fonti, L. Pace and M. Salvatore, "Nuclear imaging in cancer theranostics," The quarterly journal of

nuclear medicine and molecular imaging : official publication of the Italian Association of Nuclear Medicine (AIMN) [and] the International Association of Radiopharmacology (IAR), [and] Section of the Society of. vol. 51, No. 2, 2007, pp. 152–163.

[162] Cuenca, A. G., H. Jiang, S. N. Hochwald, M. Delano, W. G. Cance and S. R. Grobmyer, "Emerging implications of nanotechnology on cancer diagnostics and therapeutics," Cancer, vol. 107, No. 3, 2006, pp. 459–466.

[163] McCarthy, J. R., and R. Weissleder, "Multifunctional magnetic nanoparticles for targeted imaging and therapy," Adv. Drug Deliv. Rev., vol. 60, No. 11, August 2008, pp. 1241–1251.

[164] Jones, E. F., J. He, H. F. VanBrocklin, B. L. Franc and Y. Seo, "Nanoprobes for medical diagnosis: Current status of nanotechnology in molecular imaging," Curr. Nanosci., vol. 4, No. 1, February 2008, pp. 17–29.

[165] Cinteza, L. O., "Multifunctional nanosystems for cancer theragnostics" SPIE Newsroom, April 2011, pp. 1–3

[166] Gindy, M. E., and R. K. Prud'homme, "Multifunctional nanoparticles for imaging, delivery and targeting in cancer therapy," Expert Opin. Drug Deliv., vol. 6, No. 8, August 2009, pp. 865–878.

[167] Lammers, T., F. Kiessling, W. E. Hennink and G. Strom, "Nanotheranostics and Image-Guided Drug Delivery: Current Concepts and Future Directions," Molecular Pharmaceutics, vol. 7, No. 6, 2010, pp. 1899–1912.

[168] O'Brien, M. E., N. Wigler, M. Inbar, R. Rosso, E. Grischke, A. Santoro, R. Catane, D. G. Kieback, P. Tomczak, S. P. Ackland, F. Orlandi, L. Mellars, L. Alland and C. Tendler, "Reduced cardiotoxicity and comparable efficacy in a phase III trial of pegylated liposomal doxorubicin HCl (CAELYX/Doxil) versus conventional doxorubicin for first-line treatment of metastatic breast cancer," Ann. Oncol., vol. 15, No. 3, March 2004, pp. 440–449.

[169] Otsuka, H., Y. Nagasaki, and K. Kataoka, "PEGylated nanoparticles for biological and pharmaceutical applications," Adv. Drug Deliv. Rev., vol. 55, No. 3, February 2003 pp. 403–419.

[170] Dobrovolskaia, M. A., and S. E. McNeil, "Immunological properties of engineered nanomaterials," Nat Nanotechnol, vol. 2, No. 8, August 2007, pp. 469–478.

[171] Maeda, H., and Y. Matsumura, "EPR effect based drug design and clinical outlook for enhanced cancer chemotherapy," Adv. Drug. Deliv. Rev., vol. 63, No. 3, March 2011, pp. 129–130.

[172] Van vlerken, L. E., Z. Duan, S. R. Little, M. V. Seiden and M. M. Amiji, "Biodistribution and pharmacokinetic analysis of Paclitaxel and ceramide administered in multifunctional polymer-blend nanoparticles in drug resistant breast cancer model," Mol. Pharm., vol. 5, No. 4, July/August 2008, pp. 516–526.

[173] Cao, N., D. Cheng, S. Zou, H. Ai, J. Gao and X. Shuai, "The synergistic effect of hierarchical assemblies of siRNA and chemotherapeutic drugs co-delivered into hepatic cancer cells," Biomaterials, vol. 32, No. 8, March 2011, pp. 2222–2232.

[174] Bhavsar, M. D., and M. M. Amiji, "Gastrointestinal distribution and in vivo gene transfection studies with nanoparticles-in-microsphere oral system (NiMOS)," J. Control. Release, vol. 119, No. 3, Jun 2007, pp. 339–348.

[175] Uhrich, K. E., S. M. Cannizzaro, R. S. Langer, K. M. Shakesheff, "Polymeric systems for controlled drug release," Chem. Rev., vol. 99, No. 11, November 1999, pp. 3181–3198.

[176] Dillen, K., J. Vandervoort, G. Van den Mooter et al., "Evaluation of ciprofloxacin-loaded Eudragit RS100 or RL100/PLGA nanoparticles," Int. J. Pharm., vol. 314, No. 1, May 2006, pp. 72–82.

[177] Abeylath, S. C., S. Ganta, A. K. Iyer and M. Amiji, "Combinatorial-Designed Multifunctional Polymeric Nanosystems for Tumor-Targeted Therapeutic Delivery," Acc. Chem. Res., vol. 44, No. 10, July 2011, pp. 1009–1017.

[178] Abeylath, S. C., and M. M. Amiji, "'Click' synthesis of dextran macrostructures for combinatorial-designed self-assembled nanoparticles encapsulating diverse anticancer therapeutics," Bioorg Med Chem, vol. 19, No. 21, November 2011, pp. 6167–6173.

[179] Ghotbi, Z., A. Haddadi, S. Hamdy R. W. Hung, J. Samuel and A. Lavasanifar, "Active targeting of dendritic cells with mannan-decorated PLGA nanoparticles," J. Drug Target., vol. 19, No. 4, May 2011, pp. 281–292.

[180] Choi, C. H., C. A. Alabi, P. Webster, M. E. Davis, "Mechanism of active targeting in solid tumors with transferrin-containing gold nanoparticles," Proc. Natl. Acad. Sci. U. S. A., vol. 107, No. 3, January 2010, pp. 1235–1240.

[181] Choi, K. Y., H. Chung, K. H. Min, H. Y. Yoon, K. Kim, J. H. Park, I. C. Kwon and S. Y. Jeong, "Self-assembled hyaluronic acid nanoparticles for active tumor targeting," Biomaterials, vol. 31, No. 1, January 2010, pp. 106–114.

[182] Ganta, S., H. Devalapally, A. Shahiwala and M. Amiji, "A review of stimuli-responsive nanocarriers for drug and gene delivery," J. Control. Release, vol. 126, No. 3, 2008 pp. 187–204.

[183] Pecot, C. V., G. A. Calin, R. L. Coleman, G. Lopez-Berestein and A. K. Sood, "RNA interference in the clinic: challenges and future directions," Nat. Rev. Cancer, vol. 11, No. 1, January 2011, pp. 59–67.

[184] Torchilin, V. P., "Recent approaches to intracellular delivery of drugs and DNA and organelle targeting," Annu. Rev. Biomed. Eng., vol. 8, 2006, pp. 343–375.

[185] Szakacs, G., J. K. Paterson, J. A. Ludwig, C. Booth-Genthe and M. M. Gottesman, "Targeting multidrug resistance in cancer," Nat Rev Drug Discov, vol. 5, No. 3, March 2006, pp. 219–234.

[186] Devalapally, H., Z. Duan, M. V. Seiden and M. M. Amiji, "Modulation of drug resistance in ovarian adenocarcinoma by enhancing intracellular ceramide using tamoxifen-loaded biodegradable polymeric nanoparticles," Clin. Cancer. Res, vol. 14, No. 10, May 2008, pp. 3193–203.

[187] Van Vlerken, L. E., Z. Duan, M. V. Seiden and M. M. Amiji, "Modulation of intracellular ceramide using polymeric nanoparticles to overcome multidrug resistance in cancer," Cancer Res, vol. 67, No. 10, May 2007, pp. 4843–4850.

[188] Janib, S. M., A. S. Moses, and J. A. MacKay, "Imaging and drug delivery using theranostic nanoparticles," Advanced Drug Delivery Reviews, vol. 62, No. 11, 2010, pp. 1052–1063.

[189] Lee, H., E. Lee, D. K. Kim, N. K. Jang, Y. Y. Jeong and S. Jon, "Antibiofouling polymer-coated superparamagnetic iron oxide nanoparticles as potential magnetic resonance contrast agents for in vivo cancer imaging," Journal of the American Chemical Society, vol. 128, No. 22, June 2006, pp. 7383–7389.

[190] Das, M., D. Mishra, P. Dhak, S. Gupta, T. K. Maiti, A. Basak and P. Panchanan, "Biofunctionalized, Phosphonate-Grafted, Ultrasmall Iron Oxide Nanoparticles for Combined Targeted Cancer Therapy and Multimodal Imaging," Small, vol. 5, No. 24, 2009, pp. 2883–2893.

[191] Heidari M.,Majd, D. Asgari, J. Barar, H. Valizadeh, V. Kafil, G. Coukos and Y. Omidi, "Specific targeting of cancer cells by multifunctional mitoxantrone-conjugated magnetic nanoparticles," Journal of Drug Targeting, vol. 0, No. 0, May 2013, pp. 1–13.

[192] Santra, S., S. D. Jativa, C. Kaittanis, G. Normand, J. Grimm and J. M. Perez, "Gadolinium-Encapsulating Iron Oxide Nanoprobe as Activatable

NMR/MRI Contrast Agent," ACS Nano, vol. 6, No. 8, August 2012, pp. 7281–7294.

[193] Gupta, A. K., R. R. Naregalkar, V. D. Vaidya and M. Gupta, "Recent advances on surface engineering of magnetic iron oxide nanoparticles and their biomedical applications," Nanomedicine : nanotechnology, biology, and medicine, vol. 2, No. 1, February 2007, pp. 23–39.

[194] Gao, X., Y. Cui, R. M. Levenson, L. W. chung and S. Nie, "In vivo cancer targeting and imaging with semiconductor quantum dots," Nature biotechnology, vol. 22, No. 8, August 2004, pp. 969–976.

[195] Dubertret, B., P. Skourides, D. J. Norris, V. Noireaux, A. H. Brivanlou and A. Libchaber "In vivo imaging of quantum dots encapsulated in phospholipid micelles," Science, vol. 298, No. 5599, November 2002, pp. 1759–1762.

[196] Huang, P., J. Lin, X. Wang, Z. Wang, C. Zhang, M. He, K. Wang, F. Chen, Z. Li, G. Shen, D. Cui and X. Chen, "Light-Triggered Theranostics Based on Photosensitizer-Conjugated Carbon Dots for Simultaneous Enhanced-Fluorescence Imaging and Photodynamic Therapy," Advanced Materials, vol. 24, No. 37, September 2012, pp. 5104–5110.

[197] Chu, M., X. Pan, D. Zhang, Q. Wu, J. Peng and W. Hai, "The therapeutic efficacy of CdTe and CdSe quantum dots for photothermal cancer therapy," Biomaterials, vol. 33, No. 29, 2012, pp. 7071–7083.

[198] Bagalkot, V., L. Zhang, E. Levy-Nissenbaum, S. Jon, P. W. Kantoff, R. Langer and O. C. Farokhzad, "Quantum dot-aptamer conjugates for synchronous cancer imaging, therapy, and sensing of drug delivery based on bi-fluorescence resonance energy transfer," Nano Lett., vol. 7, No. 10, October 2007, pp. 3065–3070.

[199] Hu, R., W.-C. Law, G. Lin, L. Ye, J. Liu, J. L. Reynolds, K. T. Yong, "PEGylated phospholipid micelle-encapsulated near-infrared PbS quantum dots for in vitro and in vivo bioimaging," Theranostics, vol. 2, No. 7, 2012, pp. 723–733.

[200] Liu, L., K.-T. Yong, I. Roy, W. C. Law, L. Ye, J. Liu, R. Kumar, X. Zhang and P. N. Prasad, "Bioconjugated Pluronic triblock-copolymer micelle-encapsulated quantum dots for targeted imaging of cancer: in vitro and in vivo studies," Theranostics, vol. 2, No. 7, 2012 pp. 705–713.

[201] Medintz, I. L., H. T. Uyeda, E. R. Goldman and H. Mattoussi, "Quantum dot bioconjugates for imaging, labelling and sensing," Nature materials, vol. 4, No. 6, June 2005, pp. 435–446.

[202] Rana, S., A. Bajaj, R. Mout and V. M. Rotello, "Monolayer coated gold nanoparticles for delivery applications," Adv Drug Deliv Rev, vol. 64, No. 2, February 2012, pp. 200–216.

[203] Song, J., J. Zhou, and H. Duan, "Self-assembled plasmonic vesicles of SERS-encoded amphiphilic gold nanoparticles for cancer cell targeting and traceable intracellular drug delivery," J Am Chem Soc, vol. 134, No. 32, August 2012, pp. 13458–13469.

[204] Ghosh, P., G. Han, M. De, C. K. Kim and V. M. Rotello, "Gold nanoparticles in delivery applications," Adv Drug Deliv Rev, vol. 60, No. 11, August 2008 pp. 1307–1315.

[205] Boisselier, E., and D. Astruc, "Gold nanoparticles in nanomedicine: preparations, imaging, diagnostics, therapies and toxicity," Chemical Society reviews, vol. 38, No. 6, June 2009, pp. 1759–1782.

[206] Hirsch, L. R., R. J. Stafford, J. A. Bankson, S. R. Sershen, B. Rivera, R. E. Price, J. D. Hazle, N. J. Halas and J. L. West, "Nanoshell-mediated near-infrared thermal therapy of tumors under magnetic resonance guidance," Proceedings of the National Academy of Sciences of the United States of America, vol. 100, No. 23, November 2003, pp. 13549–13554.

[207] Von Maltzahn, G., J. H. Park, A. Agrawal, N. K. Bandaru, S. K. Das, M. J. Sailor and S. N. Bhatia, "Computationally guided photothermal tumor therapy using long-circulating gold nanorod antennas," Cancer research, vol. 69, No. 9, May 2009, pp. 3892–3900.

[208] Cha, E.-J., I.-C. Sun, S. C. Lee, K. Kim, I. C. Kwon and C.-H. Ahn, "Development of a pH sensitive nanocarrier using calcium phosphate coated gold nanoparticles as a platform for a potential theranostic material," Macromolecular Research, vol. 20, No. 3, 2012, pp. 319–326.

[209] Kojima, C., S.-H. Cho, and E. Higuchi, "Gold nanoparticle-loaded PEGylated dendrimers for theragnosis," Research on Chemical Intermediates, vol. 38, No. 6, April 2012, pp. 1279–1289.

[210] Heo, D. N., D. H. Yang, H.-J. Moon, J. B. Lee,M. S. Bae, S. C. Lee, W. J. Lee, I. C. Sun and I. K. Kwon, "Gold nanoparticles surface-functionalized with paclitaxel drug and biotin receptor as theranostic agents for cancer therapy," Biomaterials, vol. 33, No. 3, 2012, pp. 856–866.

[211] Day, E. S., L. R. Bickford, J. H. Slater, N. S. Riggall, R. A. Drezek, J. L. West,"Antibody-conjugated gold-gold sulfide nanoparticles as multifunctional agents for imaging and therapy of breast cancer," International journal of nanomedicine, vol. 5, August 2010, pp. 445–454.

[212] Ghosh, P., G. Han, M. De C. K. Kim, V. M. Rotello, "Gold nanoparticles in delivery applications," Advanced Drug Delivery Reviews, vol. 60, No. 11, August2008, pp. 1307–1315.

[213] Cheng, S.-H., C.-H. Lee, M.-C. Chen, J. S. Souris, F.-G. Tseng, C.-S. Yang, C.-Y. Mou, C.-T. Chen and L.-W. Lo, "Tri-functionalization of mesoporous silica nanoparticles for comprehensive cancer theranostics-the trio of imaging, targeting and therapy," Journal of Materials Chemistry, vol. 20, No. 29, 2010, pp. 6149–6157.

[214] Koole, R., M. M. van Schooneveld, J. Hilhorst, K. Castermans, D. P. Cormode, G. J. Strijkers, C. de Mello Donega, D. Vanmaekelbergh, A. W. Griffioen, K. Nicolay, Z. A. Fayad, A. Meijerink and W. J. Mulder, "Paramagnetic Lipid-Coated Silica Nanoparticles with a Fluorescent Quantum Dot Core: A New Contrast Agent Platform for Multimodality Imaging," Bioconjugate Chemistry, vol. 19, No. 12, December 2008, pp. 2471–2479.

[215] Shuba, M., S. Maksimenko, G. Y. Slepyan and G. Hanson, Nanodevices and Nanomaterials for Ecological Security, Springer Netherlands, 2012.

[216] Delogu, L. G., G. Vidili, E. Venturelli, C. Menard-Moyon, M. A. Zoroddu, G. Pilo, P. Nicolussi, C. Ligios, D. Bedognetti, F. Sgarrella, R. Manetti and A. Bianco,"Functionalized multiwalled carbon nanotubes as ultrasound contrast agents," Proceedings of the National Academy of Sciences, vol. 109, No. 41, September 2012, pp. 16612–16617.

[217] Kosuge, H., S. P. Sherlock, T. Kitagawa, R. Dash, J. T. Robinson, H. Dai and M. V. McConnell, "Near Infrared Imaging and Photothermal Ablation of Vascular Inflammation Using Single-Walled Carbon Nanotubes," Journal of the American Heart Association, vol. 1, No. 6, December 2012, pp. 1–10.

9

Nanomedicine for the Treatment of Breast Cancer

Surendra Nimesh[*,1], Nidhi Gupta[2] and Ramesh Chandra[3,4]

[1]Department of Biotechnology, School of Life Sciences,
Central University of Rajasthan, Bandarsindri, Kishangarh,
Ajmer-305801, Rajasthan, India
[2]Department of Biotechnology, IIS University, Mansarovar,
Jaipur-302020, Rajasthan, India
[3]Dr.B.R. Ambedkar Center for Biomedical Research,
University of Delhi, Delhi -110007, India
[4]Department of Chemistry, University of Delhi,
Delhi -110007, India
[*]Department of Biotechnology, School of Life Sciences,
Central University of Rajasthan, Bandarsindri,
N.H. 8, Teh.- Kishangarh, Dist., Ajmer - 305801,
Rajasthan, India

9.1 Introduction

Breast cancer is one of the most common and lethal malignancies in women worldwide. Statistically, 1.5 million females were diagnosed with breast cancer in 2010. Breast cancer accounts for almost 33% of all incident cases of cancer in women [1]. Increase in global incidence, morbidity and mortality from breast cancer have led to intensified efforts in the search for etiological factors of the disease. While increasing age and the female sex are well-recognized risk factors, reproductive characteristics such as age at menarche and menopause, menstrual irregularity, age at first and last child-birth, parity and breastfeeding have also been linked to breast carcinogenesis [2]. However, the vast majority of women who develop breast cancer (75%) have no familial or hereditary/genetic risk for breast cancer [3].

Post-genomic Approaches in Cancer and Nano Medicine, 267–286.

Reproductive hormones appear to play a critical role in initiation and promotion of breast cancer. Endocrine therapies are the cornerstone of breast cancer treatments in preventive, adjuvant, or metastatic disease settings [4]. However, up to 50% of women suffering from breast cancer are not responsive to this therapy because they lack estrogen and progesterone receptors which are the normal target for most popular endocrine therapy with tamoxifen or an aromatase inhibitor [5]. There is also a renewed interest in ovariectomy and ovarian suppression therapy in the adjuvant treatment of premenopausal women with breast cancer [6]. Chemotherapy is widely used as adjuvant therapy in women with tumors that are >1 cm in size [7]. In pre-menopausal women, chemotherapy accelerates the menopausal transition and may lead to premature menopause [6]. As a result, gonadotropin-releasing hormone agents are being studied. However, menopausal symptoms can be quite troubling for women who have abruptly stopped taking hormone therapy, with exacerbation if adjuvant endocrine therapy with tamoxifen or other inhibitor is initiated at the same time. The widespread use of adjuvant therapy for breast cancer has led to dramatic improvements in survival. However, they are associated with severe side-effects in women who had received any adjuvant therapy [8–10]. This is even higher in women who received both chemotherapy and tamoxifen or preventive oophorectomy, a common procedure to treat breast cancer [11]. In addition, chemotherapy also led to sexual dysfunction in younger women who developed amenorrhea [8, 10, 12, 13]. One of the most successful approaches involves surgical removal of primary breast tumor. But, this does not rule out relapse at local or distant sites, because of the presence of micro-metastases at the time of the diagnosis [4]. Several other strategies like radiation therapy, bisphosphonates for bone diseases, and trastuzumab for patients have been shown to improve overall survival in patients with advanced breast cancer, however, none of these treatment modalities have been able to show an increase in the cure rate [14]. Also, cancer gene therapy approaches to breast and other solid tumor treatment have been limited by the ability of the delivery vectors to achieve specific high-level expression within tumor tissues or the tumor environment following systemic administration [15].

9.2 Etiology of Breast Cancer

Tumor cells show the characteristic features of aggressive cellular differentiation and loss of proliferative control. These features are also phenotypic manifestations of many genetic alterations, mainly modifications in cell cycle regulatory genes. The morphologic alterations at tumor cells are an extension of prior molecular changes at the cellular genome level. The molecular examination of genomic DNA of tumor cells using techniques such as analysis of nucleotide sequences of short DNA fragments, and/or use of radio-labeled hybridized probes that bind to specific genomic regions have revealed a wide range of molecular alterations. The results of such genetic studies have suggested the existence of a genetic identity card with the definitive characteristics of each tumor.

Epidemically, breast cancer is one of the most prevalent and lethal malignancies in women, with greater than 1,300,000 cases and 450,000 deaths each year worldwide. In the field of molecular genetics, these lesions have provided meaningful knowledge of the genetic aberrations, with constant increase in the available information concerning molecular events in breast cancer; that are of recognized importance for the comprehension of the etiopathogenesis of the disease, and practical tools in the disease management: as information complementary to the histologic diagnosis, as prognostic markers of the disease, as possible predictors of treatment response, as an experimental basis for developing new treatment strategies, and as the molecular basis for establishing groups of hereditary cancers.

9.2.1 Histology and Diagnosis of Breast cancer

Several parameters including pathological features such as histology and size of tumor, pathologic stage, involvement of lymph nodes and eventually, metastatic spread are considered while diagnosis for breast cancer is clinically done. The oestrogen and progesterone receptor content of tumor cells is determined by radioimmunoassay or immunohistochemical (IHC) analysis, and gives necessary information on the primary tumor characteristics. The analysis along with above mentioned parameters, in addition to menopausal status and patient age, directs the mode of cancer treatment.

Breast cancer is phenotypically heterogeneous and is therapeutically categorized into three basic groups that further have subtypes. The most common and diverse is the estrogen receptor (ER) positive group with several possible genomic tests to predict outcomes for ER patients receiving endocrine therapy [16, 17]. The second group is epidermal growth factor receptor (HER2) amplified group, a big clinical success because HER2 is a potential therapeutic target and is efficaciously being targeted [18, 19]. The third group is identified as the Triple-negative breast cancers (TNBCs), lacking expression of ER, progesterone receptor (PR) and HER2, also known as basal-like breast cancers, being treated with only chemotherapy options. ER positive group is further classified into Luminal A and Luminal B breast cancers, with the former being the most extensive subtype, representing 50–60% of the total incidences of breast cancer [20].

9.2.2 Luminal A subgroup

In this case, cell proliferation genes are lowly expressed. The expression of genes in the luminal epithelium lining the mammary ducts is activated by ER transcription factor [21]. The molecular profiles of all incidences of lobular carcinoma *in situ* keep every case as luminal A type. The expression of various genes is altered with GATA3 marker expression being highest in this subgroup. IHC profile of luminal A is characterized by a low histological grade and with the expression of ER, PGR, Bcl-2 and cytokeratin CK8/18. HER2 expression is absent and a low rate of proliferation is measured by Ki67. The relapse rate in this type is significantly lower at 27.8% than that for other subtypes (10) and patients with this subtype of cancer have a good prognosis. The treatment of this subgroup of breast cancer is mainly based on selective estrogen receptor modulators (SERMs) like tamoxifen and pure selective regulators of ER like fulvestrant and aromatase inhibitors (AI) in postmenopausal patients [22].

9.2.3 Luminal B Subgroup

Based on the luminal B molecular profile, 10% to 20% of all breast cancers come under this category. These not only have worse prognosis but also have aggressive phenotype, higher histological grade and proliferative index in comparison to the luminal A subgroup. ER is expressed in both luminal A and B tumors but the prognosis of subtype B is very different, letting an emphasis on discovering new biomarkers to distinguish the two subtypes. The main biological difference between the two subtypes is an increased expression of proliferation genes, such as MKI67 and cyclin B1 in the luminal B subtype.

9.2.4 Basal-like Carcinomas, a Subgroup of Triple-negative Breast Cancers

10–20% of all breast cancers are represented by the basal-like subtype carcinomas and genes present in normal breast myoepithelial cells, including high molecular weight cytokeratins CK5 and CK17, P-cadherin, caveolin 1 and 2, nestin, CD44 and EGFR are being expressed. The cancer is quite predominant among African women and is clinically characterized by appearance at an early age, have bigger tumor size at diagnosis, a high histological grade and a high frequency of lymph node affectation [23]. Five biomarkers namely, ER, PGR, HER2, EGFR and CK5/6 have been selected by IHC profile to identify the basal-like group with 100% specificity and 76% sensitivity [24]. Despite having good response to chemotherapy, basal-like tumors have worse prognosis and higher relapse rate, raising the need to identify new therapeutic targets and designing of better treatment approaches. Mutations in BRCA1 and P53 are commonly attached with this group. BRCA1 is critical in the DNA repair and its inactivation leads to the accumulation of errors and genetic instability favoring the growth of tumors.

Poly-ADP ribosepolymerase- 1 (PARP-1) is key in DNA single strand breaks repair and PARP-1 inhibitors have emerged as the most promising strategy to treat basal-like tumors. Due to BRCA 1 mutations DNA repairing being defective, this has led to accumulation of breaks in the double-stranded DNA and to cell death. Phase I studies evaluating olaparib (AZD2281) as monotherapy in breast cancer with BRCA1 and BRCA2 mutations show a very high response rate and clinical benefit (47% and 63%, respectively) [25].

9.2.5 Normal Breast

TNs are secondly categorized as this subtype, that is rare in occurrence and grouped into the classification of intrinsic subtypes with fibroadenomas and normal breast samples and account for about 5–10% of all breast carcinomas. These tumors present an intermediate prognosis between luminal and basal-like, express genes characteristic of adipose tissue and usually show poor or no response to neo-adjuvant chemotherapy. Along with lack of expression of ER, HER2 and PR, they are also negative for CK5 and EGFR expression.

9.2.6 Claudin-low

In 2007, a new intrinsic subtype was identified and is characterized by a low expression of genes involved in tight junctions and intercellular adhesion, including claudin-3, -4, -7 cingulin, ocludin, and E-cadherin, hence the name claudin-low [26]. Similar to Basal-like, this group has low expression of HER2 yet over expresses a set of 40 immune response genes, indicating a high infiltration of tumors immune system cells [27]. These have a poor long term prognosis and inefficient response to neoadjuvant chemotherapy. Even if a tumor presents a specific drugable oncogenic dependence, tumor cells often display an unexpected resistance that allows them to escape death, leading emphasis on identification and validation of new biomarkers, especially in those entities lacking clear therapeutic targets.

9.3 Nanomedicine

Cancer treatment options are presently limited to chemotherapy, radiation and surgery. All these approaches not only remove diseased tissues but also damage the normal healthy tissues. However, nanomedicine-based treatment modalities offers the means to targeted and controlled therapies selectively and directly to cancerous cells. Traditional chemotherapy utilizes drugs that are efficient in destroying cancer cells. However, these cytotoxic drugs also damage normal healthy cells along with tumor cells, which result in adverse side effects such as nausea, neuropathy, hair-loss, fatigue, and suppressed immune system. Nanomedicine provides tools that allow delivery of these chemotherapeutic drug molecules directly to the tumor while leaving normal healthy tissue.

The term "nanomedicine" is an umbrella word that is used to collectively represent vectors such as polymeric micelles, quantum dots, liposomes, polymer-drug conjugates, dendrimers, biodegradable nanocarriers, inorganic nanoparticles and other nanomaterials with therapeutic relevance. Among these vectors, polycationic nanocarriers or nanoparticles have found numerous applications in targeted drug and gene delivery. Nanoparticles can be defined as sub-microscopic particles with at least one dimension less than 100 nm. Due to their small size and condensed nature, nanoparticles provide better tissue penetration and targeting [28]. Further, nanoparticles engineered from polycationic polymers have been largely explored to deliver DNA and siRNA. These nanoparticles can further be categorized into (1) nanospheres are spherical nanometer size particles where the drug molecules can be either entrapped inside the sphere or adsorbed on the outer surface or both, (2) nanocapsules that consist of a solid polymeric shell and an inner liquid core where the desired drug molecules can be entrapped (**Figure 9.1**). Nanoparticles bear numerous advantages over conventional chemotherapy such as:

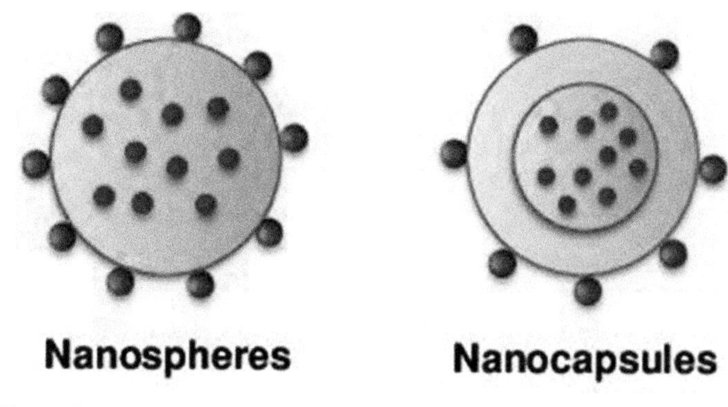

Figure 9.1 Types of nanoparticles, (1) nanospheres and (2) nanocapsules.

1. Deliver the drug molecules to the target cell in intact and active form.
2. Improve the absorption of drug molecules into tumors and by the cancerous cells.
3. Facilitates controlled and enhanced drug distribution to the tumor cells thereby resulting in reduced interaction with the non-tumorous cells and less side effects.

9.4 Targeted Nanomedicine

Successful clinical application of nanomedicine relies on target specific delivery of drug molecules to tumors. Numerous efforts have been dedicated to achieve target specific drug delivery employing nanoparticles modulated with various targeting ligands such as peptides, antibodies, and sugar molecules. The interaction of ligand modified nanoparticles occurs via the specific cell surface receptors which further favours internalization. One of the advantages of targeted drug delivery is that the desired therapeutic effect is achieved at low doses and the possible adverse effects are minimized. Drug targeting employing nanoparticles can be broadly classified as: passive and active targeting.

9.4.1 Passive Targeting

To achieve passive targeting within the body, nanoparticles are prepared without introducing much surface modifications. Intravenous administration of unmodified nanoparticles leads to rapid clearance from systemic circulation by opsonization and macrophage engulfment or accumulation in the liver and spleen [29, 30]. This rapid clearance of nanoparticles can be explored for targeting the accumulated macrophages in atherosclerosis or for the treatment of hepatic disorders such as leishmaniasis, a parasitic disease [31, 32]. Accumulation of particles up to 400 nm has been reported by passive targeting to tumors [33]. This could be achieved due to the ability of these particles to leach into the diseased tissue through the leaky vasculature network commonly observed in tumorigenesis; a phenomenon called the EPR effect (Figure 9.2) [33]. Multiplication of tumor cells leads to formation of clusters that reach a size of 2–3 mm, which induces angiogenesis to meet the ever-increasing nutrition and oxygen demands of the growing tumor [34]. This neovasculature significantly differs from that of normal tissues in microscopic anatomical architecture [35]. The blood vessels in the tumor are irregular in shape, dilated, leaky or defective, and the endothelial cells are poorly aligned or disorganized with large fenestrations (Figure 9.2). Moreover, the perivascular cells and the basement membrane, or the smooth-muscle layer, are frequently absent or abnormal in the vascular wall. Tumor vessels have a wide lumen, whereas tumor tissues have poor lymphatic drainage [35–39]. These anatomical defects along with functional abnormalities, results in extensive leakage of blood plasma components, such as macromolecules, nanoparticles and lipid particles, into the tumor tissue. Furthermore, the slow venous return in tumor tissue and the poor lymphatic clearance results in accumulation of macromolecules in the tumor, whereas extravasation into tumor interstitium continues.

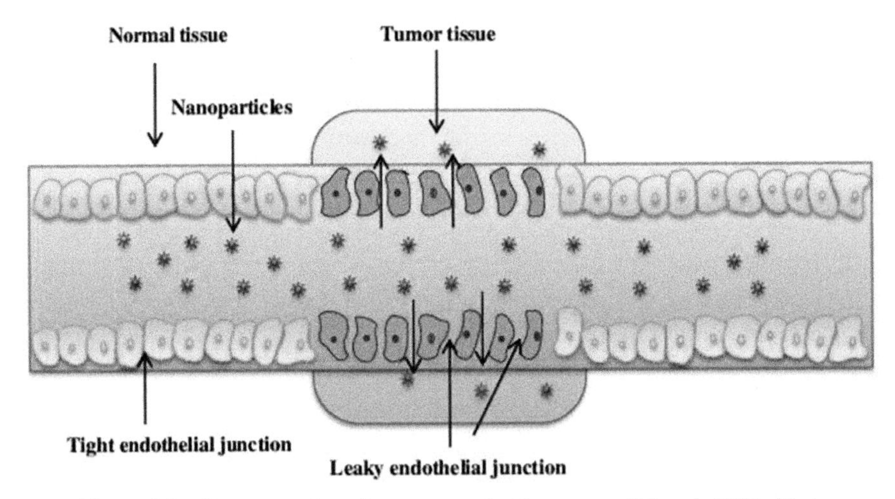

Figure 9.2 Representation of tumor targeting by nanoparticles via EPR effect.

9.4.2 Active Targeting

Nanoparticles modified with a targeting ligand are commonly employed to achieve active targeting. The surface of the nanoparticles is modulated by conjugation of a ligand, which facilitates the homing, binding and internalization of the complex to the target cells. Several strategies have been proposed to control the amount of targeting ligands on the surface of the nanoparticles. However, in the case of weak binding ligands, multivalent functionalization on the surface of the nanoparticles provides sufficient avidity. Generally, small molecule ligands such as peptides, sugars, and small molecules are more attractive than antibodies due to higher stability, purity, ease of production through synthetic routes, and non-immunogenicity.

9.5 Nanomedicince for the Treatment of Breast Cancer

Though breast cancer is a serious health challenge, better understanding of the molecular basis of disease progression has helped to decrease the mortality rate. Presently, available treatment options include surgical procedures, radiation, and/or chemotherapy, gene therapy and other methods. However, they all have some or the other limitations. Nanomedicine is a new and rapidly evolving technology that finds application in various areas of material sciences and emerging as new technology to be used in medicine; especially in drug delivery and other biomedical applications. Indeed, there has been a rapid growth in research in the use of nanoparticles in cancer therapeutics. Human epidermal growth factor receptor 2 (HER2 or ErbB2 or Neu) is a trans-membrane receptor protein that initiates the intracellular tyrosine kinase signaling cascade and mediate enhanced proliferation, oncogenesis, metastasis, and probable resistance to therapeutic agents that induce apoptosis [40]. In patients suffering from breast cancer approximately, 15% and 20% show over-expression of HER2 and henceforth an adverse prognosis [41]. Nanovectors conjugated with Trastuzumab have been reported to recognize HER2/neu receptors, followed by internalization into cells mediated through receptor-mediated endocytosis [41–43].

9.5.1 Liposomes

Liposomes are sub-microscopic vesicles fabricated from one or more bilayers of amphipathic lipids that entrap an equal number of internal aqueous compartments. They can be categorised on the basis of their size, number and arrangement of their constituting lipid bilayers (Figure 9.3). Aqueous

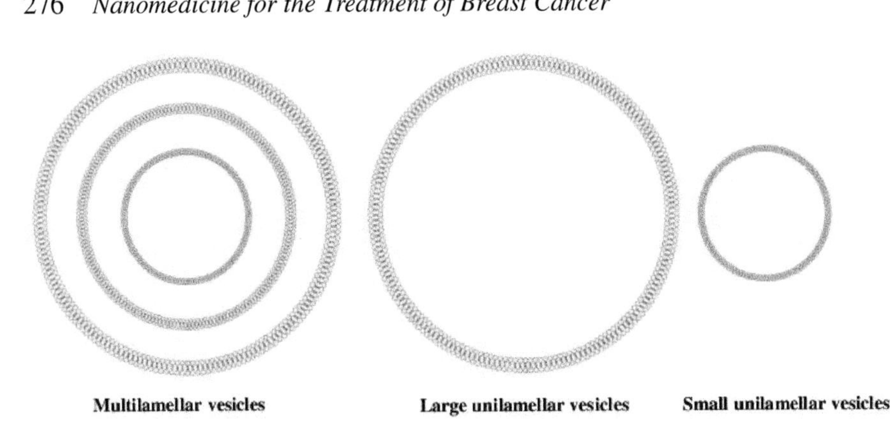

Figure 9.3 Different types of liposomes.

hydration of dried lipid films leads to formation of multilamellar vesicles (MLVs). Usually, MLVs are several hundred of nanometers in diameter, large, complex structures comprising of a series of concentric bilayers separated by small aqueous compartments. Liposomes prepared with only a single lipid bilayer are referred as unilamellar vesicles with size between 50–500 nm in diameter and are designated as large unilamellar vesicles (LUVs) while the vesicles smaller than 50 nm in diameter, called as small unilamellar vesicles (SUVs). Liposomes act as small depots for the solubilized drug encapsulated in the internal aqueous space formed by the liposomal lamellae and used as drug carriers. Further, liposomes can be used to deliver drugs with various limitations, such as limited solubility, serum stability, circulation half-life, biodistribution, and target selectivity. Owing to their structural make up, liposomes allows delivery of hydrophilic and hydrophobic drugs. Aqueous interior of the vesicles dissolves the hydrophilic compounds, while the hydrophobic compounds are entrapped in the lipid bilayer, and charged drugs can be adsorbed onto the surface of lipids [44, 45].

Recently, lipid nanocapsules (LNC) were prepared by a solvent-displacement technique, resulting in an oily core coated by a functional shell of different biocompatible molecules and surface carboxylic groups. Three different antibodies (one a specific HER2 oncoprotein antibody) were conjugated with these nanoparticles by the carbodiimide method, which allows the covalent immobilization of protein molecules through carboxylic surface groups. Immunological studies suggested that these IgG-LNC complexes showed the expected specific immuno-response. An increased uptake was observed of LNC in HER2 overexpressing breast tumoral cell lines [46].

9.5.2 Gold Nanoparticles

Gold nanoparticles (AuNP) possess strong surface plasmon fields that can effectively control fluorescence. Exploring this fluorescence manipulation property, a near-infrared (NIR) fluorophore-based contrast agent has been formulated for breast carcinoma diagnosis. The fluorophore is conjugated to AuNP with the help of a small spacer. The length of the spacer is varied so as to have the strong plasmon field to quench the fluorescence. This AuNP when it reaches the cancer tissues, the spacer would be cleaved by the enzymes secreted via the cancer cells, resulting in fluorescence and thereby detection of cancer. The nanoparticles in normal conditions are non-fluorescent and only become fluorescent when the spacer is cleaved by enzyme secreted by specific cancer cells [47, 48].

AuNPs have been investigated to destroy the targeted cancer cells by photothermal effects that facilitate conversion of laser to heat. Gold nanocages with the size $\sim$45 nm in edge length have been conjugated with monoclonal antibodies (anti-HER2) to target epidermal growth factor receptors (EGFR) that are overexpressed on the surface of breast cancer cells (SK-BR-3). Photothermal studies suggested that the nanocages strongly absorb light in the NIR region with an intensity threshold of 1.5 W/cm^2 to induce thermal destruction to the cancer cells. Furthermore, in the intensity range of 1.5–4.7 W/cm^2, the circular area of damaged cells increased linearly with the irradiation power density [49]. In another study, gold nanocages of edge length 65$\pm$7 nm as well as a powerful absorption peak at 800 nm were conjugated with anti-HER2 and targeted SK-BR-3 cells via EGFR. Cells aimed with immune Au nanocages were observed to respond instantly to laser irradiation and the cellular destruction was not reversible at power densities more than 1.6 W/cm^2. The percentage of dead cells increased with exposure time of laser irradiation up to 5 min, and then became steady [50]. A very sensitive and simple colorimetric two-photon scattering assay for selective identification of breast carcinoma SK-BR-3 cell lines in 100-cells/ml level using multifunctional (monoclonal anti- HER2/c-erb-2 antibody and S6 RNA aptamers conjugated) oval shape AuNP. On mixing assay along with breast cancer SK-BR-3 cell line, a distinctive colour change took place and two-photon scattering (TPS) intensity rises about 13 fold. This experimental result indicates a new opportunity of fast, easy and trust-worthy diagnosis of carcinoma cell lines by monitoring the colorimetric change and measuring TPS intensity from nanosystems of multifunctional gold [51].

9.5.3 Carbon Nanotubes

Carbon nanotubes (CNTs) are well defined, hollow cylindrical graphene nanomaterials with very high aspect ratios, lengths from several hundred nanometers to several micrometers and diameter of 0.4–2 nm for single walled carbon nanotubes (SWNTs) to 2–100 nm for multi walled carbon nanotubes (MWNTs). The structure of SWNTs can be considered as graphene sheet rolled up into a seamless hollow cylinder and MWNTs visualized as several co-axially arranged SWNTs of different radii with an inter-tube separation close to the inter-plane separation in graphite (0.34 –0.35 nm). CNTs have attracted much attention as a platform for the delivery of drugs and biomolecules due to their variety of size- and structure-dependent optical, thermal, electrical, and mechanical properties, as well as their unique ability to traverse across the cell membrane [52]. Although, nanotubes are insoluble in water, they can be derivatized to render them water soluble and biocompatible for a range of biological applications [53].

In a study, the dielectric properties of tissue-mimicking materials were characterized using various concentrations of SWCNTs [54]. The lower concentrations of SWCNTs were observed to considerably influence the heating response and dielectric properties of tissue-mimicking materials. At 3 GHz, SWCNTs concentrations of 0.22 % by weight showed to enhance the relative permittivity of the tissue-mimicking materials by 37% and the effective conductivity by 81%. Further, microwave heating experiments revealed significantly greater temperature increases in mixtures containing SWCNTs and the temperature increased linearly with the effective conductivity of the mixtures. These results indicated that SWCNTs might increase contrast for microwave imaging and assisted selective microwave heating for breast cancer therapeutics.

9.5.4 Human Serum Albumin Nanoparticles

Human serum albumin (HSA) is the most abundant protein in human blood plasma, synthesized in liver. It is known to be biodegradable, non-toxic and water soluble, allowing effortless delivery by injection and therefore an ideal agent for nanoparticle formation. A recent study reported the assimilation and targeted delivery of noscapine loaded HSA nanoparticles to tumor cells [55]. The nanoparticles were engineered to attain a material size in the range of 150–300 nm with 85–96 % drug-loading efficiency. Further, these nanoparticles were investigated *in vitro* for their anti-tumor activity as well as efficacy on carcinoma of breast [55]. In another study, methotrexate-HSA conjugated nanoparticles (MTX-HSA NPs) surface decorated using trastuzumab (TMAB) molecules were fabricated [56]. The size of TMAB-MTX-HSA NPs was 123–346 nm and varied according to the amount of TMAB molecules conjugated on their surfaces. The cytotoxicity studies revealed significantly higher toxicity of TMAB-MTX-HSA NPs on HER2 positive cells after 24 H as compared to that of non-targeted MTX-HSA NPs. Further, the flow cytometry analysis showed that the binding activity of TMAB molecules remained approximately unchanged after conjugation, when the number of the attached TMAB molecules on the surface of MTX-HSA NPs was not very high. Moreover, higher uptake to HER2 positive cells was observed with TMAB-MTX-HSA NPs as compared to non-targeted MTX-HSA NPs when the optimized amount of TMAB was conjugated on the surface of NPs [56].

In another report, target-oriented nanoparticles based on biodegradable HSA loaded with cytostatic drug doxorubicin were developed [41]. Trastuzumab was covalently attached to the surface of the nanoparticles. The nanoparticles were efficiently taken up by HER2 over-expressing breast cancer cells and delivered doxorubicin. The results suggested that cell-type specific drug-loaded nanoparticles could achieve an improvement in cancer therapy.

9.5.5 Other Nanomaterials

Multifunctional superparamagnetic iron oxide nanoparticle (SPION) of size 40 nm were developed for targeting metastatic breast cancer in a transgenic mouse model and imaging with magnetic resonance. SPIONs coated with a copolymer of chitosan and polyethylene glycol (PEG) were labeled with a fluorescent dye for optical detection and conjugated with a monoclonal antibody against the HER2. *In vivo* studies suggested efficient labeling of primary breast tumors with HER2 tagged SPIONs. Further, the SPIONs allowed remarkable contrast improvement in MR images of primary breast tumors; thereby suggesting the potential for MRI detection of micrometastases [57].

9.6 Conclusions

The rising global incidences of morbidity and mortality from breast cancer have led to intensified efforts in the search for effective therapeutic outcome. Progress in nanotechnology has opened up new avenues for imaging and management of breast carcinoma. Imaging of tumors through application of nanoparticles such as AuNPs or SPIONs is developing at a fast pace. Also, antibody–tagged nanomaterials facilitate simultaneous recognition of small tumors at multiple molecular targets. The mortality rate and disease progression both could be decreased to one third with timely diagnosis and treatment at an early stage. Although, improvement in disease prognosis has significantly improved the life expectancy of breast cancer patients, there is still a compelling need to look for new technologies to meet challenges of breast cancer.

References

[1] Jemal A, Murray T, Ward E, Samuels A, Tiwari RC, Ghafoor A, Feuer EJ, Thun MJ. Cancer statistics, 2005. CA: A Cancer Journal for Clinicians. 2005; 55: 10–30.

[2] Constantine GD, Pickar JH. Estrogens in postmenopausal women: Recent insights. Current Opinion in Pharmacology. 2003; 3: 626–634.

[3] Hankinson S, Colditz G, Willett W. Towards an integrated model for breast cancer etiology: The lifelong interplay of genes, lifestyle, and hormones. Breast Cancer Res. 2004; 6: 1–6.

[4] Breidenbach M, Rein DT, Schöndorf T, Khan KN, Herrmann I, Schmidt T, Reynolds PN, Vlodavsky I, Haviv YS, Curiel DT. A new targeting approach for breast cancer gene therapy using the heparanase promoter. Cancer Lett. 2006; 240: 114–122.

[5] Winer EP, Hudis C, Burstein HJ, Bryant J, Chlebowski RT, Ingle JN, Edge SB, Mamounas EP, Gelber R, Gralow J, Goldstein LJ, Pritchard KI, Braun S, Cobleigh MA, Langer AS, Perotti J, Powles TJ, Whelan TJ, Browman GP. American society of clinical oncology technology assessment working group update: Use of aromatase inhibitors in the adjuvant setting. Journal of Clinical Oncology. 2003; 21: 2597–2599.

[6] Bernhard J, Zahrieh D, Castiglione-Gertsch M, Hürny C, Gelber RD, Forbes JF, Murray E, Collins J, Aebi S, Thürlimann B, Price KN, Goldhirsch A, Coates AS. Adjuvant chemotherapy followed by goserelin compared with either modality alone: The impact on amenorrhea, hot flashes, and quality of life in premenopausal patients—the international breast cancer study group trial viii. Journal of Clinical Oncology. 2007; 25: 263–270.

[7] Panel NIoHCD. National institutes of health consensus development conference statement: Adjuvant therapy for breast cancer, november 1–3, 2000. Journal of the National Cancer Institute. 2001; 93: 979–989.

[8] Ganz PA, Greendale GA, Petersen L, Kahn B, Bower JE. Breast cancer in younger women: Reproductive and late health effects of treatment. Journal of Clinical Oncology. 2003; 21: 4184–4193.

[9] Greendale GA, Petersen L, Zibecchi L, Ganz PA. Factors related to sexual function in postmenopausal women with a history of breast cancer. Menopause. 2001; 8: 111–119.

[10] Takahashi M, Kai I. Sexuality after breast cancer treatment: Changes and coping strategies among japanese survivors. Social Science & Medicine. 2005; 61: 1278–1290.

[11] Ganz PA. Breast cancer, menopause, and long-term survivorship: Critical issues for the 21st century. The American Journal of Medicine. 2005; 118: 136–141.

[12] Ganz PA, Rowland JH, Desmond K, Meyerowitz BE, Wyatt GE. Life after breast cancer: Understanding women's health-related quality of life and sexual functioning. Journal of Clinical Oncology. 1998; 16: 501–514.

[13] Meyerowitz BE, Desmond KA, Rowland JH, Wyatt GE, Ganz PA. Sexuality following breast cancer. Journal of Sex & Marital Therapy. 1999; 25: 237–250.

[14] Hobday TJ, Perez, E. A. Molecularly targeted therapies for breast cancer. Cancer Control. 2005; 12: 73–81.

[15] Fujimori M. Genetically engineeredbifidobacterium as a drug delivery system for systemic therapy of metastatic breast cancer patients. Breast Cancer. 2006; 13: 27–31.

[16] Paik S, Shak S, Tang G, Kim C, Baker J, Cronin M, Baehner FL, Walker MG, Watson D, Park T, Hiller W, Fisher ER, Wickerham DL, Bryant J, Wolmark N. A multigene assay to predict recurrence of tamoxifen-treated, node-negative breast cancer. New England Journal of Medicine. 2004; 351: 2817–2826.

[17] van 't Veer LJ DH, van de Vijver MJ, He YD, Hart AA, Mao M, Peterse HL, van der Kooy K, Marton MJ, Witteveen AT, Schreiber GJ, Kerkhoven RM, Roberts C, Linsley PS, Bernards R, Friend SH. Gene expression profiling predicts clinical outcome of breast cancer. Nature. 2002; 415: 530–536.

[18] Slamon DJ CG, Wong SG, Levin WJ, Ullrich A, McGuire WL. Human breast cancer: Correlation of relapse and survival with amplification of the her-2/neu oncogene. Science. 1987; 235: 177–182.

[19] Chin K DS, Fridlyand J, Spellman PT, Roydasgupta R, Kuo WL, Lapuk A, Neve RM, Qian Z, Ryder T, Chen F, Feiler H, Tokuyasu T,

Kingsley C, Dairkee S, Meng Z, Chew K, Pinkel D, Jain A, Ljung BM, Esserman L, Albertson DG, Waldman FM, Gray JW. Genomic and transcriptional aberrations linked to breast cancer pathophysiologies. Cancer Cell. 2006; 10: 529–541.

[20] Perou CM. Molecular stratification of triple-negative breast cancers. Oncologist. 2011; 16: 61–70.

[21] Sørlie T PC, Tibshirani R, Aas T, Geisler S, Johnsen H, Hastie T, Eisen MB, van de Rijn M, Jeffrey SS, Thorsen T, Quist H, Matese JC, Brown PO, Botstein D, Lønning PE, Børresen-Dale AL. Gene expression patterns of breast carcinomas distinguish tumor subclasses with clinical implications. Proc Natl Acad Sci U S A. 2001; 98: 10869–10874.

[22] Kennecke H YR, Woods R, Cheang MC, Voduc D, Speers CH, Nielsen TO, Gelmon K. Metastatic behavior of breast cancer subtypes. Clinical Oncology. 2010; 28: 3271–3277.

[23] Bosch A EP, Zaragoza R, Vina JR, Lluch A. Triple-negative breast cancer: Molecular features, pathogenesis, treatment and current lines of research. Cancer Treat Rev. 2010; 36: 206–251.

[24] Nielsen TO HF, Jensen K, Cheang M, Karaca G, Hu Z, Hernandez-Boussard T, Livasy C, Cowan D, Dressler L, Akslen LA, Ragaz J, Gown AM, Gilks CB, van de Rijn M, Perou CM. Immunohistochemical and clinical characterization of the basal-like subtype of invasive breast carcinoma. Clin Cancer Res. 2004; 10: 5367–5374.

[25] Fong PC BD, Yap TA, Tutt A, Wu P, Mergui-Roelvink M, Mortimer P, Swaisland H, Lau A, O'Connor MJ, Ashworth A, Carmichael J, Kaye SB, Schellens JH, de Bono JS. Inhibition of poly(adp-ribose) polymerase in tumors from brca mutation carriers. N Engl J Med. 2009; 361: 123–134.

[26] Herschkowitz JI SK, Weigman VJ, Mikaelian I, Usary J, Hu Z, Rasmussen KE, Jones LP, Assefnia S, Chandrasekharan S, Backlund MG, Yin Y, Khramtsov AI, Bastein R, Quackenbush J, Glazer RI, Brown PH, Green JE, Kopelovich L, Furth PA, Palazzo JP, Olopade OI, Bernard PS, Churchill GA, Van Dyke T, Perou CM. Identification of conserved gene expression features between murine mammary carcinoma models and human breast tumors. Genome Biology. 2007; 8: R76.

[27] Prat A PJ, Karginova O, Fan C, Livasy C, Herschkowitz JI, He X, Perou CM. Phenotypic and molecular characterization of the claudin-low intrinsic subtype of breast cancer. Breast Cancer Research. 2010; 12: R68.

[28] Peer D, Karp JM, Hong S, Farokhzad OC, Margalit R, Langer R. Nanocarriers as an emerging platform for cancer therapy. Nat Nano. 2007; 2: 751–760.

[29] Owens DE, 3rd, Peppas NA. Opsonization, biodistribution, and pharmacokinetics of polymeric nanoparticles. Int J Pharm. 2006; 307: 93–102.

[30] Panagi Z, Beletsi A, Evangelatos G, Livaniou E, Ithakissios DS, Avgoustakis K. Effect of dose on the biodistribution and pharmacokinetics of plga and plga-mpeg nanoparticles. Int J Pharm. 2001; 221: 143–152.

[31] Durand R, Paul M, Rivollet D, Houin R, Astier A, Deniau M. Activity of pentamidine-loaded methacrylate nanoparticles against leishmania infantum in a mouse model. Int J Parasitol. 1997; 27: 1361–1367.

[32] Ruehm SG, Corot C, Vogt P, Kolb S, Debatin JF. Magnetic resonance imaging of atherosclerotic plaque with ultrasmall superparamagnetic particles of iron oxide in hyperlipidemic rabbits. Circulation. 2001; 103: 415–422.

[33] Kim JH, Kim YS, Park K, Lee S, Nam HY, Min KH, Jo HG, Park JH, Choi K, Jeong SY, Park RW, Kim IS, Kim K, Kwon IC. Antitumor efficacy of cisplatin-loaded glycol chitosan nanoparticles in tumor-bearing mice. J Control Release. 2008; 127: 41–49.

[34] Folkman J. Angiogenesis in cancer, vascular, rheumatoid and other disease. Nat Med. 1995; 1: 27–31.

[35] Skinner SA, Tutton PJ, O'Brien PE. Microvascular architecture of experimental colon tumors in the rat. Cancer Res. 1990; 50: 2411–2417.

[36] Matsumura Y, Maeda H. A new concept for macromolecular therapeutics in cancer chemotherapy: Mechanism of tumoritropic accumulation of proteins and the antitumor agent smancs. Cancer Res. 1986; 46: 6387–6392.

[37] Suzuki M, Hori K, Abe I, Saito S, Sato H. A new approach to cancer chemotherapy: Selective enhancement of tumor blood flow with angiotensin ii. Journal of the National Cancer Institute. 1981; 67: 663–669.

[38] Maeda H, Matsumura Y. Tumoritropic and lymphotropic principles of macromolecular drugs. Crit Rev Ther Drug Carrier Syst. 1989; 6: 193–210.

[39] Iwai K, Maeda H, Konno T. Use of oily contrast medium for selective drug targeting to tumor: Enhanced therapeutic effect and x-ray image. Cancer Res. 1984; 44: 2115–2121.

[40] Steinhauser I, Spänkuch B, Strebhardt K, Langer K. Trastuzumab-modified nanoparticles: Optimisation of preparation and uptake in cancer cells. Biomaterials. 2006; 27: 4975–4983.
[41] Anhorn MG, Wagner S, Kreuter Jr, Langer K, von Briesen H. Specific targeting of her2 overexpressing breast cancer cells with doxorubicin-loaded trastuzumab-modified human serum albumin nanoparticles. Bioconjug Chem. 2008; 19: 2321–2331.
[42] Yang H-M, Park CW, Woo M-A, Kim MI, Jo YM, Park HG, Kim J-D. Her2/neu antibody conjugated poly(amino acid)-coated iron oxide nanoparticles for breast cancer mr imaging. Biomacromolecules. 2010; 11: 2866–2872.
[43] Goldstein D, Sader O, Benita S. Influence of oil droplet surface charge on the performance of antibody–emulsion conjugates. Biomedicine & Pharmacotherapy. 2007; 61: 97–103.
[44] Malam Y, Loizidou M, Seifalian AM. Liposomes and nanoparticles: Nanosized vehicles for drug delivery in cancer. Trends Pharmacol Sci. 2009; 30: 592–599.
[45] Drulis-Kawa Z, Dorotkiewicz-Jach A. Liposomes as delivery systems for antibiotics. Int J Pharm. 2010; 387: 187–198.
[46] Sánchez-Moreno P, Ortega-Vinuesa JL, Boulaiz H, Marchal JA, Peula-Garcia JM. Synthesis and characterization of lipid immuno-nanocapsules for directed drug delivery. Selective anti-tumor activity against her2 positive breast-cancer cells. Biomacromolecules. 2013; 14: 4248–4259
[47] Wang J, O'Toole M, Massey A, Biswas S, Nantz M, Achilefu S, Kang K. Highly specific, nir fluorescent contrast agent with emission controlled by gold nanoparticle. In: LaManna JC, Puchowicz MA, Xu K, Harrison DK, Bruley DF, eds. Oxygen transport to tissue xxxii. Springer US; 2011: 149–154.
[48] Kumar A, Boruah BM, Liang X-J. Gold nanoparticles: Promising nanomaterials for the diagnosis of cancer and hiv/aids. J. Nanomaterials. 2011; 2011: 22–22.
[49] Chen J, Wang D, Xi J, Au L, Siekkinen A, Warsen A, Li Z-Y, Zhang H, Xia Y, Li X. Immuno gold nanocages with tailored optical properties for targeted photothermal destruction of cancer cells. Nano Lett. 2007; 7: 1318–1322.
[50] Au L, Zheng D, Zhou F, Li Z-Y, Li X, Xia Y. A quantitative study on the photothermal effect of immuno gold nanocages targeted to breast cancer cells. ACS Nano. 2008; 2: 1645–1652.

[51] Lu W, Arumugam SR, Senapati D, Singh AK, Arbneshi T, Khan SA, Yu H, Ray PC. Multifunctional oval-shaped gold-nanoparticle-based selective detection of breast cancer cells using simple colorimetric and highly sensitive two-photon scattering assay. ACS Nano. 2010; 4: 1739–1749.

[52] Kam NWS, Liu Z, Dai H. Functionalization of carbon nanotubes via cleavable disulfide bonds for efficient intracellular delivery of sirna and potent gene silencing. J Am Chem Soc. 2005; 127: 12492–12493.

[53] Hirsch A. Functionalization of single-walled carbon nanotubes. Angew Chem Int Ed Engl. 2002; 41: 1853–1859.

[54] Mashal A SB, Li X, Avti PK, Sahakian AV, Booske JH. Toward carbon-nanotube-based theranostic agents for microwave detection and treatment of breast cancer: Enhanced dielectric and heating response of tissue-mimicking materials. IEEE Trans Biomed Eng 2010; 57: 1831–1834.

[55] Sebak S, Mirzaei M, Malhotra M, Kulamarva A, Prakash S. Human serum albumin nanoparticles as an efficient noscapine drug delivery system for potential use in breast cancer: Preparation and in vitro analysis. Int J Nanomedicine. 2010; 5: 525–532.

[56] Taheri A, Dinarvand R, Atyabi F, Ghahremani MH, Ostad SN. Trastuzumab decorated methotrexate–human serum albumin conjugated nanoparticles for targeted delivery to her2 positive tumor cells. European Journal of Pharmaceutical Sciences. 2012; 47: 331–340.

[57] Kievit FM, Stephen ZR, Veiseh O, Arami H, Wang T, Lai VP, Park JO, Ellenbogen RG, Disis ML, Zhang M. Targeting of primary breast cancers and metastases in a transgenic mouse model using rationally designed multifunctional spions. ACS Nano. 2012; 6: 2591–2601.

10

Nanoparticle-Based Drug Delivery Systems: Associated Toxicological Concerns and Solutions

James Lyons[1,2], Aniruddha Bhati[3], Jaimic Trivedi[3] and Arati Sharma[2]

[1]Harrisburg University of Science and Technology,
Department of Biotechnology, 326 Market Street
Harrisburg, PA 17101
[2]The Pennsylvania State University, College of Medicine,
Department of Pharmacology, 500 University Drive,
Hershey, PA 17033, USA
[3]PD Patel Institute of Applied Sciences,
Charotar University of Science and technology,
Gujarat, India

10.1 Introduction

The advent of no other technology has created so much hype and expectation as the introduction of nanotechnology in the last few decades. Although nanotechnology could potentially hold the key to overcoming a number of traditional issues regarding therapeutics, diagnostics, and research; the safety of this technology is not fully defined and the technology in itself raises concern. As nanomedicine grows in popularity and becomes more of a commercially viable technology, major questions regarding the toxicological issues associated with nanotechnology will arise. In this chapter, we discuss nanoparticles and their use as onco-diagnostics. We then discuss the use of nanotechnology for drug delivery and some of the factors, which limit the use of nanotechnology. Finally, some of the methods being used to assess nanotoxicology and techniques being explored to help these

Post-genomic Approaches in Cancer and Nano Medicine, 287–314.

effective therapeutic and diagnostic tools to reach their clinical potential are discussed.

10.1.1 Nanotechnology and Nanoparticles (NP)

The National Nanotechnology Initiative (NNI) defines nanotechnology as "The understanding and control of matter at the nanoscale, at dimensions between approximately 1 and 100 nanometers, where unique phenomena enable novel applications." A nanometer, in actual parlance is one billionth of a meter (http://www.nano.gov). Nanotechnology has found applications in almost all fields of biology, physics, chemistry or an amalgam of all of these [1]. Therefore it is an interdisciplinary field of technological innovation on the nanometer scale offering comprehensive applicability. Nanotechnology has brought together scientists, engineers and physicians to revolutionize the fields of life sciences and healthcare by enabling interventions at cellular and molecular levels [1]. However, our main focus here will be on biology with special reference to oncology.

Kreuter has defined a nanoparticle as a solid colloidal particle ranging in size from 1 to 1000 nm [2]. In nanomedicine, nanoparticles can be elaborated as materials or surfaces that are intentionally manipulated at nanometer scale imparting it some novel properties [3]. Specific characteristics of nanoparticles which have captured attention of scientists include their small size, large surface area, and their unusual electronic, optical, and magnetic properties which make them vulnerable to desirable manipulations [4]. Nanomaterials have the potential to revolutionize the field of medicine because of their ability to interact with organs and tissues at the molecular and cellular levels[5]. Various classes of nanoparticles have been listed in Table 10.1.

With the recent development of nanobiotechnology, NPs have gained increasing attention for use in biomedical applications such as magnetic

Table 10.1 Various classes of nanoparticles

Classes	Examples	References
Lipid-based	Liposomes	(9, 10, 50, 89, 90)
Polymer-based	Dextran, PLGApoly(lactic-co-glycolic acid), Dendrimers	(91–99)
Non-metal-based	Carbon nanotubes	(100,101)
Metal- and Metal oxide-based	Gold and Silver, Supramagnetic	(102–104)
Semiconductor-based	Quantum dots	(105–107)

resonance imaging (MRI), detection of pathogens, fluorescent bio-labeling, detection of biomolecules with high specificity, tissue engineering, phagokinetic studies, magnetic separation of cells and biomolecules, enzyme catalysis, gene therapy, and targeted chemotherapy and radiotherapy (4, 6).

10.1.2 Nanotechnology and Cancer

Cancer has been a prodigal foe to mankind for centuries. Large amounts of money and manpower have been directed towards understanding, prognosis, diagnosis, and possible cure of cancer. Traditional cancer therapies such as chemotherapy, radiation, and surgery possess certain limitations, mainly due to undesirable drug associated side effects, low drug concentrations at the tumor site, and/or development of resistance by the tumor cells [7]. Novel targeted therapeutic approaches possessing improved efficacy and negligible toxicity are required to translate the drug discoveries into clinical applications more effectively (8, 9). Recently, nanotechnology has come up with particles that would be capable of detecting oncogenesis at early stages, specifically targeting cancer cells without causing any deleterious effect to normal cells, and facilitating combinatorial interplay of functionally linked agents by putting them together in a single nanoparticle to treat this disease more efficiently (8, 9). Properties of an ideal nanoparticle-based cancer therapeutic and/or imaging agent include (i) rapid detection, (ii) capacity to act as a vector for single or multiple anti-cancer drugs, (iii) rapid delivery to target cancer cells, and (iv) real time assay of treatment efficacy [9]. Advantages of using nanoparticles over classical drugs include (i) prevention of degradation of the entrapped content like drugs and siRNA, (ii) targeted delivery, (iii) ability to deliver higher drug concentrations, (iv) controlled, time dependent release of encapsulated molecules, and (v) minimized compound toxicity by masking effect (10, 11).

10.1.2.1 Nanoparticles in diagnostics

The primary goal in cancer diagnostics is the detection of precancerous and neoplastic lesions at early stages. Classical cancer imaging technologies used clinically are inefficient to achieve this goal. An ideal device which could serve this purpose should be able to identify cancers based on their molecular expression profiles, convert these molecular signals into amplified detectable data, and possess contrast enhancing agents for visualization and targeting; such as an antibody [12]. Nanoparticles also provide an added advantage of non-invasive diagnosis; eliminating the painful and cumbersome procedures involved in biopsies of cancerous tissues.

10.1.2.2 MRI (Magnetic Resonance Imaging)

Magnetic properties of magnetic nanoparticles can be desirably altered by controlling their size and surface coatings which enhances their applicability in tumor imaging as contrast enhancing agents and signal amplifiers [13]. MRI has facilitated high resolution imaging of cellular and subcellular events by its ability to image cross linked iron oxide and related NPs. These probes can be used for early diagnosis, risk assay, disease progression monitoring, and therapeutic efficacy analysis [4]. The superiority of dextran-coated ultra-small paramagnetic iron-oxide nanoparticles over the conventional gadolinium MRI contrast in the surgical treatment of brain tumors was demonstrated by the virtue of their intraoperative imaging enhancement, inflammatory targeting, and detect ability at low magnet strength [14]. Cancerous cells possess remarkable sustained telomerase activity, which was detected by MRI assays with the help of nanoparticles that switch their magnetic state on annealing with telomeric TTAGGG sequences [15]. Sustained angiogenesis, an early-to-mid stage event in human cancers, has been successfully imaged by various formulations of derivatized nanoparticles using MRI in animal models, targeted by $\alpha v\beta 3$-integrin [12].

10.1.2.3 Implantable nanosensors

Implantable nanosensors are designed such that they sense specific signals *in vivo* and transmit the information extra corporeally. Development of these devices have not been realized clinically mainly because of biofouling (non-specific, undesirable adsorption of serum proteins on the surfaces of sensors), leading to decreased sensitivity of the sensors to desired signals corresponding to the protein of interest against the background signal of adsorbed serum proteins [16]. More realistically, nanotechnology can be exploited to design novel sensing strategies resistant to bio fouling.

10.1.2.4 Cantilevers

Cantilevers, a new dimension to molecular recognition, was brought into light by James Gimzewski *et al.* with a concept involving detection of forces and deformations generated due to bio-molecular binding events by selective sensing nanostructures [17]. Majumdar *et al.* applied micro cantilevers to quantify clinically significant concentrations of Prostate Specific Antigen [18]. The use of nanoparticle probes might facilitate discrimination of individual single-pair mismatches in DNA[19]. Nanocantilevers have enormous diagnostic potential due to their multiplexing capability translated as DNA microarrays and proteomic profiling arrays [20].

10.1.2.5 Bio-barcode

Nam *et al.* introduced a novel two particle DNA-detection technology, the 'bio-barcode', demonstrating 500 zeptomolar sensitivity [21]. It involves oligonucleotide-modified gold nanoparticles and magnetic particles carrying a predetermined nucleotide sequence serving as an identification probe. Further, gold-nanoparticle-modified probes in conjunction with micro cantilevers have been developed for DNA assay with single mismatch discrimination [14].

10.2 Nanoparticles as Onco-therapeutics

In recent years, numerous nanoparticles designed as multifunctional diagnostic and therapeutic devices specifically for cancer drug delivery and cancer imaging are applied or are under development for the treatment of cancer [22]. Drug carriers such as liposomes, dendrimers, synthetic polymers, microcapsules, and several others have been used widely (23–25).

10.2.1 Drug Delivery Systems (DDSs)

Currently available anticancer drugs have distinct mechanisms of action which vary in their effects on different types of normal and cancerous cells. Due to the fact that there are very few biochemical differences between normal cells and cancerous cells, the effectiveness of many anticancer drugs is limited by their toxicity to normal rapidly growing cells. Additionally, cancerous cells that are normally suppressed by a specific drug may develop a resistance to that drug [26]. There are three main common goals to any anticancer therapeutic; [1] damage the DNA of affected cancer cells, [2] inhibit the synthesis of new DNA strands to stop the cell from replicating, or [3] stop mitosis. The most popular cancer drug, chemotherapy, is traditionally feared by patients due to its high toxicity level. Unfortunately, their cause for concern is not without reason. Many anticancer drugs are very non-specific; therefore, they subsequently cause serious side effects in bodily systems that naturally have a rapid turnover rate including skin, hair, gastrointestinal, and bone marrow [26].

The most important aspect of drug delivery is the accurate targeting of the drug to cells or tissues of choice [27]. Although today's technology is still far from the proposed "magic bullet" design envisioned by Paul Ehrlich in the beginning of the twentieth century, nanotechnology offers a way to bring this vision closer to reality [27]. By attaching the drug to the surface of a nanoparticle or the drug being integrated into the matrix of the nanoparticle, researchers can overcome a number of the shortcomings of traditional anticancer therapeutics. Applicability of nanoparticles as potent drug delivery systems (DDSs) is realized by their ability to enhance the efficacy of "free" drugs by improving their critical properties including solubility, stability *in vivo*, pharmacokinetics and biodistribution, and by eliminating the major problem of resistance [28].

10.2.1.1 Polymeric biodegradable nanoparticles

Polymeric biodegradable nanoparticles offer a drug delivery system which provides a way to sustain a localized drug therapeutic agent for weeks. This allows for a controlled release mechanism so that a specific amount of the drug is released into the targeted area at a specific time. These systems can carry a number of therapeutic agents including plasmid DNA, proteins, peptides, and other low molecular weight compounds [29]. Olivier *et al.* explored the use of polybutylcyanoacrylate nanoparticles for delivery of doxorupolysorbate 80 to the brain, a potential therapy for brain tumor treatment [30].

10.2.1.2 Metallic nanoparticles

Due to their extremely small size and vast surface area, metallic nanoparticles are able to carry very high doses of drugs. Some of the things these tiny particles can carry include anticancer therapeutic agents, proteins, and DNA [29]. Priyabrata *et al.* demonstrated this capability in a study where they looked at gold nanoparticles bearing functional anti-cancer drugs and anti-angiogenic agents [31].

10.2.1.3 Ceramic nanoparticles

The advantages of ceramic nanoparticles include easy preparation, water solubility, and stability in biological environments. These nanoparticles have been shown to be carriers for proteins, DNA, anticancer therapeutic agents, and other high molecular weight compounds. They have been used in a number of studies involving such therapies as photodynamic therapies and diabetes therapies [29]. However, they have also been explored for possible cancer therapies. Roy *et al.* [32] demonstrated the potential of ceramic-based nanoparticles as drug carriers for photodynamics therapies by using these particles to encapsulate insoluble photosensitizing anti-cancer agents [32].

10.2.1.4 Polymeric micelles

Polymeric micelles have an advantage much like that of ceramic nanoparticles. Due to a hydrophobic core, polymeric micelles make a suitable carrier for water-insoluble drugs. They can carry such compounds as DNA, anticancer therapeutic agents and proteins. Although they have seen some promise in antifungal treatment, they are mainly recognized for their potential application in solid tumor therapies [29]. Some of the different polymeric micelle formulations of anti-cancer drugs that are already in clinical trials [33] are summarized in Table 10.2.

10.2.1.5 Dendrimers

Dendrimers are repetitively branched molecules. They are typically symmetric around their core and often adopt a spherical three dimensional morphology [34]. Dendrimers are often utilized as drug delivery systems because they are perfect for the encapsulation of the drug being used [35]. The unique advantage of using these systems for drug delivery is that they can be modified to carry either hydrophobic or hydrophilic drugs [29]. Therefore, the potential agents that can be carried by these nanoparticles are abundant. These nanoparticles have been explored as potential carriers for DNA, anticancer therapeutic agents, antibacterial therapeutic agents, antiviral

Table 10.2 Issues surrounding nanotoxicology

Formulation	Polymer	Drug	Incorporation Method	Issues Solved
Genexol§-PM	mPEG-PDLLA	Paclitaxel	Physical entrapment	Solubilization
NK105	PEG-P(Asp)^{d}	Paclitaxel	Physical entrapment	Solubilization/Targeting
NC-6004	PEG-P(Glu) (Cisplatin)	Cisplatin	Coordinate bonding	Targeting
NC-4016	PEG-P(Glu) (DACHPt)	DACHPt	Coordinate bonding	Targeting
NK012	PEG-P(Glu)(SN-38)	SN-38	Chemical conjugation	Solubilization/Targeting
NK911	PEG-P(Asp)(DOX)	Doxorubicin	Physical entrapment	Targeting
SP1049C	PluronicL61, F127	Doxorubicin	Physical entrapment	Anti-MDR

therapeutic agents, and other high molecular weight compounds [29]. The versatility of these drug carriers has shown promise in a vast array of applications. For example, Chen and Cooper showed their potential for bacterial infection treatment[36]. Additionally, Witvrouw *et al.* demonstrated the promise of dendrimers in treatment of the human immunodeficiency virus [37].

The most promising application for this system however is in tumor therapies. Latallo *et al.* demonstrated their potential application in delivery of anticancer drugs using an animal model of human epithelial cancer [38]. Likewise, Medina *et al.* showed the potential of dendrimers for use as carriers for anticancer drugs against hepatic cancer. They used N-acetylgalactosamine-functionalized dendrimers as cell targeted carriers *in vitro* with hepatic cancer cells [39]. Finally, Wang *et al.* explored the possibilities of using generation 4 polymidoaminedendrimers as a novel candidate of a nanocarrier for gene delivery agents in breast cancer treatment [40].

10.2.1.6 Liposomes

Liposomes are nanoscale vesicles composed of membrane-like lipid bilayers surrounding an aqueous compartment. The unique feature of liposomes is the ability to encapsulate both lipophilic and hydrophilic compounds. This makes them fantastic candidates for drug carrier systems [41]. Furthermore they have facilitated the targeting of specific agents. Thus, they have been of extreme interest for cancer researchers as a cancer drug delivery system. Yang *et al.* explored the use of a liposome delivery system in treatment of pancreatic cancer. Yang states in his article, “a liposome-based delivery system can concomitantly improve the pharmacokinetics, reduce the side effects, and potentially increase selective tumor uptake...... ”[42]. Likewise, Zavaleta *et al.* used a liposome delivery system for enhanced peritoneal drug delivery in an ovarian cancer model. Again, they concluded that there was increase in selective tumor uptake as well as reduced side effects [43]. Furthermore, liposome delivery systems have even been explored for use against advanced breast cancer [44].

While liposome delivery systems have been of great promise to the delivery of cancer therapeutics, there has also been a multitude of studies conducted using liposomes as a system for gene delivery. This is mainly due to the fact that gene delivery systems follow many similar strategies of drug delivery systems [45]. A study done by Dowty *et al.* showed the use of liposomes for delivery of plasmid DNA into postmitotic nuclei of primary rat myotubes [46]. Likewise, Wolff *et al.* demonstrated the use of liposomes for direct gene transfer into mouse muscle *in vivo* [47]. Liposomes have also been studied extensively for delivery systems in treating infectious diseases and other therapeutic agents (48, 49). Additionally, Tran *et al.* combined nanoliposomal-ceramide with sorafenib as a novel therapeutics agent for melanoma and breast cancer. The formula synergistically inhibited cell survival and decreased tumour development (49, 50).

10.2.2 Nucleic acid carriers

With the advent in antisense therapeutics, gene therapy, and rDNA technology, the applicability of nanoparticles as carriers of nucleic acids such as plasmids and siRNA has come into picture. This approach involves encapsulation of nucleic acids with in nanoparticles, generally liposomes, and thereby enhancing their internalization by membrane fusion or endocytosis[3]. For *in vitro* delivery, siRNA has also been conjugated with variety of nanoparticles, such as gold, quantum dots, and iron oxide [51].

10.2.3 Magnetofection

The principle underlying magnetofection is binding of nucleic acids to magnetic nanoparticles, subjecting the desired area to a magnetic field, and thereby resulting in the concentration and transfection of nucleic acids to the targeted area. For cancer therapy, a high-field, high-gradient, rare earth permanent magnet is placed above the solid tumor in order to retain administrated magnetic nanoparticles with bound nucleic acids *in situ* until they internalize and transfect malignant cells [52].

As it is evident, nanotechnology has made a significant impact on drug delivery systems. Unfortunately, although much effort has been devoted to the development of nanoparticle drug delivery systems, a fully functional therapy based on nanoparticle delivery has yet to be in clinical use [29]. However, novel nanoparticle structures and materials are continuously being reported and many of which have reached clinical trials. Thus, a future of fully functional and commercially available nanoparticle drug delivery systems is not out of question [29]. Unfortunately, despite the promising future outlook of nanotechnology, there is still great concern regarding the toxicity associated with these therapeutic and diagnostic technologies.

10.3 Nanotoxicology

Donaldson *et al.* defined nanotoxicology as "a new branch of toxicology which addresses the gaps in knowledge, specifically, the adverse effects on health likely to be caused by nanomaterials" [53]. The administration of nanoparticles in medicine involves deliberate direct ingestion or injection into the body. Nanoparticles are potent vectors for delivering cancer therapeutics into tumors [12], however, nanoparticles themselves possess inherent toxicity which can limit their use in therapeutics. Prime determinants of nanoparticle toxicity include size, shape, charge,composition, constituent leaching, and triggering of immune reactions [8].

10.3.1 Important physiochemical properties, which causes toxicity

10.3.1.1 Particle size

Nanoparticle size is one of the most critical factors that can influence toxicity. Nanoparticles which are less than 100 nm in size posses desirable mechanical, electrical, and chemical properties for drug delivery; however smaller sizes can cause issues, such as passing through the blood-brain barrier, triggering immune reactions, and damaging cell membranes [54].

10.3.1.2 Particle composition and charge

The composition and charge of nanoparticles can affect organ accumulation and toxicty. Nanoparticles with a zeta potential above (+/–) 30 nV are usually stable in suspension, as the surface charge prevents the aggregation of the particles. However, when being considered for a drug delivery system, it is important to address the issue of organ accumulation when designing the system as potential organ-related toxicities may require alternative formulations for safety [55].

10.3.1.3 Particle surface area

Nanoparticles possess the unique property of having a very large surface area to size ratio. Nanoparticles interact with biological systems through this surface area. This means that controlling the surface properties such as composition, charge, and porosity are critical factors which influence nanotoxicity [56]. While the increased surface area of nanoparticles has the advantage of providing opportunities for antibody conjugation and material delivery, it also causes an inherent disadvantage of making them toxic if not rationally designed [9].

10.3.2 Nanodrug delivery systems and toxicity

After relinquishment of the initial aura that surrounded the advent of nanotechnology in drug delivery systems; the cases pointing toward toxicity of these famed nanocarriers started surfacing in the scientific world. In this section, some of the drug delivery systems and problems associated with them are discussed.

10.3.2.1 Liposome toxicity

There are a number of aspects of liposomal drug delivery systems that one must be concerned with when analyzing toxicity; including lipid composition, particle size, and charge. Therefore, the lipid composition, formulation, and charge of the liposomes should be developed such that the potential adverse side effects can be minimized [9]. For example, cationic liposomes have been shown to interact with serum proteins, lipoproteins, and the extra cellular matrix. This leads to aggregation or release of agents before reaching the targeted cell, resulting insystemic toxicity [57]. Likewise, they have been shown to cause liver damage due to their positive charge [58]. Cationic liposomes have also been shown to cause cellular influx and inflammation of the lungs through reactive oxygen species induction [57]. Finally, studies have shown that cationic liposomes can lead to macrophage-mediated toxicity following exposure of longer than three hours [59].

10.3.2.2 Dendrimer-based drug delivery system toxicity

One major area of concern when using a dendrimer-based delivery system is the charge. For example, cationic dendrimers can destabilize cell membranes and cause cell lysis [9]. Jevprasesphant *et al.* reported that amino-terminated Poly(amidoamine) dendrimers cause cytotoxicty on human intestinal adenocarcinoma Caco-2 cells [60].

10.3.2.3 Metallic nanoparticle toxicity

Although metallic nanoparticles show great promise as effective drug delivery systems, there is great concern surrounding toxicity of these particles. For example, gold nanoparticle could potentially cross a mother's placenta and cause deletorious toxic effects to the developing fetus [61]. Additionally, gold nanoparticles interact with cellular proteins that could cause free radical-induced cell death or modify protein structure leading to auto-immune related toxicity [62]. Likewise, silver nanoparticles can cause blood-brain destruction, by producing reactive oxygen species, as well as neuronal degeneration and brain edema [63]. Also, silver nanoparticles have been linked to mitochondrial targeting, reactive oxygen species production, and glutathione depletion in liver cells [64]. Finally, silver nanoparticles have been shown to release silver ions, which can be toxic [9].

Table 10.3 Issues surrounding nanotoxicology

Issues	Examples
Physicochemical determinants	Size, Shape, Surface area, Surface chemistry
Routes of exposure	Skin, Respiratory tract, Gastrointestinal tract
Biodistribution	Clearance, Opsonization
Molecular determinants	Oxidative stress, Inflammation
Genotoxicity	Mutagenesis, Chromosomal aberrations
Regulatory aspects	Government, Industries, Academia

10.3.2.4 Quantum dot toxicity

Quantum dots have shown great promise in medical diagnostics due to their unique optical properties. Quantum dots are capable of emitting a broad spectrum of fluorescence, a property which has been exploited mainly for use in biomedical imaging [9]. However, researchers have also shown interest in using these semiconductor nanoparticles for possible drug delivery and gene therapy. Unfortunately, as with all nanoparticles, toxicity is still of great concern. First and foremost, the metal core constituents of quantum dots can be toxic following removal or dissolution of their coating (9). Exposure of quantum dots to acidic or oxidative environments, such as those found in endosomes, can cause decomposition and subsequent release of their metalloid core into the cytoplasm causing toxicity [65]. However, this is not the only concern for quantum dots as the surface coating itself can be toxic [66]. Various issues surrounding Nanotoxicology are summarized in Table 10.3.

10.4 Assessment of Nanotoxicology

To overcome the risks posed by a nanoparticulate system, it is a primary prerequisite to assess the magnitude, type and the reason behind the suspected toxicity. This assessment can be at the cellular or at the genetic level depending upon the system which is under study. The knowledge about this enables future experimentations and modifications needed to overcome the toxicity concerns. In this section, such available techniques are discussed.

10.4.1 *In vitro* cell culture assays

In vitro cell culture assays are essential to evaluate mechanisms underlying toxicity of nanoparticles in biological systems. Despite the inadequacy of these assays to mimic the actual, they can be used for determination of preliminary toxicity [67].

10.4.1.1 Biocompatibility assays

Biocompatibility of nanoparticles can be assessed by viability assays such as MTT, sulforhodamine B, BUDR incorporation, etc. which sort for membrane integrity, metabolic status and adherence of the cells as basic parameters for analysis [68]. Changes in any of these parameters or a drop in viability could indicate potential toxicity.

10.4.1.2 Hemolytic and platelet aggregation tests

Hemolytic and platelet aggregation tests hold relevance due to the possibility of triggering of hemolysis or platelet aggregation. This leads to a variety of issues uponintravenous administration of nanoparticles [9].

10.4.1.3 Reactive Oxygen Species (ROS) and oxidative stress detection assays

The generation of ROS can be detected by techniques such as Electroparamagnetic resonance [69], 2, 7-dichlorofluorescin assays [70], cytochrome C assays[71], Glutathione levels quantification, and measurement of mRNA expressionchanges of oxidative stress-dependent genes [72].

10.4.1.4 Genotoxicity assays

A very few assays such as the salmonellareverse mutation assay (Ames test) for mutagenesis detection [73], the micronucleus test, and the alkalinecomet assay for clastogenic effect [74], have been used for the assessment of genotoxicity of nanoparticles.

10.4.2 *In vivo* animal assays

In vivo analysis provides a clear picture of the actual interactions of the nanoparticles in a living system, and hence, are of prime importance [9]. However, these studies are limited due to time, cost, or ethical issues.

10.4.2.1 Dose-range determination

Mice serve as the most convenient animal models for *in vivo* studies following intravenous administration of nanoparticles. Experimentally, animals are usually monitored for parameters suchas mortality, body weight, organ weight, clinical chemistry, hematology, gross pathology, and histology of control versus treatment groups (http://ncl.cancer.gov).

10.4.2.2 Pharmacokinetics

Pharmacokinetics, or organ distribution and clearance studies, of nanoparticles is essential for evaluating potential toxic effects [75]. Degradation, excretion, and clearance of nanoparticles can be assessed with the use of radiolabels as well as scintillation counting, inductively coupled plasma mass spectroscopy, electron microscopy in conjunction with energy dispersive spectroscopy, or high-performance liquid chromatography (http://ncl.cancer.gov).

10.4.2.3 Immunotoxicity

Nanoparticles, depending on their compositions and characteristics, are known to interact with the immune system causing either enhanced or, suppressed immune system function [76]. The lymph node proliferation assay can be used to estimate nanoparticle-associated toxicity, but it has inherent disadvantages. The U.S. Food and Drug Administration recommendsexamining immunotoxicity through changes in gross pathology hematology, clinicalchemistry, immune organ weights, or histological evaluation (http://ncl.cancer.gov) and [75].

10.5 Toxicity Modulation to Optimally Exploit Clinical Potential of Nanoparticles as a Therapeutics

As nanotechnology becomes indespensible for modern therapeutics it is the need of the hour to modulate them in such a way that it nullifies the associated risks. This modulation is to be seen through changes in various physical parameters in the nanoparticles. Some of these strategies are enlisted in the this section and the summary is presented in a schematic form in Figure 10.1.

10.5.1 Size

The body's immune system is designed to recognize and entrap foreign particles. Nanoparticles that are too large or too small in size get entrapped in the reticuloendothelial system or interstitial spaces of the body hindering their optimum delivery, triggering immune responses, and decreasing clearance from the body. Particles smaller than 80 nm are nontoxic; however, particles smaller than 10 nm can cause significant toxicity. The smaller the particle size, the higher the surface area to volume ratio and the higher the surface reactivity and chances of aggregation. Therefore, too small or too large particles might end up causing irreversible deposition of particles in tissues such as the liver and spleen, thereby affecting their normal biochemical or enzymatic functions [12]. Yet, smaller nanoparticles like quantum dots which are of sizes less than 5.5 nm are rapidly subjected to renal clearance from the body. Studies have demonstrated 4-fold more effective accumulation of nanoparticles with sizes

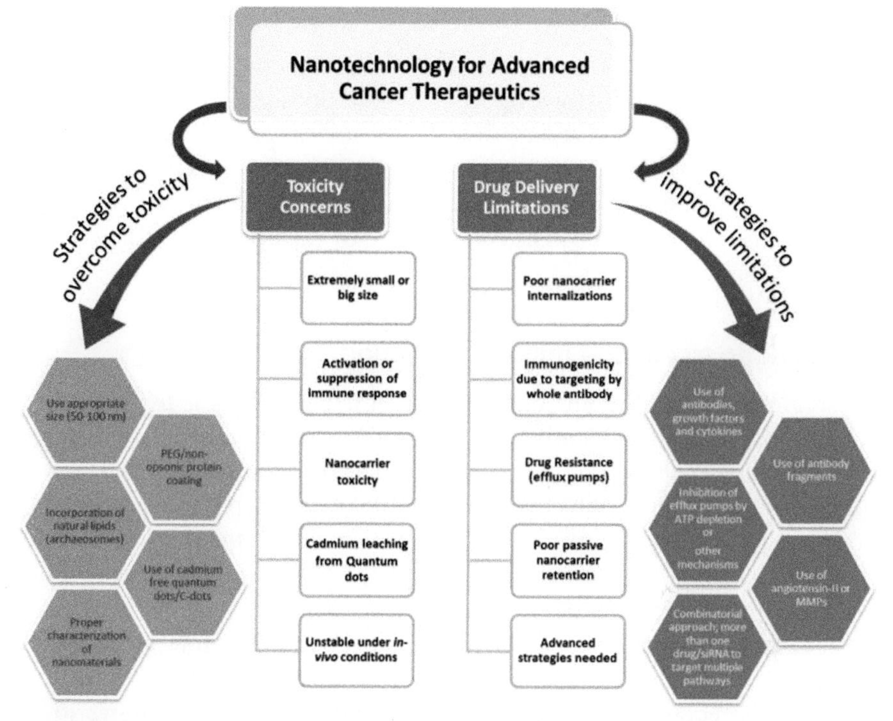

Figure 10.1 Various approaches to address the toxicological concerns assoiciated with nanoparticledbased drug delivery systems.

ranging from 100 to 200 nm in tumors as compared to those less than 50 nm or greater than 300 nm [77]. Therefore, a nanoparticle with an average size of 100 nm can be considered as optimum for onco-therapeutics as it is not so small as would cause toxicity and yet, large enough to enter tumors and deliver a desirable effect.

10.5.2 Shape

The aspect ratio (i.e. the ratio of width to length) of the nanoparticles is intimately associated with their cellular uptake, thereby playing a crucial role in toxicity modulation. Spherical gold particles (aspect ratio 1:1) cause minimal toxicity. However, studies have shown that lower aspect (1:3) ratio gold nanorods demonstrate an increase in their endocytic uptake as compared to the ones with a higher (1:5) aspect ratio [78]. This issue is addressed by addition of capping agents [79].

10.5.3 Charge

Positively charged liposomes are more toxic to cells as compared to negatively charged or neutralones [10]. However, to use these liposomes as vectors for negatively charged genetic therapeutic molecules such as plasmidsor siRNA, electrostatic interaction needs to be established. This can be achieved by the use of cationic liposomes. The toxicity caused due to using cationic, positively charged, liposomes can be counteracted by incorporation of neutral or negatively charged lipids [80].

10.5.4 Masking to escape immune rejection

The most relevant way of modulating toxicity of nanoparticles is by surface protection [50]. It involves coating of nanoparticles with biocompatible or biodegradable hydrophilic polymers, surfactants, or copolymers; such as polyethylene glycol (PEG), polyethylene oxide, polyoxamer, poloxamine, and polysorbate [4]. This coating minimizes the interactions of the particle with body fluids and cells so as to minimize any probable side-effect that might be caused due to their interactions.

Opsonization is a process of recognizing and covering foreign particles with complement proteins C3, C4, C5, and immunoglobulins, thereby enhancing their recognition by the reticuloendothelial system, leading to phagocytosis and removal from the body [81]. Mechanistically, a protective hydrophilic layer is formed around the nanoparticles by PEGylation, which, by virtue of steric hinderance repels the absorption of opsonin proteins. Thus, it eliminates the problem of opsonization [82]. PEG molecules also increase the retention time of a drug in the body by reducing the circulation clearance [50].

10.5.5 Leaching of the constituents entrapped in nanoparticles

On exposure to body fluids, nanovectors composed of metals or semiconductors get oxidized to corresponding ions. Gold nanoparticles, despite their biocompatiblity, pose the threat of toxicity orthwarting normal enzymatic functions by complex formation with biologic components in its oxidized state [83]. A solution to this issue could be surface functionalization to curtail the direct contact of particles with body fluids, thereby circumventing any possibility of oxidation into ions [84].

Quantum dots made of heavy metals such as cadmium could release ions causing toxicity. Therefore, use of cadmium-free quantum dots (CFQD) containing phosphorescent materials and fluorescent materials conjugated with tumor-targeting ligands are being synthesized using rare earth metal-doped oxide in phosphor. C-dots (Cornell-dots) are such examples which are silica-based fluorescent nanoparticles recently approved by the FDA in human phase I trials for clinical evaluation of cancer imaging as an Investigational New Drug (IND) [85].

10.5.6 Adding targeting moieties to nanoparticles

Targeting of nanoparticles is clinically significant so as to protect the normal non-cancerous cells from getting affected by the siRNA or anticancer drug, in antisense and chemotherapy, respectively. Additionally, targeting of nanoparticles is significant to maximize delivery of drugs to a specific site; resulting in high efficacy at low dosage concentrations. This active targeting is achieved by tagging the nanoparticles with targeting moieties such as antibodies or peptides bound to the surface, which interact with receptors and facilitate recepto rmediated endocytosis of encapsulated drugs [86]. An ideal targeting ligand should possess properties like stability and easy conjugability with nanoparticles, without affecting the properties of either the nanoparticles or the ligand [10].

Among the plethora of available ligands, antibodies are most sensitive, stable, and easy to produce; using hybridoma technology. However, the potentially immunogenic Fc-receptor of the antibodies can lead to accumulation of immunoliposomes in the liver and the spleen. Instead, the use of antibody fragments possessing only F(ab')2, Fab', and scFv has circumvented this problem, but raised an issue of stability[10].

However, with the growing understanding of nanotechnology and the continued collaboration between disciplines, such as biology, chemistry, bioengineering, physics, etc., we will continue to see advancement in the modulation of nanotoxicity; eventually allowing these novel technologies to reach their full clinical potential.

10.6 Future Implication

The future of nanotechnology is very bright. Nanotechnology could help to resolve a variety of issues related to medicine and therapeutics. The future holds the key to development of "smart drugs" which can be used as "Theranostics" which are agents that could serve as both therapeutic and imaging agents [87]. This type of combination therapy could potentially be revolutionary; allowing nanotechnology to provide platforms designed to provide a diverse set of simultaneous multifunctional preventative, diagnostic, and therapeutic approaches (Figure 10.2) [88]. Such devices appear promising in revolutionizing the field of oncology by providing non-invasive means to diagnose cancer at early stage and deliver effective therapeutics in a targeted, dose, and time-dependent manner; while simultaneously aiding in the visualization of tumor regression.

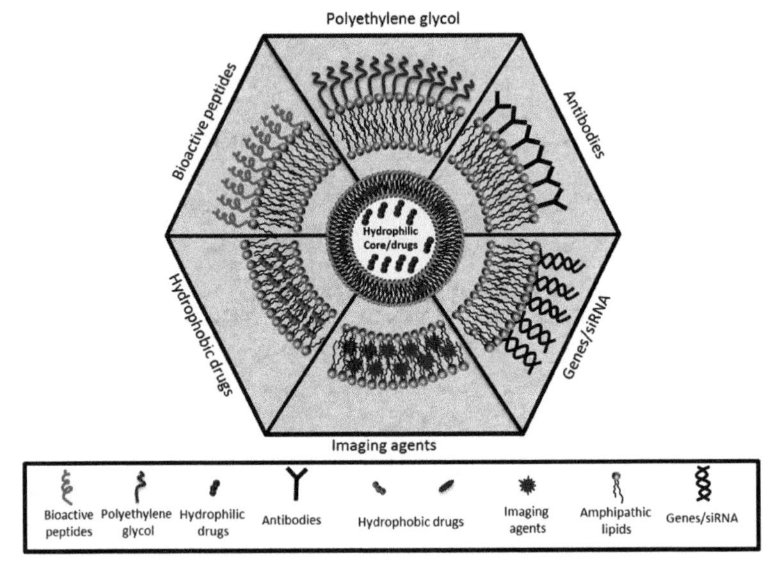

Figure 10.2 Schematic Theranostic, a multifunctional particle.

10.7 Conclusions

Onco-nanotechnology is an enormously promising approach for advancing prognostic, diagnostic, preventative, and therapeutic efficacy and imaging in cancer, and therefore has been proved to be superior to traditional cancer drugs. However, to exploit their true clinical potential optimally, many obstacles need to be addressed. Prime concerns to this approach involve nano-toxicity and targeted delivery to cancer cells. Better understanding of the properties, behavior, and interactions of nanoparticles with human tissues, fluids, and immune system would help scientists to circumvent these issues. Moreover, the methods used for toxicity testing of nanoparticles should also be improved. Nevertheless, nanotechnology is an unparalleled approach that opens new roads to the field of oncology.

References

[1] Porter, A., and Youtie, J. (2009) How Interdisciplinary is Nanotechnology? Journal of Nanoparticle Research Volume 11, Issue 5, pp 1023–1041. Journal of Nanoparticle Research **11**, 1023–1041.

[2] Kreuter, J. (1994) Drug delivery and nanoparticles: Applications and hazards. Eur J Drug Metab Pharmacokinet **3**, 133.

[3] Prijic, S., and Sersa, G. (2011) Magnetic nanoparticles as targeted delivery systems in oncology. Radiology and oncology **45**, 1–16.

[4] Yang, H. W., Hua, M. Y., Liu, H. L., Huang, C. Y., and and Wei, K. C. (2012) Potential of magnetic nanoparticles for targeted drug delivery. Nanotechnology, Science and Applications **5**, 73–86.

[5] Linkov, I., Satterstrom, F. K., and Corey, L. M. (2008) Nanotoxicology and nanomedicine: making hard decisions. Nanomedicine **4**, 167–171.

[6] Salata, O. (2004) Applications of nanoparticles in biology and medicine. Journal of nanobiotechnology **2**, 3.

[7] Langer, R. (1998) Drug delivery and targeting. Nature **392**, 5–10.

[8] Singh, S., Sharma, A., and Robertson, G. P. (2012) Realizing the clinical potential of cancer nanotechnology by minimizing toxicologic and targeted delivery concerns. Cancer research **72**, 5663–5668.

[9] Sharma, A., Madhunapantula, S. V., and Robertson, G. P. (2012) Toxicological considerations when creating nanoparticle-based drugs and drug delivery systems. Expert opinion on drug metabolism & toxicology **8**, 47–69.

[10] Puri, A., Loomis, K., Smith, B., Lee, J. H., Yavlovich, A., Heldman, E., and Blumenthal, R. (2009) Lipid-based nanoparticles as pharmaceutical drug carriers: from concepts to clinic. Crit Rev Ther Drug Carrier Syst **26**, 523–580.

[11] Youns, M., Hoheisel, J. D., and Efferth, T. (2010) Therapeutic and Diagnostic Applications of Nanoparticles. Curr Drug Targets.

[12] Ferrari, M. (2005) Cancer nanotechnology: opportunities and challenges. Nat Rev Cancer **5**, 161–171.

[13] Rogers, W. J., and Basu, P. (2005) Factors regulating macrophage endocytosis of nanoparticles: implications for targeted magnetic resonance plaque imaging. Atherosclerosis **178**, 67–73.

[14] Su, M., Li, S., and Dravid, V. P. (2003) Microcantilever resonance-based DNA detection with nanoparticle probes.. Appl. Phys. Lett. **82**.

[15] Voura, E. B., Jaiswal, J. K., Mattoussi, H., and Simon, S. M. (2004) Tracking metastatic tumor cell extravasation with quantum

dot nanocrystals and fluorescence emission-scanning microscopy. Nat Med **10**, 993–998.

[16] Desai, T. A., Hansford, D. J., Leoni, L., Essenpreis, M., and Ferrari, M. (2000) Nanoporous anti-fouling silicon membranes for biosensor applications. Biosens Bioelectron **15**, 453–462.

[17] Fritz, J., Baller, M. K., Lang, H. P., Rothuizen, H., Vettiger, P., Meyer, E., -J. Güntherodt, H., Gerber, C., and Gimzewski, J. K. (2000) Translating Biomolecular Recognition into Nanomechanics. Science **288**, 316–318.

[18] Wu, G., Datar, R. H., Hansen, K. M., Thundat, T., Cote, R. J., and Majumdar, A. (2001) Bioassay of prostate-specific antigen (PSA) using microcantilevers. Nature biotechnology **19**, 856.

[19] Hansen, K. M., Ji, H. F., Wu, G., Datar, R., Cote, R., Majumdar, A., and Thundat, T. (2001) Cantilever-based optical deflection assay for discrimination of DNA single-nucleotide mismatches. Anal Chem **73**, 1567–1571.

[20] Yue, M., Lin, H., Dedrick, D., Satyanarayana, S., Majumdar, A., Bedekar, J., Jenkins, S., and Sundaram, A. (2004) 2-D microcantilever array for multiplexed biomolecular analysis. J. Microelectromech. Syst. **13**, 290–299.

[21] Nam, J. M., Stoeva, S. I., and Mirkin, C. A. (2004) Bio-bar-code-based DNA detection with PCR-like sensitivity. Journal of the American Chemical Society **126**, 5932–5933.

[22] Liu, Y., Solomon, M., and Achilefu, S. (2013) Perspectives and potential applications of nanomedicine in breast and prostate cancer. Medicinal research reviews **33**, 3–32.

[23] Zhang, G., Zeng, X., and Li, P. (2013) Nanomaterials in cancer-therapy drug delivery system. J Biomed Nanotechnol **9**, 741–750.

[24] Peiris, P. M., Toy, R., Doolittle, E., Pansky, J., Abramowski, A., Tam, M., Vicente, P., Tran, E., Hayden, E., Camann, A., Mayer, A., Erokwu, B. O., Berman, Z., Wilson, D., Baskaran, H., Flask, C. A., Keri, R. A., and Karathanasis, E. (2012) Imaging metastasis using an integrin-targeting chain-shaped nanoparticle. ACS nano **6**, 8783–8795.

[25] Nishimura, Y., Mieda, H., Ishii, J., Ogino, C., Fujiwara, T., and Kondo, A. (2013) Targeting cancer cell-specific RNA interference by siRNA delivery using a complex carrier of affibody-displaying bio-nanocapsules and liposomes. Journal of nanobiotechnology **11**, 19.

[26] Ophardt, C. E. (2003) "Anti-Cancer Drugs I." in Virtual Chembook Elmhusrt College..
[27] Kayser, O., Lemke, A., and Hernandez-Trejo, N. (2005) The impact of nanobiotechnology on the development of new drug delivery systems. Curr Pharm Biotechnol **6**, 3–5.
[28] Allen, T. M., and Cullis, P. R. (2004) Drug Delivery Systems: Entering the Mainstream. Science **303**, 1818–1822.
[29] Yih, T. C., and Al-Fandi, M. (2006) Engineered nanoparticles as precise drug delivery systems. Journal of Cellular Biochemistry **97**, 1184–1190.
[30] Olivier, J. C., Fenart, L., Chauvet, R., Pariat, C., Cecchelli, R., and Couet, W. (1999) Indirect evidence that drug brain targeting using polysorbate 80-coated polybutylcyanoacrylate nanoparticles is related to toxicity. Pharmaceutical research **16**, 1836–1842.
[31] Mukherjee, P., Bhattacharya, R., and Mukhopadhyay, D. (2005) Gold Nanoparticles Bearing Functional Anti-Cancer Drug and Anti-Angiogenic Agent: A "2 in 1" System with Potential Application in Cancer Therapeutics. Journal of Biomedical Nanotechnology **1**, 224–228.
[32] Roy, I., Ohulchanskyy, T. Y., Pudavar, H. E., Bergey, E. J., Oseroff, A. R., Morgan, J., Dougherty, T. J., and Prasad, P. N. (2003) Ceramic-based nanoparticles entrapping water-insoluble photosensitizing anticancer drugs: a novel drug-carrier system for photodynamic therapy. Journal of the American Chemical Society **125**, 7860–7865.
[33] Gong, J., Chen, M., Zheng, Y., Wang, S., and Wang, Y. (2012) Polymeric micelles drug delivery system in oncology. Journal of controlled release : official journal of the Controlled Release Society **159**, 312–323.
[34] Astruc, D., Boisselier, E., and Ornelas, C. (2010) Dendrimers designed for functions: from physical, photophysical, and supramolecular properties to applications in sensing, catalysis, molecular electronics, photonics, and nanomedicine. Chemical reviews **110**, 1857–1959.
[35] Morgan, M. T., Nakanishi, Y., Kroll, D. J., Griset, A. P., Carnahan, M. A., Wathier, M., Oberlies, N. H., Manikumar, G., Wani, M. C., and Grinstaff, M. W. (2006) Dendrimer-encapsulated camptothecins: increased solubility, cellular uptake, and cellular retention affords enhanced anticancer activity in vitro. Cancer research **66**, 11913–11921.
[36] Chen, C. Z., and Cooper, S. L. (2002) Interactions between dendrimer biocides and bacterial membranes. Biomaterials **23**, 3359–3368.

[37] Witvrouw, M., Fikkert, V., Pluymers, W., Matthews, B., Mardel, K., Schols, D., Raff, J., Debyser, Z., De Clercq, E., Holan, G., and Pannecouque, C. (2000) Polyanionic (i.e., polysulfonate) dendrimers can inhibit the replication of human immunodeficiency virus by interfering with both virus adsorption and later steps (reverse transcriptase/integrase) in the virus replicative cycle. Molecular pharmacology **58**, 1100–1108.

[38] Kukowska-Latallo, J. F., Candido, K. A., Cao, Z., Nigavekar, S. S., Majoros, I. J., Thomas, T. P., Balogh, L. P., Khan, M. K., and Baker, J. R., Jr. (2005) Nanoparticle targeting of anticancer drug improves therapeutic response in animal model of human epithelial cancer. Cancer research **65**, 5317–5324.

[39] Medina, S. H., and El-Sayed, M. E. (2009) Dendrimers as carriers for delivery of chemotherapeutic agents. Chemical reviews **109**, 3141–3157.

[40] Wang, P., Zhao, X. H., Wang, Z. Y., Meng, M., Li, X., and Ning, Q. (2010) Generation 4 polyamidoamine dendrimers is a novel candidate of nano-carrier for gene delivery agents in breast cancer treatment. Cancer letters **298**, 34–49.

[41] Alexis, F., Pridgen, E. M., Langer, R., and Farokhzad, O. C. (2010) Nanoparticle technologies for cancer therapy. Handb Exp Pharmacol, 55–86.

[42] Yang, F., Jin, C., Jiang, Y., Li, J., Di, Y., Ni, Q., and Fu, D. (2011) Liposome based delivery systems in pancreatic cancer treatment: from bench to bedside. Cancer treatment reviews **37**, 633–642.

[43] Zavaleta, C. L., Phillips, W. T., Soundararajan, A., and Goins, B. A. (2007) Use of avidin/biotin-liposome system for enhanced peritoneal drug delivery in an ovarian cancer model. International journal of pharmaceutics **337**, 316–328.

[44] Treat, J., Greenspan, A., Forst, D., Sanchez, J. A., Ferrans, V. J., Potkul, L. A., Woolley, P. V., and Rahman, A. (1990) Antitumor activity of liposome-encapsulated doxorubicin in advanced breast cancer: phase II study. Journal of the National Cancer Institute **82**, 1706–1710.

[45] Cullis, P. R., and Chonn, A. (1998) Recent advances in liposome technologies and their applications for systemic gene delivery. Advanced drug delivery reviews **30**, 73–83.

[46] Dowty, M. E., Williams, P., Zhang, G., Hagstrom, J. E., and Wolff, J. A. (1995) Plasmid DNA entry into postmitotic nuclei of primary rat

myotubes. Proceedings of the National Academy of Sciences of the United States of America **92**, 4572–4576.

[47] Wolff, J. A., Malone, R. W., Williams, P., Chong, W., Acsadi, G., Jani, A., and Felgner, P. L. (1990) Direct gene transfer into mouse muscle in vivo. Science **247**, 1465–1468.

[48] Stover, T. C., Sharma, A., Robertson, G. P., and Kester, M. (2005) Systemic delivery of liposomal short-chain ceramide limits solid tumor growth in murine models of breast adenocarcinoma. Clin Cancer Res **11**, 3465–3474.

[49] Tran, M. A., Smith, C. D., Kester, M., and Robertson, G. P. (2008) Combining nanoliposomal ceramide with sorafenib synergistically inhibits melanoma and breast cancer cell survival to decrease tumor development. Clin Cancer Res **14**, 3571–3581.

[50] Tran, M. A., Watts, R. J., and Robertson, G. P. (2009) Use of liposomes as drug delivery vehicles for treatment of melanoma. Pigment Cell Melanoma Res **22**, 388–399.

[51] De, M., Ghosh, P. S., and Rotello, V. M. (2008) Applications of Nanoparticles in Biology.. Adv. Mater. **20**, 4225–4241.

[52] Scherer, F., Anton, M., Schillinger, U., Henke, J., Bergemann, C., Kruger, A., Gansbacher, B., and Plank, C. (2002) Magnetofection: enhancing and targeting gene delivery by magnetic force in vitro and in vivo. Gene Ther **9**, 102–109.

[53] Donaldson, K., Stone, V., Tran, C., Kreyling, W., and Borm, P. J. A. (2004) Nanotoxicology. Occup Environ Med **61**, 727–728.

[54] Dhawan, A., and Sharma, V. (2010) Toxicity assessment of nanomaterials: methods and challenges. Anal Bioanal Chem **398**, 589–605.

[55] Lv, H., Zhang, S., Wang, B., Cui, S., and Yan, J. (2006) Toxicity of cationic lipids and cationic polymers in gene delivery. Journal of controlled release : official journal of the Controlled Release Society **114**, 100–109.

[56] Powers, K. W., Brown, S. C., Krishna, V. B., Wasdo, S. C., Moudgil, B. M., and Roberts, S. M. (2006) Research strategies for safety evaluation of nanomaterials. Part VI. Characterization of nanoscale particles for toxicological evaluation. Toxicol Sci **90**, 296–303.

[57] Dokka, S., Toledo, D., Shi, X., Castranova, V., and Rojanasakul, Y. (2000) Oxygen radical-mediated pulmonary toxicity induced by some cationic liposomes. Pharmaceutical research **17**, 521–525.

[58] Akhtar, S., and Benter, I. (2007) Toxicogenomics of non-viral drug delivery systems for RNAi: potential impact on siRNA-mediated gene

silencing activity and specificity. Advanced drug delivery reviews **59**, 164–182.

[59] Filion, M. a. P., NC. (1998) Major limitations in the use of cationic liposomes for DNA delivery International journal of pharmaceutics **162**, 159–170.

[60] Jevprasesphant, R., Penny, J., Jalal, R., Attwood, D., McKeown, N. B., and D'Emanuele, A. (2003) The influence of surface modification on the cytotoxicity of PAMAM dendrimers. International journal of pharmaceutics **252**, 263–266.

[61] Keelan, J. A. (2011) Nanotoxicology: nanoparticles versus the placenta. Nat Nanotechnol **6**, 263–264.

[62] Chang, C. (2010) The immune effects of naturally occurring and synthetic nanoparticles. J Autoimmun **34**, J234–246.

[63] Rutberg, F. G., Dubina, M. V., Kolikov, V. A., Moiseenko, F. V., Ignat'eva, E. V., Volkov, N. M., Snetov, V. N., and Stogov, A. Y. (2008) Effect of silver oxide nanoparticles on tumor growth in vivo. Dokl Biochem Biophys **421**, 191–193.

[64] Hussain, S. M., Hess, K. L., Gearhart, J. M., Geiss, K. T., and Schlager, J. J. (2005) In vitro toxicity of nanoparticles in BRL 3A rat liver cells. Toxicology in vitro : an international journal published in association with BIBRA **19**, 975–983.

[65] Hoshino, A., Hanaki, K., Suzuki, K., and Yamamoto, K. (2004) Applications of T-lymphoma labeled with fluorescent quantum dots to cell tracing markers in mouse body. Biochem Biophys Res Commun **314**, 46–53.

[66] Hardman, R. (2006) A toxicologic review of quantum dots: toxicity depends on physicochemical and environmental factors. Environ Health Perspect **114**, 165–172.

[67] Arora, S., Rajwade, J. M., and Paknikar, K. M. (2012) Nanotoxicology and in vitro studies: the need of the hour. Toxicology and applied pharmacology **258**, 151–165.

[68] Mickuviene, I., Kirveliene, V., and Juodka, B. (2004) Experimental survey of non-clonogenic viability assays for adherent cells in vitro. Toxicology in vitro: an international journal published in association with BIBRA **18**, 639–648.

[69] Schins, R. P., Duffin, R., Hohr, D., Knaapen, A. M., Shi, T., Weishaupt, C., Stone, V., Donaldson, K., and Borm, P. J. (2002) Surface modification of quartz inhibits toxicity, particle uptake, and oxidative

DNA damage in human lung epithelial cells. Chem Res Toxicol **15**, 1166–1173.

[70] Wilson, M. R., Lightbody, J. H., Donaldson, K., Sales, J., and Stone, V. (2002) Interactions between ultrafine particles and transition metals in vivo and in vitro. Toxicology and applied pharmacology **184**, 172–179.

[71] Stone, V., Johnston, H., and Schins, R. P. (2009) Development of in vitro systems for nanotoxicology: methodological considerations. Crit Rev Toxicol **39**, 613–626.

[72] Xiao, G. G., Wang, M., Li, N., Loo, J. A., and Nel, A. E. (2003) Use of proteomics to demonstrate a hierarchical oxidative stress response to diesel exhaust particle chemicals in a macrophage cell line. J Biol Chem **278**, 50781–50790.

[73] Johnston, H. J., Hutchison, G. R., Christensen, F. M., Peters, S., Hankin, S., Aschberger, K., and Stone, V. (2010) A critical review of the biological mechanisms underlying the in vivo and in vitro toxicity of carbon nanotubes: The contribution of physico-chemical characteristics. Nanotoxicology **4**, 207–246.

[74] Mroz, R. M., Schins, R. P., Li, H., Jimenez, L. A., Drost, E. M., Holownia, A., MacNee, W., and Donaldson, K. (2008) Nanoparticle-driven DNA damage mimics irradiation-related carcinogenesis pathways. Eur Respir J **31**, 241–251.

[75] Hall, J. B. D., MA, Patriak, McNeil, SE. (2007) Characterization of nanoparticles for therapeutics. Nanomedicine (Lond) **2**, 789–803.

[76] Elsabahy, M., and Wooley, K. L. (2013) Cytokines as biomarkers of nanoparticle immunotoxicity. Chemical Society reviews **42**, 5552–5576.

[77] Moreira, J. N., Gaspar, R., and Allen, T. M. (2001) Targeting Stealth liposomes in a murine model of human small cell lung cancer. Biochimica et biophysica acta **1515**, 167–176.

[78] Chithrani, B. D., Ghazani, A. A., and Chan, W. C. (2006) Determining the size and shape dependence of gold nanoparticle uptake into mammalian cells. Nano letters **6**, 662–668.

[79] Hauck, T. S., Ghazani, A. A., and Chan, W. C. (2008) Assessing the effect of surface chemistry on gold nanorod uptake, toxicity, and gene expression in mammalian cells. Small **4**, 153–159.

[80] Patel, H. M., and Moghimi, S. M. (1998) Serum-mediated recognition of liposomes by phagocytic cells of the reticuloendothelial system - The concept of tissue specificity. Advanced drug delivery reviews **32**, 45–60.

[81] Owens, D. E., 3rd, and Peppas, N. A. (2006) Opsonization, biodistribution, and pharmacokinetics of polymeric nanoparticles. International journal of pharmaceutics **307**, 93–102.

[82] Dadashzadeh, S., Mirahmadi, N., Babaei, M. H., and Vali, A. M. (2010) Peritoneal retention of liposomes: Effects of lipid composition, PEG coating and liposome charge. Journal of controlled release : official journal of the Controlled Release Society **148**, 177–186.

[83] Murphy, C. J., Gole, A. M., Stone, J. W., Sisco, P. N., Alkilany, A. M., Goldsmith, E. C., and Baxter, S. C. (2008) Gold nanoparticles in biology: beyond toxicity to cellular imaging. Acc Chem Res **41**, 1721–1730.

[84] Richards, D., and Ivanisevic, A. (2012) Inorganic material coatings and their effect on cytotoxicity. Chemical Society reviews **41**, 2052–2060.

[85] Ow, H., Larson, D. R., Srivastava, M., Baird, B. A., Webb, W. W., and Wiesner, U. (2005) Bright and stable core-shell fluorescent silica nanoparticles. Nano letters **5**, 113–117.

[86] Immordino, M. L., Dosio, F., and Cattel, L. (2006) Stealth liposomes: review of the basic science, rationale, and clinical applications, existing and potential. Int J Nanomedicine **1**, 297–315.

[87] LaVan, D. A., McGuire, T., and Langer, R. (2003) Small-scale systems for in vivo drug delivery. Nature biotechnology **21**, 1184–1191.

[88] De Jong, W. H., and Borm, P. J. (2008) Drug delivery and nanoparticles:applications and hazards. Int J Nanomedicine **3**, 133–149.

[89] Tran, M. A., Gowda, R., Sharma, A., Park, E. J., Adair, J., Kester, M., Smith, N. B., and Robertson, G. P. (2008) Targeting V600EB-Raf and Akt3 using nanoliposomal-small interfering RNA inhibits cutaneous melanocytic lesion development. Cancer research **68**, 7638–7649.

[90] Felgner, P. L., Gadek, T. R., Holm, M., Roman, R., Chan, H. W., Wenz, M., Northrop, J. P., Ringold, G. M., and Danielsen, M. (1987) Lipofection: a highly efficient, lipid-mediated DNA-transfection procedure. Proceedings of the National Academy of Sciences of the United States of America **84**, 7413–7417.

[91] Farokhzad, O. C., Cheng, J., Teply, B. A., Sherifi, I., Jon, S., Kantoff, P. W., Richie, J. P., and Langer, R. (2006) Targeted nanoparticle-aptamer bioconjugates for cancer chemotherapy in vivo. Proceedings of the National Academy of Sciences of the United States of America **103**, 6315–6320.

[92] Nagpal, K., Singh, S. K., and Mishra, D. N. (2010) Chitosan nanoparticles: a promising system in novel drug delivery. Chem Pharm Bull (Tokyo) **58**, 1423–1430.

[93] Bisht, S., and Maitra, A. (2009) Dextran-doxorubicin/chitosan nanoparticles for solid tumor therapy. Wiley Interdiscip Rev Nanomed Nanobiotechnol **1**, 415–425.

[94] Caminade, A. M., and Majoral, J. P. (2010) Dendrimers and nanotubes: a fruitful association. Chemical Society reviews **39**, 2034–2047.

[95] Jain, R., Shah, N. H., Malick, A. W., and Rhodes, C. T. (1998) Controlled drug delivery by biodegradable poly(ester) devices: different preparative approaches. Drug Dev Ind Pharm **24**, 703–727.

[96] Panyam, J., and Labhasetwar, V. (2003) Biodegradable nanoparticles for drug and gene delivery to cells and tissue. Advanced drug delivery reviews **55**, 329–347.

[97] Musumeci, T., Vicari, L., Ventura, C. A., Gulisano, M., Pignatello, R., and Puglisi, G. (2006) Lyoprotected nanosphere formulations for paclitaxel controlled delivery. J Nanosci Nanotechnol **6**, 3118–3125.

[98] Hoshino, A., Fujioka, K., Oku, T., Nakamura, S., Suga, M., Yamaguchi, Y., Suzuki, K., Yasuhara, M., and Yamamoto, K. (2004) Quantum dots targeted to the assigned organelle in living cells. Microbiol Immunol **48**, 985–994.

[99] Gardikis, K., Hatziantoniou, S., Bucos, M., Fessas, D., Signorelli, M., Felekis, T., Zervou, M., Screttas, C. G., Steele, B. R., Ionov, M., Micha-Screttas, M., Klajnert, B., Bryszewska, M., and Demetzos, C. (2010) New drug delivery nanosystem combining liposomal and dendrimeric technology (liposomal locked-in dendrimers) for cancer therapy. J Pharm Sci **99**, 3561–3571.

[100] Shen, C., Brozena, A. H., and Wang, Y. (2010) Double-walled carbon nanotubes: Challenges and opportunities. Nanoscale.

[101] Thakare, V. S., Das, M., Jain, A. K., Patil, S., and Jain, S. (2010) Carbon nanotubes in cancer theragnosis. Nanomedicine (Lond) **5**, 1277–1301.

[102] Hainfeld, J. F., Slatkin, D. N., and Smilowitz, H. M. (2004) The use of gold nanoparticles to enhance radiotherapy in mice. Phys Med Biol **49**, N309–315.

[103] Zwiorek, K., Kloeckner, J., Wagner, E and Coester, C. (2004) Gelatin nanoparticles as a new and simple gene delivery system. J Pharm Pharm Sci **7**, 22–28.

[104] Hirsch, L. R., Stafford, R. J., Bankson, J. A., Sershen, S. R., Rivera, B., Price, R. E., Hazle, J. D., Halas, N. J., and West, J. L. (2003)

Nanoshell-mediated near-infrared thermal therapy of tumors under magnetic resonance guidance. Proceedings of the National Academy of Sciences of the United States of America **100**, 13549–13554.

[105] Surendiran, A., Sandhiya, S., Pradhan, S. C., and Adithan, C. (2009) Novel applications of nanotechnology in medicine. Indian J Med Res **130**, 689–701.

[106] Smith, A. M., Dave, S., Nie, S., True, L., and Gao, X. (2006) Multicolor quantum dots for molecular diagnostics of cancer. Expert Rev Mol Diagn **6**, 231–244.

[107] Tang, M., Xing, T., Zeng, J., Wang, H., Li, C., Yin, S., Yan, D., Deng, H., Liu, J., Wang, M., Chen, J., and Ruan, D. Y. (2008) Unmodified CdSe quantum dots induce elevation of cytoplasmic calcium levels and impairment of functional properties of sodium channels in rat primary cultured hippocampal neurons. Environ Health Perspect **116**, 915–922.

11

Biodegradable Carrier Systems for Drug and Vaccine Delivery

Anil Mahapatro[1] and Dinesh Singh[2]

[1]Bioengineering Program & Department of Industrial and Manufacturing Engineering, Wichita State University, 1845 Fairmount Street, Wichita, KS 67260, USA
[2]Department of Life Sciences, Winston-Salem State University, 601 S MLK Jr. Drive Winston Salem, NC 27110, USA

11.1 Introduction

The role of nanotechnology in material science and engineering has seen extensive development in the last decade [1–4]. Nanostructured materials refer to materials with sizes in the 1–100 nm range which demonstrate unique properties and functions due to their "size effect"[5, 6]. Nanostructured materials have the flexibility and capability to be modified and integrated in to biomedical devices since most biological systems, including viruses, membranes and protein complex are natural nanostructures [7]. Rapid developments in nanomaterials have been made in a wide variety of biomedical sectors including cardiovascular and orthopedic with specific applications in tissue engineered scaffolds and devices, site specific drug delivery systems, cancer therapy and clinical bioanalytical diagnostics and therapeutics [8–13]. In recent years, significant effort has been devoted to develop nanotechnology for drug delivery. Table 11.1 describes examples of nanotechnology-based products that are being tested or approved for commercial use [9]. Nanosize drug carriers offers a suitable means of delivering small molecular weight drugs as well as macromolecules such as proteins, peptides or genes by either localized or targeted delivery to the tissue of interest [8]. Table 11.2 depicts examples of use of nanoparticles for drug delivery and the subsequent therapeutic enhancement

Post-genomic Approaches in Cancer and Nano Medicine, 315–340.

[9]. Nanotechnology is being currently leveraged to formulate therapeutic agents in biocompatible nanocomposites such as nanoparticles, nanocapsules, micellar systems and conjugates [9]. In this chapter, we will review polymeric-based biodegradable nanoparticles and their applications in drug delivery.

Polymeric-based nanoparticles for drug delivery are submicron size polymeric colloidal particles with a therapeutic agent of interest embedded or encapsulated within their polymeric matrix or adsorbed or conjugated onto the surface [14]. During the 1980's and 1990's, several drug delivery systems were developed to improve the efficiency of drugs and minimize toxic side-effects [15]. The early nanoparticles (NPs) and microparticles were mainly formulated from poly(alkylcyanoacrylate) [16]. Initial promise for microparticles was dampened by the fact that there was a size limit for the particle to cross the intestinal lumen into the lymphatic system following oral drug delivery. Similarly the therapeutic effect of drug loaded nanoparticles was relatively poor due to rapid clearance of the particles by phagocytosos post intravenous administration. In recent years, this problem has been solved by surface modification of the nanoparticles [15]. Nanoparticles have a further advantage over larger microparticles because they are better suited for intravenous (i.v) delivery [17]. A wide variety of drugs can be delivered using nanoparticulate carriers via a number of routes. NPs can be used to deliver hydrophilic, hydrophobic drugs, proteins, vaccines, biological macromolecules etc. They can be designed for targeted delivery to brain, arterial walls, lungs, tumor cells, liver spleen or long term systemic circulation. In addition, nanoparticles with imaging agents offer opportunities to exploit optical imaging or MRI in cancer imaging and guided hyperthermia therapy [18]. Figure 11.1 illustrates the possibility of using a multimodal approach and integrated systems that combine differing properties such as tumor targeting, therapy and imaging in an all in one system [18]. Numerous techniques exist for synthesizing nanoparticles which may differ based on the type of drug used and the targeted organ and delivery mechanism selected. Based on the protocol to be followed, the parameters must be tailored to create the best possible characteristics for the nanoparticles. In this chapter, we will review the different biodegradable nanoparticles currently used and the techniques for preparing them. We will also discuss advances in surface modifications and specific end applications of these NPs.

Table 11.1 Examples nanotechnology based products that are being tested or approved for commercial use. Adapted from ref [9] Copyright 2012 Elsevier

Product	Type Nanome-trial	Indicator	Phase	Advantages	Company
Doxil	PEGylated liposome	Ovarian cancer and multiple myeloma	On Market	Enhanced circulation time and is up to six limes more effective (liar free DOX.	Jansscn
Ahraxanc	Albumin NPs	Lung Cancer	On	Enhanced cytoloxicily, shorter infusion lime.	Cclgene Corporation.'
		Breast Cancer	Market	[in. Jose rcqimvd	Abraxis Bioseicnees
Aunmniuiie (CYT-6091)	AuNPs coupled to TNF and PEG-Thiol	Solid Tumors	Phase II	Selectively destroys cancer cells without harming healtbv tissues	Cvtlmmune Sciences
Auto She II	Gold-coated silica NPs	Solid Tumors	Phase I	Highly selective and rapid rumor destruction with minimal damage to surrounding liullbv tissues	Nanospcctra Biosciences
Cnmbide.x	Iron oxide NPs	Tumor Imaging	NDA FM	Efficient for ihe detection ol merastatic lyniph nodes in various cancers	Advanced Maundies
Cycloscrt	Cycledextrin NPs	Mctastatic solid tumor	1ND Filed	Very effective in preventing tumor progression	Insert therapeutics

(Continued)

Table 11.1 Continued

Product	Type Nanometrial	Indicator	Phase	Advantages	Company
ING N-401	Liposome	Mctestalk lung	Phase I	Suppresses lumor enmlh and inhibits metastasis of lung cancer	Intnogen
MRX-952	Formulation of irinolecan metabolite	onocology	preclinical		lmaRx Therapeu-tics therapeu tics
Nanoxel	Nanoparticulate delivery	Breast Cancer	On mark el	Cremnphnr fn-e filter-soluble formulation.	Dabur Pharma
	system for paclitaxel	Ovarian Cancer		Greater etTieaev and decreased cviukrucitv	
TNT	Polymer-coated	Solid Tumors	Preclimcal	Provides targeted therapy. Selectively kills	Triton BioSystems
AnuEpCAM	iron oxide			EpCAM positive cells	
Verigene	DNA func-tionalized	Diagnostics	On market	AuNPs enable efficient diagnosis for	Nanosphcre Triton
platform	AuNPs			melhici II in-resistant *Stapbylococ-cus turnus*	BioSy stems

11.2 Nanoparticle Fabrication

Biodegradable nanoparticles can be fabricated from a variety of materials such as proteins, polysaccharides and synthetic biodegradable polymers. The selection of the base polymer is based on various design and end application criterion and depends on many factors such as (1) size of nanoparticle

Table 11.2 Nanoparticles for drug delivery and the subsequent therapeutic improvement. Adapted from ref [9] Copyright 2012 Elsevier

Type of nanomaterial	Encapsulant	Indicator	Therapeutic improvement
Polyisohexyl cyanoacrylate NPs	DOX	Hepatocellular Carcinoma	Higher antitumor efficacy than native doxorubicin and can overcome multiple drug resistance phenotype.
PLGA NPs	Paclitaxel	Various cancers	Effective in chemotherapeutic and photothermal destruction of cancer cells
Gold NPs (AuNPs)	–	Various cancers	Effective as radiation sensitizers for cancer therapy
Chitosan NP (CNP)	siRNA	Ovarian cancer	Increased selective intratumoral delivery and significant inhibition of tumor growth compared to controls
Cetyl alcohol/polysorbate NPs	Paclitaxel	Brain tumor	Higher brain and tumor cell uptake, thus leading to greater cytotoxicity; also effective towards p-glycoprotein expressing tumor cells.
Lipid nanocapsules	Etoposide	Glioma	Greater cytotoxicity. Can overcome p-glycoprotein dependent multidrug resistance.
P (4-vinyl pyridine) particles	–	Antimicrobial agent	These particles can be used to inhibit bacterial growth for various bacteria as biocolloids
Chitosan-alginate NPs	Carboplatin	Retinoblastoma	Enhanced antiproliferative activity and cytotoxicity of NPs in comparison with native carboplatin
Poly (3-hydroxybutyrate-co-3-hydroxyoctanoate) NPs	DOX	Various cancers	Effective in selective delivery of anticancer drug to the folate receptor-overexpressed cancer cells

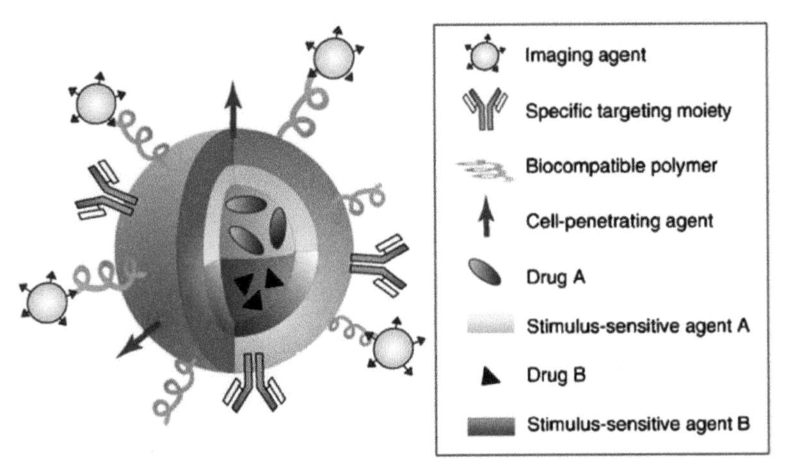

Figure 11.1 Multifunctional nanoparticles. Multifunctional nanoparticles can combine a specific targeting agent (usually with an antibody or peptide) with nanoparticles for imaging (such as quantum dots or magnetic nanoparticles), a cell-penetrating agent (e.g., the polyArg peptide TAT), a stimulus-selective element for drug release, a stabilizing polymer to ensure biocompatibility polyethylene glycol most frequently), and the therapeutic compound. Development of novel strategies for controlled released of drugs will provide nanoparticles with the capability to deliver two or more therapeutic agents. Adapted from ref [18] Copyright 2009 Wiley interscience.

desired, (2) properties of the drug (aqueous solubility, stability etc) to be encapsulated in the polymer, (3) surface characteristics and functionality, (4) degree of biodegradability and biocompatibility and (5) drug release profile of the final product. Contingent of appropriate selection of desired criterion for the preparation of the nanoparticle the methods of preparing the NPs can be classified as following(1) dispersion of preformed polymers, (2) polymerization of monomers and (3) ionic gelation method for hydrophilic polymers. Figure 11.2 schematically describes the various techniques for the preparation of polymer nanoparticles [5]. The general advantages and disadvantages of individual methods are summarized in Table 11.3[19]. The various methods of nanoparticle fabrication are briefly reviewed below.

11.2.1 Dispersion of preformed polymers

this technique is a common technique used to prepare biodegradable nanoparticles from poly (lactic acid) (PLA); poly (D, L-glycolide) PLG; poly (D, L-lactide-co-glycolide) PLGA and poly (cyanoacrylate) PCA. This technique can be used in several ways as described below.

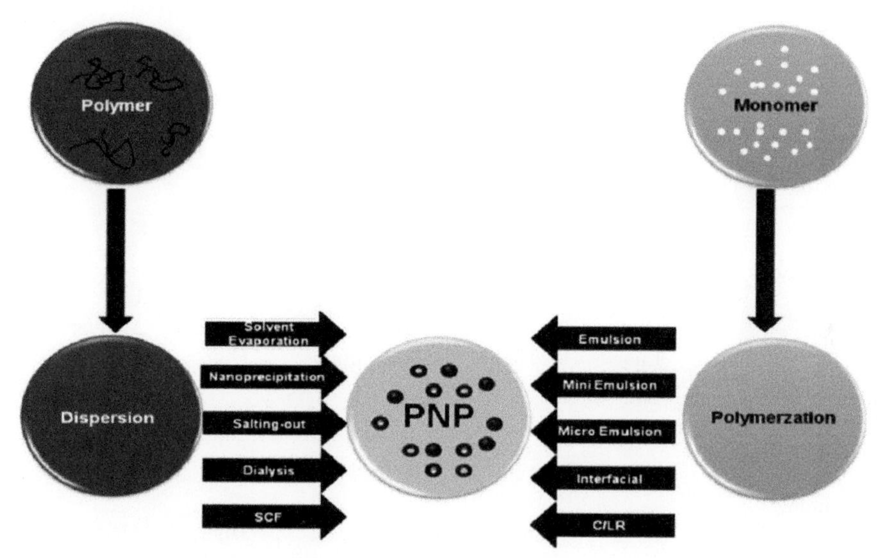

Figure 11.2 Schematic representation of various techniques for the preparation of polymer nanoparticles. SCF: supercritical Fluid technology, C/LR: controlled/livingradical. Adapted from ref [5] Copyright 2011 Elsevier.

(a) *Solvent evaporation method:* In this technique, the polymer is dissolved in an organic solvent such as dichloromethane, chloroform or ethyl acetate. The drug is dissolved or dispersed in this preformed polymer solution and then this mixture is emulsified into an aqueous solution to form an oil-water emulsion, using an appropriate surfactant/emulsifying agent such as gelatin, polyvinyl alcohol etc. After formation of a stable emulsion, the organic solvent is evaporated by increasing the temperature or pressure and by continuous stirring. Figure 11.3 shows a schematic representation of this method [19]. Process parameters such as stabilizer and polymer concentration and stirring speed have a great influence on the particle size of the NP formed [17, 20].

(b) *Spontaneous emulsification/solvent diffusion method:* this is a modified solvent diffusion method where a water miscible solvent such as acetone and methanol along with a water insoluble organic solvent dichloromethane or chloroform are used as the oil phase [21]. Due to the spontaneous diffusion of solvents, an interfacial turbulence is created between the two phases leading to the formation of smaller particles. As the concentration of water soluble solvent increases smaller particle sizes of NP can be achieved [19, 21].

Table 11.3 Polymeric nanoparticles: general advantages and drawbacks of the preparation methods

Method	Simplicily of procedure	Need for purification	Facility Scaling-up	EE(%)	Safely of compounds
Polymerization of monomers					
Emulsion polymerization					
Organic	Low	High	NR	Low	Low
Aqueous	High	low	High	Medium	Medium
Intefacial Polymerization	Luw	High	Medium	High	Low
Preformed polymers					
Synthetic					
Emulsifcation/Solvent evaporation	High	Low	Low	Medium	Medium
Solvent displacement and interfecia] deposition	High	NR	NR	High	Medium
Salting out	High	High	High	High	Low
Emulsion/solverl diffusion	Medium	Medium	High	High	Medium
Natural					
Albumin	NR	High	NR	Medium	Low
Gelatin	NR	High	NR	Medium	Low
Polysaccharides					
Alginate	High	Medium	High	High	High
Coitosan.	High	Medium	High	High	High
Agnrae	Medium	High	NR	NR	High
Desolvation	NR	High	NR	Low	Low

(c) *Nanoprecipitation method:* Typically, this method is used for hydrophobic drug entrapment, but it has been adapted for hydrophilic drugs as well. Polymer and drug are dissolved in a polar, water miscible solvent such as acetone, acetonitrile, ethanol, or methanol. The solution is then poured in a controlled manner (i.e. drop-wise addition) into an aqueous solution with surfactant. Nanoparticles are formed instantaneously by rapid solvent diffusion. Finally, the solvent is removed under reduced pressure [22].

(d) *Salting out method:* In this synthesis method, the polymer is dissolved in the organic phase, which should be water-miscible, like acetone or tetrahydrofuran (THF). The organic phase is emulsified in an aqueous

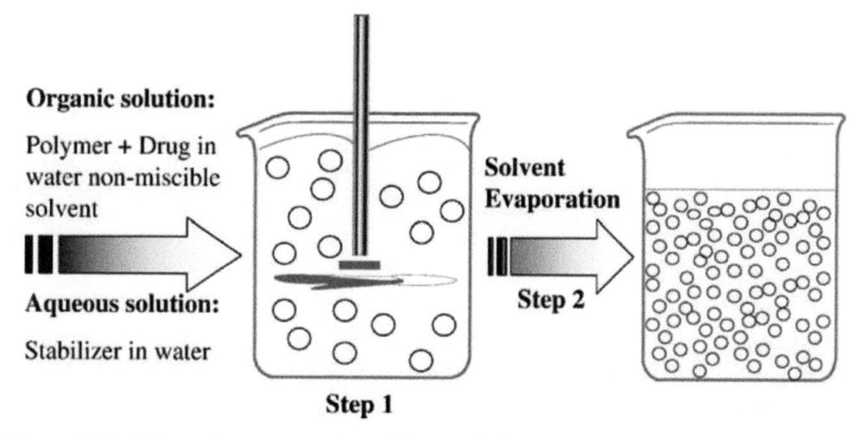

Figure 11.3 Schematic representation of the emulsification-evaporation technique. Adapted from ref [19] Copyright 2006 Elsevier.

phase, under strong mechanical shear stress. The aqueous phase contains the emulsifier and a high concentration of salts which are not soluble in the organic phase. Typically, the salts used are 60% w/w of magnesium chloride hexahydrate [23] or magnesium acetate tetrahydrate in a ratio of 1:3 polymer to salt [24]. Contrary to the emulsion diffusion method, there is no diffusion of the solvent due to the presence of salts. The fast addition of pure water, to the o/w emulsion, under mild stirring, reduces the ionic strength and leads to the migration of the water-soluble organic solvent to the aqueous phase inducing nanosphere formation. The final step is purification by cross flow filtration or centrifugation to remove the salting out agent [23, 24].

11.2.1.1 Polymerization Methods

NPs are prepared from monomers that are polymerized to form NPs in an aqueous solution. Drug/therapeutic agents are incorporated in the NP either by dissolving the drug in the polymerization medium or by adsorption/attachment of the drug onto the NPs after the polymerization and when NPs formation has been completed. The NP suspension is then purified to remove stabilizers and surfactants that may be recycled for subsequent polymerization by ultracentrifugation and by resuspending the particles in an isotonic surfactant medium. This technique of forming NPs has been reported for making polybutylcyanoacrylate or poly (alkylcyanoacrylate) NPs [25, 26]. Concentration of surfactant and the stabilizer used determines the final size of the NPs formed [27].

11.2.1.2 Ionic gelation method for hydrophilic polymers

Some of the natural macromolecules used to prepare NPs include gelatin, alginate, chitosan and agarose NPs. These are hydrophilic natural polymers and have been used to synthesize biodegradable NPs by ionic gelation method which involves the materials undergoing transition from liquid to gel due to ionic interaction at room temperature. Examples of formation of gelatin NPs includes formation of gelatin NPs by hardening the emulsified gelatin solution droplets by cooling the emulsion below the gelation point in an ice bath resulting in gelation of the gelatin droplets [28]. Alginate NPs are reported to be produced by drop-wise extrusion of sodium alginate solution into the calcium chloride solution [29]. Sodium alginate is a water-soluble polymer that gels in the presence of multivalent cations such as calcium [30]. Chitosan NPs are prepared based on the spontaneous formation of complexes between chitosan and polyanions or the gelation of a chitosan solution dispersed in an oil emulsion [31].

11.2.1.3 Biodegradable polymer matrix for nanoparticle fabrication

Examples of some extensively studied biodegradable polymer matrixes used to form nanoparticles are described below:

- Poly-D-L- lactide-co-glycolide (PLGA): Poly-D-L- lactide-co-glycolide (PLGA) is one of the most successfully used biodegradable polymer because it undergoes hydrolysis in the body to produce the biodegradable metabolite monomers, lactic acid and glycolic acid. Since the body effectively deals with these two monomers, there is very minimal systemic toxicity associated by using PLGA for drug delivery or biomaterial applications. PLGA NPs have been mostly prepared by emulsification-diffusion, solvent evaporation and nanoprecipitation method [32]. PLGA nanoparticles have been used to develop the proteins and peptides nanomedicine, nanovaccines, nanoparticles-based gene delivery systems [32, 33].
- Polylactic acid (PLA): PLA is a biocompatible and biodegradable polymer which undergoes scission in the body to monomeric units of lactic acid as a natural intermediate. PLA nanoparticles have been mostly prepared by solvent evaporation, solvent displacement salting out and solvent diffusion [19, 34]. The salting out procedure is based on the separation of a water miscible solvent from aqueous solution by adding salting out agent like magnesium chloride, calcium chloride, etc. The

main advantage of salting out procedure is that it minimizes stress to protein encapsulants [32]

- Poly-ϵ-caprolactone (PCL): poly-ϵ-caprolactone is degraded by hydrolysis of its ester linkages in physiological conditions (such as in the human body) and has therefore received a great deal of attention for use in drug delivery. In particular, it is interesting for the preparation of long-term implantable devices, owing to its degradation slower than that of polylactide. PCL nanoparticles have been prepared mostly by nanoprecipitation, solvent displacement and solvent evaporation [32, 35, 36].
- Chitosan: Chitosan is a modified natural carbohydrate polymer prepared by the partial N-deacetylation of crustacean derived natural biopolymer chitin. There are at least four methods reported for the preparation of chitosan nanoparticles as ionotropic gelation, microemulsion, emulsification solvent diffusion and polyelectrolyte complex [32, 37, 38].
- Gelatin: Gelatin is extensively used in food and medical products and is attractive for use in controlled release due to its nontoxic, biodegradable, bioactive and inexpensive properties. It is a poly-ampholyte having both cationic and anionic groups along with a hydrophilic group. It is known that the mechanical properties, swelling behavior and thermal properties of gelatin NP depend significantly on the crosslinking degree of gelatin. Gelatin nanoparticles can be prepared by desolvation/coacervation or emulsion method [32, 39, 40].
- Poly-alkyl-cyano-acrylates (PAC): The biodegradable as well as biocompatible poly-alkylcyanoacrylates are degraded by esterases in biological fluids and produce some toxic products that may stimulate or damage the central nervous system. Thus, this polymer is not authorized for application in humans. However, PAC nanoparticles are prepared mostly by emulsion polymerization, interfacial polymerization and nanoprecipitation [19, 32].

11.2.1.4 Nanoparticle functionalization

Surface modification of the NPs plays a critical role for their successful application since the mononuclear phagocytic system (MPS), one of the body's innate modes of defense, eliminates any injected NPS from the blood stream unless the particles are modified to escape recognition [41]. Longer circulation time increases the probability for the nanoparticles to reach their target. Small particles ($<$100 nm) with a hydrophilic surface have the greatest ability to evade the MPS [42, 43]. Several methods have been developed for surface modification of the NP. Of these methods, the most preferred method is the adsorption or grafting of poly-ethylene glycol (PEG) to the surface of nanoparticles. Figure 11.4 schematically represents the surface functionalization of biodegradable nanoparticles with PEG which results in low protein binding and thus leading to increased nanoparticle circulation time within the body [44]. Addition of PEG and PEG-containing copolymers to the surface of nanoparticles results in an increase in the blood circulation half-life of the particles. The mechanisms involved in prolonged circulation time of the PEG surface modified NPs are still not well understood. It is generally thought that the increased residency of the nanoparticles in blood is mainly due to the suppression of surface opsonization of serum or plasma protein caused by the steric repulsion of PEG hydrated barriers on nanoparticle surfaces.

Studies have shown that the degree to which proteins adsorb onto particle surfaces can be minimized by increasing the PEG density on the particle surface and increasing the molecular weight of the PEG chains used [42]. For

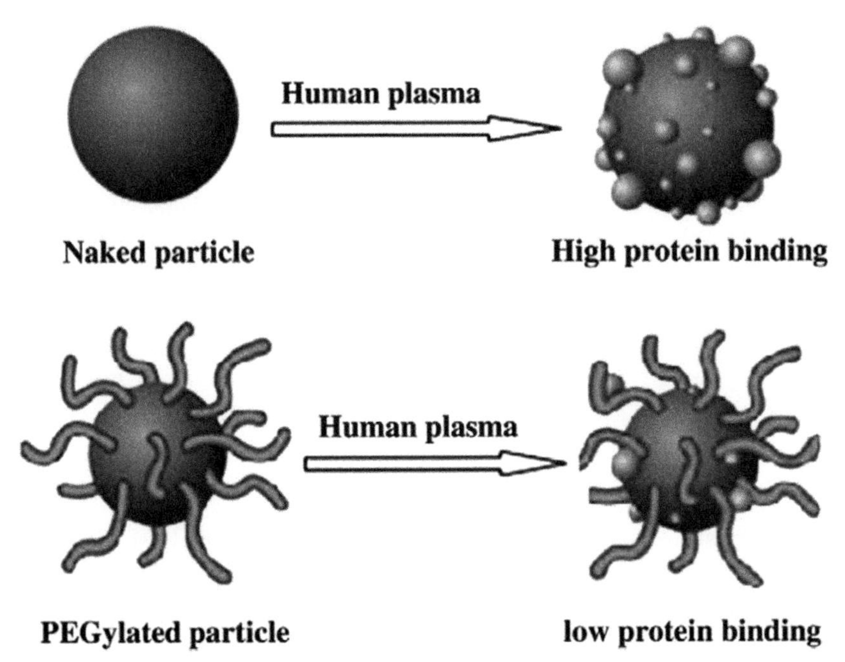

Figure 11.4 3PEGylated nanoparticles are able to avoid clearance from the blood stream by repelling protein adsorption, thus prolonging nanoparticle circulation time within the body. Adapted from ref [44] Copyright 2013 Elsevier.

example, Leroux *et al.* [45] showed that an increase in PEG molecular weight in PLGA nanoparticles was associated with less interaction with the MPS, and longer systemic circulation. PEG has been shown to impart stability on PLA particles submerged in simulated gastric fluid (SGF). Tobio *et. al.* showed that after 4 hours in SGF, 9% of PLA nanoparticles converted to lactic acid versus 3% conversion for PEG-PLA particles [46]. PEG is also believed to facilitate mucoadhesion and consequently transport through the Peyers patches of the GALT (gut associated lymphoid tissue) [47]. In addition, PEG may benefit nanoparticle interaction with blood constituents. Thus, the presence of PEG allows for important number of functions for the use of polymeric NPs.

Apart from PEG, other hydrophilic polymers such as poloxamers, polysorbate 80, TPGS, polysorbate 20, polysaccharides like dextran and different type of copolymers can be used to efficiently coat conventional nanoparticles surface leading to variation in the surface properties [48, 49]. These coatings provide a dynamic cloud of hydrophilic and neutral chains at the particle surface, which repel plasma proteins. Surface modification by TPGS increases the adhesion of nanoparticles to tumor cells surface. It also provides milder environments to the encapsulated proteins. IgG coating on the surface of nanoparticles increases the immunoresponse of the nanoparticles. Hydrophilic polymers can be introduced at the surface in two ways, either by adsorption of surfactants or by use of block or branched copolymers [48–50]. Figure 11.5 schematically elaborates on the different surface functionalization approaches for drug targeting [9].

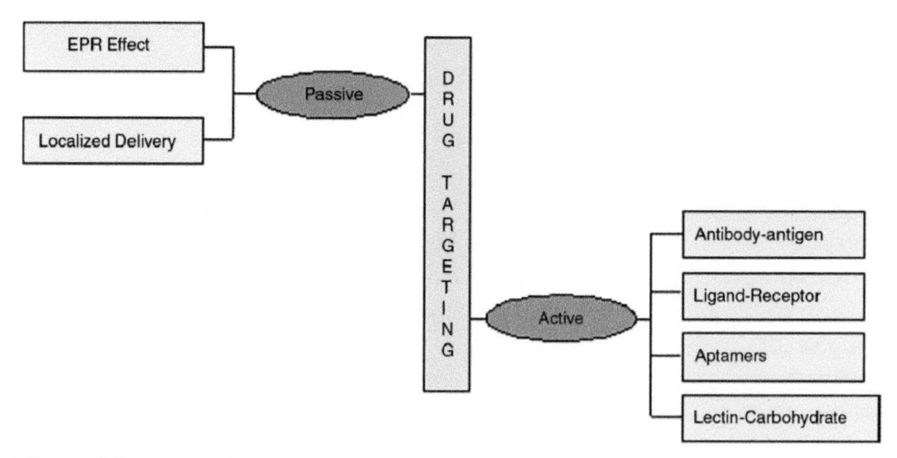

Figure 11.5 Schematic representation of different drug-targeting approaches. Adapted from ref [9] Copyright 2012 Elsevier.

A successful nanoparticle also needs to have a high loading capacity of the drug so as to reduce the quantity of the polymer carrier required for administration. Drug loading into nanoparticles is achieved by two methods: (1) by incorporating the drug at the time of nanoparticle production or (2) by adsorbing the drug after the formation of nanoparticles by incubating the NPs in a concentrated drug solution [17]. These two methods provides ways by which the drug is adsorbed/attached to the NPs; namely, encapsulation of the drug in the polymer, dispersion of the drug in the polymer, adsorption of the drug onto the surface of the nanoparticles and chemical binding of the drug to the polymer. The amount of bound drug can be determined by subtracting the drug content in the supernatant from the primary amount of drug present in the suspension. The drug release mechanisms are equally important as the drug polymer formulation because of the proposed application in sustained drug delivery. In general, drug release rate depends on solubility of the drug, desorption of the surface bound/adsorbed drug, drug diffusion through the polymer matrix, NP matrix erosion/degradation and combination of the erosion diffusion process [32]. One way to modify the drug release profiles is by appropriate choice of the polymer matrices. Drug release kinetics also depends upon the size of the NPs and the loading efficiency of drug. The drug loading efficiency will determine the initial burst and the sustained release rate of nanoencapsulated drug molecule. The rapid initial release or burst of drug seen in drug release profiles is mainly attributed to weakly bound or adsorbed drug on to the surface [15, 51].

11.3 Specific Applications of Biodegradable Nps

11.3.1 Tumor Targeting

The rationale of using nanoparticles for tumor targeting is based on the following factors (1) NPs will be able to deliver the requisite dose load of drug in the vicinity of the tumor due to the enhanced permeability and retention effect or active targeting by ligands on the surface of NPs; (2) NPs will reduce the drug exposure of healthy tissues by limiting drug distribution to target organ. Active tumor targeting of NPs may be achieved with either direct targeting or the pretargeting method. In direct targeting, NPs is covalently coupled with the ligand and the resulting drug carrier is administered at once. In the pretargeting approach, the therapeutic molecule is not coupled with the ligand and is administered after an appropriate delay time following the targeting ligand. Nobs *et al.* [52] explored both the approaches to target PLA

NPs to tumor cells. In the direct approach, NPs with mAbs exposed on their surface were incubated with the two tumor cells, while in the pretargeting protocol, tumor cells were pretargeted with biotinylated mABs prior to the administration of avidin-labeled NPs [52].

Verdun *et al.* [53] demonstrated in mice treated with doxorubicin incorporated into poly (isohexylcyanoacrylate) nanopsheres that higher concentrations of doxorubicin manifested in the liver, spleen and lungs than in mice treated with free doxorubicin [53]. Studies show that the polymeric composition of nanoparticles such as type, hydrophobicity and biodegradation profile of the polymer along with the associated drug's molecular weight, its localization in the nanospheres and mode of incorporation technique, adsorption or incorporation, have a great influence on the drug distribution pattern *in vivo* [54].

Extensive efforts have been devoted to achieving "active targeting" of nanoparticles in order to deliver drugs to the right targets, based on molecular recognition processes such as ligand-receptor or antigen–antibody interaction. Considering that fact that folate receptors are over–expressed on the surface of some human malignant cells and the cell adhesion molecules such as selectins and integrins are involved in metastatic events, nanoparticles bearing specific ligands such as folate may be used to target ovarian carcinoma while specific peptides or carbohydrates may be used to target integrins and selectins [55]. Oyewumi *et al.* [56] demonstrated that the benefits of folate ligand coating were to facilitate tumor cell internalization and retention of Gd-nanoparticles in the tumor tissue [56]. Targeting with small ligands appears more likely to succeed since they are easier to handle and manufacture. Furthermore, it could be advantageous when the active targeting ligands are used in combination with the long-circulating nanoparticles to maximize the likelihood of the success in active targeting of nanoparticles.

11.3.1.1 Nanoparticles for Oral delivery

In recent years, significant research has been done using nanoparticles as oral drug delivery vehicles. Oral delivery using nanoparticles has been shown to be far superior to free drug in terms of bioavailability, residence time, and biodistribution [57]. Advances in biotechnology and biochemistry have led to the discovery of a large number of bioactive molecules and vaccines based on peptides and proteins. Development of suitable carriers remains a challenge due to the fact that bioavailability of these molecules is limited by the epithelial barriers of the gastrointestinal tract and their susceptibility to gastrointestinal degradation by digestive enzymes. Polymeric nanoparticles allow encapsulation of bioactive molecules and protect them against enzymatic and hydrolytic degradation. For instance, it has been found that insulin-loaded nanoparticles have preserved insulin activity and produced blood glucose reduction in diabetic rats for up to 14 days following the oral administration [58].

Another study showed antifungal drug encapsulated in particles of less than 300 nm in diameter was detected in the lungs, liver, and spleen of mice seven days post oral administration, whereas oral-free formulations were cleared within 3 hours post administration [57]. For this application, the major interest is in lymphatic uptake of the nanoparticles by the Peyers Patches in the GALT (gut associated lymphoid tissue). There have been many reports as to the optimum size for Peyers Patch uptake ranging from less than 1 μm to 5 μm [59, 60]. However, it has been shown that microparticles remain in the Peyers Patches while nanoparticles are disseminated systemically [61, 62]

Nanoparticles can be engineered not only for oral absorption themselves, but can be used to deliver a drug directly to the source for gastrointestinal uptake, thereby protecting the drug from low pH and enzymes in the stomach. pH sensitive nanoparticles made from a poly(methylacrylic acid and methacyrlate) copolymer can increase the oral bioavailability of drugs like cyclosporin A by releasing their load at a specific pH within the gastrointestinal tract. The pH sensitivity allows this to happen as close as possible to the drug's absorption window through the Peyer's patches [63].

11.3.1.2 Nanoparticles for vaccine adjuvants and gene delivery

Polynucleotide vaccines work by delivering genes encoding relevant antigens to host cells where they are expressed, producing the antigenic protein within the vicinity of professional antigen presenting cells to initiate immune response. Such vaccines produce both humoral and cell-mediated immunity because intracellular production of protein, as opposed to extracellular deposition, stimulates both arms of the immune system [64]. The key ingredient of polynucleotide vaccines, DNA, can be produced cheaply and has much better storage and handling properties than the ingredients of the majority of protein-based vaccines. Hence, polynucleotide vaccines are set to supersede many conventional vaccines particularly for immunotherapy. However, there are several issues related to the delivery of polynucleotides which limit their application. These issues include efficient delivery of the polynucleotide to the target cell population and its localization to the nucleus of these cells, and ensuring that the integrity of the polynucleotide is maintained during delivery to the target site [7]. Nanoparticles loaded with plasmid DNA could also serve as an efficient sustained release gene delivery system due to their rapid escape from the degradative endo-lysosomal compartment to the cytoplasmic compartment [65]. Hedley *et al.* [66] reported that following their intracellular uptake and endolysosomal escape, nanoparticles could release DNA at a sustained rate resulting in sustained gene expression. This gene delivery strategy could be applied to facilitate bone healing by using PLGA nanoparticles containing therapeutic genes such as bone morphogenic protein. Antigen loaded polymeric nanoparticles represent an exciting approach to the enhancement of antigen- specific humeral and cellular responses via selective targeting of the antigen to the antigen presenting cells [67]. Akagi *et al.* reported fabrication of poly(amino acids) derivatives as vaccine delivery and adjuvants [68].

11.3.1.3 Nanoparticles for drug delivery into the brain

The blood-brain barrier (BBB) is the most important factor limiting the development of new drugs for the central nervous system [69]. The BBB is characterized by relatively impermeable endothelial cells with tight junctions, enzymatic activity and active efflux transport systems. It effectively prevents the passage of water-soluble molecules from the blood circulation into the CNS, and consequently only permits selective transport of molecules that are essential for brain function [70]. Strategies for nanoparticle targeting to the brain rely on the presence of and nanoparticle interaction with specific receptor-mediated transport systems in the BBB. For example, polysorbate 80/LDL, transferrin receptor binding antibody (such as OX26), lactoferrin, cell penetrating peptides and melanotransferrin have been shown capable of delivery of a self non-transportable drug into the brain via the chimeric construct that can undergo receptor-mediated transcytosis [71–74]. It has been reported that poly(butylcyanoacrylate) nanoparticles was able to deliver hexapeptide dalargin, doxorubicin and other agents into the brain which is significant because of the great difficulty for drugs to cross the BBB [73]. Despite some reported success with polysorbate 80 coated NPs, this system does have many shortcomings including desorption of polysorbate coating, rapid NP degradation and toxicity caused by presence of high concentration of polysorbate 80 [75]. OX26 MAbs (anti-transferrin receptor MAbs), the most studied BBB targeting antibody, have been used to enhance the BBB penetration of lipsosomes [76].

Another study by Kreuter *et. al.*[77] demonstrates the delivery of several drugs successfully through the blood brain barrier using polysorbate 80 coated PACA nanoparticles [77]. It is thought that after administration of the polysorbate 80-coated particles, apolipoprotein E (ApoE) adsorbs onto the surface. The ApoE protein mimics low density lipoprotein (LDL) causing the particles to be transported across the blood brain barrier via the LDL receptors. The effects of polysorbate-80 on transport through the blood brain barrier were confirmed by Sun et. al. with PLA nanoparticles [78]. Nanoparticles were also functionalized with a thiamine surface ligand and these particles, with an average diameter of 67 nm, were able to associate with the blood brain barrier thiamine transporters and thereby increase the unidirectional transfer coefficient for the particles into the brain [79].

11.4 Conclusions

In summary, NPs are a potentially viable drug delivery system capable of delivering a multitude of therapeutic agents. To optimize NPs as a drug delivery system, greater understanding of the different mechanisms of biological interactions and particle engineering is still required. However, biodegradable NPs because of their versatility for formulation, sustained release properties, sub-cellular size and biocompatibility with tissue and cells appear to be a promising drug delivery carrier system.

References

[1] Liu X, Liu C, Zhang W, Xie C, Wei G, Lu W: Oligoarginine-modified biodegradable nanoparticles improve the intestinal absorption of insulin. International Journal of Pharmaceutics, 2013, 448(1): 159–167.

[2] Mandal B, Bhattacharjee H, Mittal N, Sah H, Balabathula P, Thoma LA, Wood GC: Core–shell-type lipid–polymer hybrid nanoparticles as a drug delivery platform. Nanomedicine: Nanotechnology, Biology and Medicine, 2013, 9(4): 474–491.

[3] Silva JM, Videira M, Gaspar R, Préat V, Florindo HF: Immune system targeting by biodegradable nanoparticles for cancer vaccines. Journal of Controlled Release, 2013, 168(2): 179–199.

[4] Ye Z, Squillante E: The development and scale-up of biodegradable polymeric nanoparticles loaded with ibuprofen. Colloids and Surfaces A: Physicochemical and Engineering Aspects, 2013, 422(0): 75–80.

[5] Rao JP, Geckeler KE: Polymer nanoparticles: Preparation techniques and size-control parameters. Progress in Polymer Science, 2011, 36(7): 887–913.
[6] Xu T, Zhang N, Nichols HL, Shi D, Wen X: Modification of nanostructured materials for biomedical applications. Materials Science and Engineering: C, 2007, 27(3): 579–594.
[7] Mohanraj VJ, Chem Y: Nanoparticles - A Review. Tropical Journal of Pharmaceutical Research, 2006, 5(1): 561–573.
[8] Couvreur P: Nanoparticles in drug delivery: Past, present and future. Advanced Drug Delivery Reviews, 2013, 65(1): 21–23.
[9] Parveen S, Misra R, Sahoo SK: Nanoparticles: a boon to drug delivery, therapeutics, diagnostics and imaging. Nanomedicine: Nanotechnology, Biology and Medicine, 2012, 8(2): 147–166.
[10] Nagarwal RC, Kant S, Singh PN, Maiti P, Pandit JK: Polymeric nanoparticulate system: A potential approach for ocular drug delivery. Journal of Controlled Release, 2009, 136(1): 2–13.
[11] Liu Y, Miyoshi H, Nakamura M: Nanomedicine for drug delivery and imaging: A promising avenue for cancer therapy and diagnosis using targeted functional nanoparticles. International Journal of Cancer, 2007, 120(12): 2527–2537.
[12] van Vlerken LE, Amiji MM: Multi-functional polymeric nanoparticles for tumour-targeted drug delivery. Expert Opinion on Drug Delivery, 2006, 3(2): 205–216.
[13] Vasir JK, Labhasetwar V: Biodegradable nanoparticles for cytosolic delivery of therapeutics. Advanced Drug Delivery Reviews, 2007, 59(8): 718–728.
[14] Labhasetwar V, Song C, Levy RJ: Nanoparticle drug delivery system for restenosis. Advanced Drug Delivery Reviews, 1997, 24(1): 63–85.
[15] Hans ML, Lowman AM: Biodegradable nanoparticles for drug delivery and targeting. Current Opinion in Solid State and Materials Science, 2002, 6(4): 319–327.
[16] Mahapatro A, Singh DK: Biodegradable nanoparticles are excellent vehicle for site directed in-vivo delivery of drugs and vaccines. Journal of Nanobiotechnology, 2011, 9(55): 1–11.
[17] Soppimath KS, Aminabhavi TM, Kulkarni AR, Rudzinski WE: Biodegradable polymeric nanoparticles as drug delivery devices. Journal of Controlled Release, 2001, 70(1–2): 1–20.

[18] Park K, Lee S, Kang E, Kim K, Choi K, Kwon IC: New Generation of Multifunctional Nanoparticles for Cancer Imaging and Therapy. Advanced Functional Materials, 2009, 19(10): 1553–1566.

[19] Pinto Reis C, Neufeld RJ, Ribeiro AJ, Veiga F: Nanoencapsulation I. Methods for preparation of drug-loaded polymeric nanoparticles. Nanomedicine: Nanotechnology, Biology and Medicine, 2006, 2(1): 8–21.

[20] Scholes PD, Coombes AGA, Illum L, Daviz SS, Vert M, Davies MC: The preparation of sub-200 nm poly(lactide-co-glycolide) microspheres for site-specific drug delivery. Journal of Controlled Release, 1993, 25(1–2): 145–153.

[21] Niwa T, Takeuchi H, Hino T, Kunou N, Kawashima Y: Preparations of biodegradable nanospheres of water-soluble and insoluble drugs with D,L-lactide/glycolide copolymer by a novel spontaneous emulsification solvent diffusion method, and the drug release behavior. Journal of Controlled Release, 1993, 25(1–2): 89–98.

[22] Govender T, Stolnik S, Garnett MC, Illum L, Davis SS: PLGA nanoparticles prepared by nanoprecipitation: drug loading and release studies of a water soluble drug. Journal of Controlled Release, 1999, 57(2): 171–185.

[23] Zweers MLT, Engbers GHM, Grijpma DW, Feijen J: In vitro degradation of nanoparticles prepared from polymers based on DL-lactide, glycolide and poly(ethylene oxide). Journal of Controlled Release, 2004, 100(3): 347–356.

[24] Eley JG, Pujari VD, McLane J: Poly (Lactide-co-Glycolide) Nanoparticles Containing Coumarin-6 for Suppository Delivery: In Vitro Release Profile and In Vivo Tissue Distribution, 2004, 11(4): 255–261.

[25] Zhang Q, Shen Z, Nagai T: Prolonged hypoglycemic effect of insulin-loaded polybutylcyanoacrylate nanoparticles after pulmonary administration to normal rats. International Journal of Pharmaceutics, 2001, 218(1–2): 75–80.

[26] Boudad H, Legrand P, Lebas G, Cheron M, Duchêne D, Ponchel G: Combined hydroxypropyl-[beta]-cyclodextrin and poly(alkylcyanoacrylate) nanoparticles intended for oral administration of saquinavir. International Journal of Pharmaceutics, 2001, 218(1–2): 113–124.

[27] Puglisi G, Fresta M, Giammona G, Ventura CA: Influence of the preparation conditions on poly(ethylcyanoacrylate) nanocapsule formation. International Journal of Pharmaceutics, 1995, 125(2): 283–287.

[28] Toshio Y, Mitsuru H, Shozo M, Hitoshi S: Specific delivery of mitomycin c to the liver, spleen and lung: Nano- and m1crospherical carriers of gelatin. International Journal of Pharmaceutics, 1981, 8(2): 131–141.
[29] Kwok KK, Groves M, Burgess D: Production of 5–15 μm Diameter Alginate-Polylysine Microcapsules by an Air-Atomization Technique. Pharmaceutical Research, 1991, 8(3): 341–344.
[30] Aslani P, Kennedy RA: Studies on diffusion in alginate gels. I. Effect of cross-linking with calcium or zinc ions on diffusion of acetaminophen. Journal of Controlled Release, 1996, 42(1): 75–82.
[31] Calvo P, Remuñan-Lόpez C, Vila-Jato JL, Alonso MJ: Chitosan and Chitosan/Ethylene Oxide-Propylene Oxide Block Copolymer Nanoparticles as Novel Carriers for Proteins and Vaccines. Pharmaceutical Research, 1997, 14(10): 1431–1436.
[32] Kumari A, Yadav SK, Yadav SC: Biodegradable polymeric nanoparticles based drug delivery systems. Colloids and Surfaces B: Biointerfaces, 2010, 75(1): 1–18.
[33] Carrasquillo KG, Stanley AM, Aponte-Carro JC, De Jésus P, Costantino HR, Bosques CJ, Griebenow K: Non-aqueous encapsulation of excipient-stabilized spray-freeze dried BSA into poly(lactide-co-glycolide) microspheres results in release of native protein. Journal of Controlled Release, 2001, 76(3): 199–208.
[34] Fessi H, Puisieux F, Devissaguet JP, Ammoury N, Benita S: Nanocapsule formation by interfacial polymer deposition following solvent displacement. International Journal of Pharmaceutics, 1989, 55(1): R1–R4.
[35] Choi C, Chae SY, Nah J-W: Thermosensitive poly(N-isopropylacry lamide)-b-poly([epsilon]-caprolactone) nanoparticles for efficient drug delivery system. Polymer, 2006, 47(13): 4571–4580.
[36] Kim SY, Lee YM: Taxol-loaded block copolymer nanospheres composed of methoxy poly(ethylene glycol) and poly(caprolactone) as novel anticancer drug carriers. Biomaterials, 2001, 22(13): 1697–1704.
[37] Sinha VR, Singla AK, Wadhawan S, Kaushik R, Kumria R, Bansal K, Dhawan S: Chitosan microspheres as a potential carrier for drugs. International Journal of Pharmaceutics, 2004, 274(1–2): 1–33.
[38] Gan Q, Wang T: Chitosan nanoparticle as protein delivery carrier–Systematic examination of fabrication conditions for efficient loading and release. Colloids and Surfaces B: Biointerfaces, 2007, 59(1): 24–34.

[39] Zillies JC, Zwiorek K, Hoffmann F, Vollmar A, Anchordoquy TJ, Winter G, Coester C: Formulation development of freeze-dried oligonucleotide-loaded gelatin nanoparticles. European Journal of Pharmaceutics and Biopharmaceutics, 2008, 70(2): 514–521.

[40] Ofokansi K, Winter G, Fricker G, Coester C: Matrix-loaded biodegradable gelatin nanoparticles as new approach to improve drug loading and delivery. European Journal of Pharmaceutics and Biopharmaceutics, 76(1): 1–9.

[41] Kim D, El-Shall H, Dennis D, Morey T: Interaction of PLGA nanoparticles with human blood constituents. Colloids and Surfaces B: Biointerfaces, 2005, 40(2): 83–91.

[42] Gref R, Lück M, Quellec P, Marchand M, Dellacherie E, Harnisch S, Blunk T, Müller RH: 'Stealth' corona-core nanoparticles surface modified by polyethylene glycol (PEG): influences of the corona (PEG chain length and surface density) and of the core composition on phagocytic uptake and plasma protein adsorption. Colloids and Surfaces B: Biointerfaces, 2000, 18(3–4): 301–313.

[43] Gref R, Couvreur P, Barratt G, Mysiakine E: Surface-engineered nanoparticles for multiple ligand coupling. Biomaterials, 2003, 24(24): 4529–4537.

[44] Naahidi S, Jafari M, Edalat F, Raymond K, Khademhosseini A, Chen P: Biocompatibility of engineered nanoparticles for drug delivery. Journal of Controlled Release, 2013, 166(2): 182–194.

[45] Leroux J-C, Allémann E, De Jaeghere F, Doelker E, Gurny R: Biodegradable nanoparticles – From sustained release formulations to improved site specific drug delivery. Journal of Controlled Release, 1996, 39(2–3): 339–350.

[46] Tobío M, Sánchez A, Vila A, Soriano I, Evora C, Vila-Jato JL, Alonso MJ: The role of PEG on the stability in digestive fluids and in vivo fate of PEG-PLA nanoparticles following oral administration. Colloids and Surfaces B: Biointerfaces, 2000, 18(3–4): 315–323.

[47] Vila A, Sánchez A, Toblo M, Calvo P, Alonso MJ: Design of biodegradable particles for protein delivery. Journal of Controlled Release, 2002, 78(1–3): 15–24.

[48] Torchilin VP, Trubetskoy VS: Which polymers can make nanoparticulate drug carriers long-circulating? Advanced Drug Delivery Reviews, 1995, 16(2–3): 141–155.

[49] Stolnik S, Illum L, Davis SS: Long circulating microparticulate drug carriers. Advanced Drug Delivery Reviews, 1995, 16(2–3): 195–214.

[50] Storm G, Belliot SO, Daemen T, Lasic DD: Surface modification of nanoparticles to oppose uptake by the mononuclear phagocyte system. Advanced Drug Delivery Reviews, 1995, 17(1): 31–48.
[51] Mahapatro A, Johnson DM, Patel DN, Feldman MD, Ayon AA, Agrawal CM: Drug Delivery from Therapeutic Self-Assembled Monolayers (T-SAMs) on 316L Stainless Steel Current Topics in Medicinal Chemistry, 2008, 8(4): 281–289.
[52] Nobs L, Buchegger F, Gurny R, AllÃ©mann E: Biodegradable Nanoparticles for Direct or Two-Step Tumor Immunotargeting. Bioconjugate Chemistry, 2005, 17(1): 139–145.
[53] Verdun C, Brasseur F, Vranckx H, Couvreur P, Roland M: Tissue distribution of doxorubicin associated with polyisohexylcyanoacrylate nanoparticles. Cancer Chemotherapy and Pharmacology, 1990, 26(1): 13–18.
[54] Couvreur P, Kante B, Lenaerts V, Scailteur V, Roland M, Speiser P: Tissue distribution of antitumor drugs associated with polyalkylcyanoacrylate nanoparticles. Journal of Pharmaceutical Sciences, 1980, 69(2): 199–202.
[55] Stella B, Arpicco S, Peracchia MT, Desmaële D, Hoebeke J, Renoir M, D'Angelo J, Cattel L, Couvreur P: Design of folic acid-conjugated nanoparticles for drug targeting. Journal of Pharmaceutical Sciences, 2000, 89(11): 1452–1464.
[56] Oyewumi MO, Yokel RA, Jay M, Coakley T, Mumper RJ: Comparison of cell uptake, biodistribution and tumor retention of folate-coated and PEG-coated gadolinium nanoparticles in tumor-bearing mice. Journal of Controlled Release, 2004, 95(3): 613–626.
[57] Pandey R, Ahmad Z, Sharma S, Khuller GK: Nano-encapsulation of azole antifungals: Potential applications to improve oral drug delivery. International Journal of Pharmaceutics, 2005, 301(1–2): 268–276.
[58] Damgé C, Michel C, Aprahamian M, Couvreur P, Devissaguet JP: Nanocapsules as carriers for oral peptide delivery. Journal of Controlled Release, 1990, 13(2–3): 233–239.
[59] Lemoine D, Préat V: Polymeric nanoparticles as delivery system for influenza virus glycoproteins. Journal of Controlled Release, 1998, 54(1): 15–27.
[60] Torché A-M, Jouan H, Le Corre P, Albina E, Primault R, Jestin A, Le Verge R: Ex vivo and in situ PLGA microspheres uptake by pig ileal Peyer's patch segment. International Journal of Pharmaceutics, 2000, 201(1): 15–27.

[61] Jenkins PG, Howard KA, Blackball NW, Thomas NW, Davis SS, O'Hagan DT: Microparticulate absorption from the rat intestine. Journal of Controlled Release, 1994, 29(3): 339–350.
[62] Eldridge JH, Hammond CJ, Meulbroek JA, Staas JK, Gilley RM, Tice TR: Controlled vaccine release in the gut-associated lymphoid tissues. I. Orally administered biodegradable microspheres target the peyer's patches. Journal of Controlled Release 1990, 11(1–3): 205–214.
[63] Dai J, Nagai T, Wang X, Zhang T, Meng M, Zhang Q: pH-sensitive nanoparticles for improving the oral bioavailability of cyclosporine A. International Journal of Pharmaceutics, 2004, 280(1–2): 229–240.
[64] Gurunathan S, Wu C-Y, Freidag BL, Seder RA: DNA vaccines: a key for inducing long-term cellular immunity. Current Opinion in Immunology, 2000, 12(4): 442–447.
[65] Panyam J, Zhou W-Z, Prabha S, Sahoo SK, Labhasetwar V: Rapid endo-lysosomal escape of poly(DL-lactide-co-glycolide) nanoparticles: implications for drug and gene delivery. The FASEB Journal, 2002, 16(10): 1217–1226.
[66] Hedley ML, Curley J, Urban R: Microspheres containing plasmid-encoded antigens elicit cytotoxic T-cell responses. Nat. Med., 1998, 4(3): 365–368.
[67] Akagi T, Baba M, Akashi M: Biodegradable Nanoparticles as Vaccine Adjuvants and Delivery Systems: Regulation of Immune Responses by Nanoparticle-Based Vaccine. In: Polymers in Nanomedicine. Volume 247, edn. Edited by Kunugi S, Yamaoka T: Springer Berlin Heidelberg; 2012: 31–64.
[68] Akagi T, Wang X, Uto T, Baba M, Akashi M: Protein direct delivery to dendritic cells using nanoparticles based on amphiphilic poly(amino acid) derivatives. Biomaterials, 2007, 28(23): 3427–3436.
[69] Chakraborty C, Sarkar B, Hsu C, Wen Z, Lin C, Shieh P: Future prospects of nanoparticles on brain targeted drug delivery. Journal of Neuro-Oncology, 2009, 93(2): 285–286.
[70] Chen Y, Dalwadi G, Benson HAE: Drug Delivery Across the Blood-Brain Barrier Current Drug Delivery, 2004, 1(4): 361–376.
[71] Gabathuler R, Arthur G, Kennard M, Chen Q, Tsai S, Yang J, Schoorl W, Vitalis TZ, Jefferies WA: Development of a potential protein vector (NeuroTrans) to deliver drugs across the blood-brain barrier. International Congress Series, 2005, 1277: 171–184.

[72] Ji B, Maeda J, Higuchi M, Inoue K, Akita H, Harashima H, Suhara T: Pharmacokinetics and brain uptake of lactoferrin in rats. Life Sciences, 2006, 78(8): 851–855.

[73] Pardridge WM: Drug and gene targeting to the brain with molecular trojan horses. Nat Rev Drug Discov, 2002, 1(2): 131–139.

[74] Scherrmann J-M, Temsamani J: The use of Pep: Trans vectors for the delivery of drugs into the central nervous system. International Congress Series, 2005, 1277: 199–211.

[75] Olivier J-C: Drug Transport to Brain with Targeted Nanoparticles. NeuroRx: the journal of the American Society for Experimental NeuroTherapeutics, 2005, 2(1): 108–119.

[76] Pardridge WM: Drug and gene targeting to the brain via blood-brain barrier receptor-mediated transport systems. International Congress Series, 2005, 1277: 49–62.

[77] Kreuter J: Nanoparticulate systems for brain delivery of drugs. Advanced Drug Delivery Reviews, 2001, 47(1): 65–81.

[78] Sun W, Xie C, Wang H, Hu Y: Specific role of polysorbate 80 coating on the targeting of nanoparticles to the brain. Biomaterials, 2004, 25(15): 3065–3071.

[79] Lockman PR, Oyewumi MO, Koziara JM, Roder KE, Mumper RJ, Allen DD: Brain uptake of thiamine-coated nanoparticles. Journal of Controlled Release, 2003, 93(3): 271–282.

About the Editors

Kishore R Sakharkar is an adjunct Professor at the Department of Biotechnology, Pune, India. He is also a research Director at OmicsVista, Singapore.

Meena K Sakharkar is an Associate Professor at the Department of Pharmacy and Nutrition, University of Saskatchewan, Canada. She was a Professor at the Graduate School of Life and Environmental Sciences, University of Tsukuba, Japan from 2010.03–2014.02.

Ramesh Chandra did his PhD. in Chemistry at Department of Chemistry, University of Delhi in 1982. He held several scientific and academic positions which includes, Assistant Research Professor at Pharmacology Dept. School of Medicine, SUNY, Stonybrook, New York (1986), Research Scientist-C (UGC Professor) at University of Delhi (1993–96), Founder Director ACBR (1991 onwards), Professor at Department of Chemistry, University of Delhi, Delhi (1993 onwards), Vice-Chancellor, Bundelkhand University, Jhansi (1999–2005). His research interests include metalloporhyrins, natural products and nanomedicine.

About the Contributors

Rashmi M. Bhande is working as a Women Scientist and Principal Investigator in the Department of Science and Technology Government of India supported major research project related to resistance of Urinary tract infections causing bacteria and development of new antimicrobial therapy to treat Infectious Diseases. She is having about five years research experience as a Women Scientist. Her research interest is in Medical Microbiology specially Treatment of Infectious Disease using nonmaterial. She has participated in various national and international Conferences/Workshops on nanotechnology and published research articles on the nanotechnology based subjects in Journals of National and International repute.

Aniruddha Bhati presently is a PhD student in the Department of Nanoscience and technology, PD Patel Institute of Applied Sciences, Charotar University of Science and technology, Gujarat. His current research area is Magnetic nanoparticles-Protein interaction and Magnetic nanoparticles based bioanalytical extraction of small molecules.

In the era of Nanoscience and technology, understanding the interaction of nanoparticles with the biological system is one of the important aspects for various applications like nanotoxicology. His research work also include enhancement of the total protein extraction efficiency using magnetic nanoparticles and developing an alternative method for bio-analytical extraction of drugs using magnetic nanoparticles. Recently, both the methods were successfully developed and patented.

He has completed his Masters in Biotechnology from VIT University, Vellore, India with dissertation on studying activity of common plant peroxidases.

A.F. Fonseca is currently working at the Institute of Bioinformatics and Biotechnology, Natal, Brazil, and Brain Institute, UFRN, Natal, Brazil.

Ran Gao received her MSc and PhD degrees from Tsinghua University, Peking Union Medical College, China, and University of Tsukuba, Japan, respectively. She is currently doing her post-doctoral research at the National

Institute of Advanced Industrial Science and Technology (AIST), Tsukuba, Japan. Her main research interest is to understand the molecular mechanism of aging and cancer metastasis using different techniques.

Nidhi Gupta received her M.S. in Biomedical Science from Dr. B.R. Ambedkar Center for Biomedical Research (ACBR), University of Delhi, Delhi, India in 2004. She did her PhD. with Dr. A.C. Banerjea at National Institute of Immunology (NII), New Delhi, India in 2008. After completing her doctoral research she joined Clinical Research Institute of Montreal, Montreal, Canada as Postdoctoral Fellow in 2008 and later worked at McGill University, Montreal. She joined The IIS University, India as Research Scientist at Department of Biotechnology in 2014. Her field of interest includes catalytic nucleic acids, siRNA and antisense.

Arun Iyer is an Assistant Professor of Pharmaceutical Sciences in the Eugene Applebaum College of Pharmacy and Health Sciences at Wayne State University in Detroit, MI and serves as a Principal Investigator on a National Cancer Institute (NCI/NIH) funded research grant on designing drug and gene delivery systems for targeting lung cancers. He completed his Bachelor's degree in Chemistry and Master's degree in Polymer Science from the University of Pune, India and received his Ph.D. in Polymer Chemistry under the mentorship of Prof. Hiroshi Maeda from the Graduate School of Engineering at Sojo University in Japan. He completed his postdoctoral training at the Department of Radiology and Biomedical Imaging at the University of California San Francisco (UCSF), School of Medicine. In 2012 He was awarded the prestigious CRS T. Nagai Postdoctoral Research Achievement award co-sponsored by the Controlled Release Society (CRS), USA, and the Nagai Foundation, Japan, for his outstanding contribution to the controlled release science and technology. He has a broad research interest in the area of nanomedicine and drug/gene delivery systems, biomedical imaging and development of use-inspired bio- and nano-medical technologies aimed towards clinical translation.

Yoshio Kato is a Senior Researcher at the Biomedical Research Institute, National Institute of Advanced Industrial Science and Technology (AIST), Tsukuba, Japan. He received his B.S. in Chemistry from University of Tsukuba in 1999, and Ph.D. in Biotechnology from the University of Tokyo in 2004. He won Young Scientist Award of Japanese Gene-Delivery organization in 2003. He has been engineering the reactive biomolecules that detect or control the gene expression in living cells. Recently, he devised the cell-penetrating zinc-finger nuclease proteins that modify genomic DNA sequence specifically for genome editing.

Sunil Kaul is a Chief Senior Research Scientist at the National Institute of Advanced Industrial Science & Technology (AIST). He had his M. Phil. and Ph. D. degrees from the University of Delhi, India. After his initial post-doctoral training, he was appointed as a researcher at the AIST in 1992. His major research interest is to study the molecular mechanism of stress, aging and cancer. Specific interests include high-resolution imaging and cancer interventional studies. He has been merging the traditional knowledge with modern technologies like gene silencing and imaging to understand the mechanism of action of Ayurvedic herb, Ashwagandha.

C. N. Khobragade is a Professor and Head (Biotechnology) in the School of Life Sciences. He is having about 20 years teaching and research experience in the area of Therapeutic enzymology and Nanobiotechnology. He has organized national and international conferences related to his area of interest and published several research articles in National and International journals.

J.E. Kroll is currently working at the Institute of Bioinformatics and Biotechnology, Natal, Brazil, and Brain Institute, UFRN, Natal, Brazil.

James Lyons James has a bachelor's of science in Biotechnology with a concentration in Nanobiotechnology from Harrisburg University of Science and Technology. He has spent time working on onco-therapeutics both in industry and academia. He has a thorough understanding of nanotechnology and drug delivery systems. His work has focused on nanotechnology as it applies to cancer therapeutics and diagnostics. Much of his research has been in the development of nanotechnology based therapeutics for Melanoma. James currently serves as a Research Technologist in the Penn State Hershey Melanoma Center at the Penn State Hershey Cancer Institute. He plans to pursue a PhD. next fall.

Anil Mahapatro is an assistant professor of Bioengineering in the Department of Industrial and Manufacturing Engineering at Wichita State University, Wichita, KS. His lab is working on development and application of biodegradable implants and drug delivery systems. His lab is developing biodegradable nanoparticles for gene and drug delivery and is also participating in a collaborative project with Dinesh K. Singh's lab on the use of biodegradable nanoparticles in delivering a HIV- DNA vaccine within the cervical and vaginal mucosa.

Faryal Mir received her BS degree in Biochemistry from Northeastern University in Boston, MA. She is currently a Research Assistant in the Laboratory of Biomaterials and Advanced Nano-Delivery Systems at Northeastern University, headed by Prof. Mansoor Amiji. She is interested in oncology and clinical chemistry.

Yukio Nagasaki was born in 1959. He received a B.S. and Ph.D. degrees in Engineering School of Science University of Tokyo in 1982, and 1987. Since 1987, he was working Science University of Tokyo as Research Associate, Assistant Professor, Associate Professor and Professor. In 2004, he moved to Graduate School of Pure and Applied Sciences, University of Tsukuba. He hold a concurrent posts of Adjunct Professor, Master's School of Medical Sciences, Graduate School of Comprehensive Human Sciences, University of Tsukuba, Principal Investigator, International Center for Materials Nanoarchitectonics Satellite (WPI-MANA), and National Institute for Materials Science (NIMS). During last 30 years, he was engaged in materials science especially in the field of biology, pharmaceutics and medical sciences. Especially, he was focusing on biointerface, drug delivery system and nanomedicine. He published more than 200 scientific papers. He received the excellent Ph.D. thesis award from Inoue Foundation of Science in 1989, Young Researcher Award from Polymer Society, Japan in 1993 and SPSJ Mitsubishi Chemical Award from Polymer Society, Japan in 2010.

Surendra Nimesh is an internationally recognized expert of nanotechnology for biological applications with specialization in drug and gene delivery. He received his M.S. in Biomedical Science from Dr. B.R. Ambedkar Center for Biomedical Science Research (ACBR), University of Delhi, Delhi, India in 2001. He completed his PhD. in Nanotechnology at ACBR and Institute of Genomics and Integrative Biology (CSIR), Delhi, India in 2007. After completing his postdoctoral studies at Ecole Polyetchnique of Montreal, Montreal in 2009, he joined Clinical Research Institute of Montreal, Montreal, Canada as Postdoctoral Fellow. He worked for a short duration at McGill University, Montreal and thereafter joined Health Canada, Canada as NSERC visiting fellow in 2012. He joined Central University of Rajasthan, India as UGC-Assistant Professor at School of life Sciences in 2013. He has authored more than 14 research papers, 7 review articles in international peer reviewed journal, 7 book chapters and 3 books. His research interests include nanoparticles mediated gene, siRNA and drug delivery for therapeutics.

Alan Prem Kumar earned his Ph.D. from University of North Texas, USA. From his Ph.D. work, he discovered a novel regulatory protein, PyrR for the pyrimidine biosynthetic pathway in Pseudomonas. Because pyrimidine biosynthesis is an essential step in the progression of secondary Pseudomonas infections, PyrR presents an attractive anti-pseudomonal drug target. Dr. Kumar then pursued Postdoctoral training in Cancer Research at Sidney Kimmel Cancer Center, California, USA. He was awarded a Postdoctoral Fellowship for his work on the role of nuclear receptors in the transcriptional regulation of human myeloperoxidase, a leukocyte enzyme implicated as causative agent in atherosclerosis and Alzheimer's disease. Dr. Kumar relocated back to Singapore to join the Faculty of Medicine, National University of Singapore as an independent Principal Investigator to continue on his expertise on nuclear receptor signaling in cancer biology. His current research focus in the areas of signaling by nuclear receptors and oncogenes in breast tumor cells as well as the development of molecular therapeutics and biomarkers of drug action in breast cancer. Dr. Kumar's group also focuses on developing new drugs for the treatment of breast cancer. Over the years, Dr. Kumar and his group have forged relationships with scientists and oncologists in cancer research and with cancer advocacy groups in Singapore.

Kishore R Sakharkar is an adjunct Professor at the Department of Biotechnology, Pune, India. He is also a research Director at OmicsVista, Singapore.

Meena K Sakharkar is an Associate Professor at the Department of Pharmacy and Nutrition, University of Saskatchewan, Canada. She was a Professor at the Graduate School of Life and Environmental Sciences, University of Tsukuba, Japan from 2010.03–2014.02.

Shikha Satendra Singh, a self-motivated student to pursue biological science and to have a better understanding, attained both Bachelors and Master's in Biotechnology (University of Mumbai, India). She is currently a PhD student at Cancer Science Institute of Singapore and Department of Pharmacology at the National University of Singapore.

Gautam Sethi After completion of his Ph.D research in Cellular Immunology from India in 2004, Dr. Sethi joined MD Anderson Cancer Center (# 1 Cancer Hospital in US) and worked for three years in the area of inflammation and cancer research for his post-doctoral studies. Later in 2008, he accepted the position of Assistant Professor in the Department of Pharmacology at National University of Singapore. The focus of his research over the past few years has been to elucidate the mechanism(s) of activation of transcription factors namely NF-κB and STAT3 by carcinogens and inflammatory agents and the identification of novel pharmacological inhibitors of these proteins for prevention of and therapy for cancer. From natural sources, his group has identified numerous small molecules that can suppress NF-κB/STAT3 activation and inhibit TNF/IL-6 signaling. Thus, these agents have potential treatment properties for numerous inflammatory diseases, including cancer and generally quite safe. Also, the results his research work have so far resulted in more than hundred ten scientific publications in high impact factor peer reviewed journals and several international research awards.

Arati Sharma, Ph.D. is an Assistant Professor at the Department of Pharmacology, Pennsylvania State University, USA. She obtained her Ph.D. in Life Sciences from the Gujarat University, India with specialization in Reproductive Toxicology followed by Postdoctoral studies at Pennsylvania State University, USA. Current research efforts focus on melanoma, the most aggressive and metastatic form of skin cancer. Her interest lies in understanding the biology of melanoma, proteins and signaling pathways involved in melanoma progression and development. Her research interests have expanded to include chemopreventive and chemotherapeutic agent development and elucidating the mechanism of action as well as in the development of novel drug delivery approaches using Nanotechnology based approaches. She has authored more than 50 papers, several patents book chapter. She has been the recipient of several prestigious awards including the by the National Institutes of Health, as well as a number of generous foundations.

Karun Sharma is a PhD student at the Graduate School of Life and Environmental Sciences, University of Tsukuba, Japan and holds a MEXT scholarship. He has a Masters in Pharmacy T.M.U Moradabad, India.

Babita Shashni is a PhD student at the Graduate School of Life and Environmental Sciences, University of Tsukuba, Japan and holds a Mitsubishi scholarship. She has a Masters in Biotechnology from Guru Nanak Dev University, Amritsar, India.

Sakshi Sikka has a diverse background, with Indian roots, with a high school education at Shanghai, China and an undergraduate education at National University of Singapore. She earned a degree in Biomedical Science from the National University of Singapore, graduating in 2011. She served as a Research Assistant at Cancer Science Institute of Singapore and is involved in research projects on cancer and natural compounds.

Dinesh K. Singh is a professor of microbiology at the Winston Salem State University. His lab is working on development of a DNA vaccine for HIV/AIDS. His other research interest involves prevention of HIV-1 transmission at the cervical/vaginal mucosal surfaces, use of nanoparticles in prevention of transmission of HIV at the mucosal surfaces. He is also participating in a collaborative project with Anil Mahapatro's lab on the use of biodegradable nanoparticles in delivering a HIV- DNA vaccine within the cervical and vaginal mucosa.

S. J. de Souza is currently working at Brain Institute, UFRN, Natal, Brazil.

Jaimic Trivedi is a biotechnopreneur and a research enthusiast. Currently he is working to bridge the gaps between academia and industry and more importantly research world and industries. His endeavours enable him to understand basic research developments at the grassroots and his work is to nurture them for market needs. Mr. Jaimic is keeping himself updated with recent research trends as he is also pursuing PhD in Microbiology from Charotar University of Science and technology.

His other research interest includes Nanotoxicology, Industrial Microbiology, Diagnostics and tissue culture. He has previously worked at Pharmaceutical education and Research development centre Ahmedabad. He completed Masters in Sciences from Gujarat University

Ganesan Venkatesan currently works as a Manufacturing Associate II for Seracare Life Sciences Inc., in Milford, MA. He received his MS degree in Pharmaceutical Sciences from Northeastern University in Boston, MA and Bachelors degree in Biotechnology form PSG College of Technology, Coimbatore, India. Prior to joining Seracare, he worked as a Graduate Research

Assistant under the mentorship of Dr. Arun Iyer in the Laboratory of Biomaterials and Advanced Nano-Delivery Systems at Northeastern University, headed by Prof. Mansoor Amiji. His area of research interests includes diagnostic and therapeutic biopolymer purification.

Long Binh Vong was born in 1984. He received his B.S. in Biochemistry at the Fonseca is currently working at the Institut in 2006, as well as a Master Degree in Biochemistry in 2010. He received the research fellowship of The Japan-East Asia Network of Exchange for Students & Youths (JENESYS, 2010–2011) to study in University of Tsukuba, Japan. Since 2012, he has been studying the Ph.D. in Materials Science, Graduate School of Pure and Applied Sciences, University of Tsukuba and receiving the fellowship of Japan Society for the Promotion of Science (JSPS) for his Ph.D. course. His research activities focus on the development of novel nanomedicine for treatment of oxidative stress-induced diseases in gastrointestinal tract, especially ulcerative colitis, as well as colitis-associated colon cancer.

Renu Wadhwa had her first Ph. D. from the Guru Nanak Dev University, India and the second Ph. D. from the University of Tsukuba, Japan. She had her post-doctoral training at the University of Newcastle, England and RIKEN Japan. At present, she is a Prime Senior Researcher and has been leading a research team working on the mechanisms of cell proliferation controls at the Biomedical Research Institute, National Institute of Advanced Industrial Science and Technology (AIST), Japan. Her major research interest is to understand the molecular mechanism of aging and cancer using normal and cancer cells as model systems. She had originally cloned a novel member of hsp70 family protein in 1993 and named it “mortalin”. Since then she has made several original findings describing the functional characteristics of this protein and its role in cancer and age-related disorders.

Toru Yoshitomi was born in 1981. He received a B.S. in Faculty of Industrial Science and Technology, Tokyo University of Science in 2005 and Ph.D. degrees in Graduate School of Pure and Applied Sciences, University of Tsukuba in 2010. Since 2010, he worked in Graduate School of Pure and Applied Sciences, University of Tsukuba as postdoctoral fellow. In 2013, he moved to Department of Chemistry, Graduate School of Science, University of Tokyo as postdoctoral fellow of Japan Society for the Promotion of Science (JSPS). He was focusing on nanomedicine for treatment of oxidative stress injuries.

Index

A
Active targeting 205, 221, 275, 303
Animal 9, 85, 98, 201, 299
Anticancer reagents 38, 87, 220, 319
Antimicrobial Nanotechnology Based Drug Delivery 171
Appications of Nanoparticles 189

B
Biodegradable Nanoparticles 292, 316, 326
Breast cancer 8, 61, 242, 267

C
Cancer 1, 29, 51, 86, 107
Carbon nanotubes (CNTs) 181, 243, 278
Chitosan Nanoparticles 175, 283, 313
CNS delivery of drugs 278, 333
Colitis 199, 213
Copper Nanoparticle 176, 194

D
Dendrimers 174, 232, 293
Differentiation 10, 28, 122, 268
Drug delivery system 169, 190, 224, 287
Drug-infused nanoparticles 175

F
Fluorescent Nanoparticles 183, 303

G
Glycolysis 66, 69, 245

H
Hepatocellular carcinoma 51, 101, 123, 231

I
Immunomodulatory effects of nanotechnology-based drug delivery systems 179
Infection 49, 85, 99, 127, 167
Inflammation 61, 127, 205, 298
Inflammatory bowel disease 209
In-vivo 87, 333
Ischemia-reperfusion injuries 205

L
Liposome for antimicrobial drug delivery 171
Liposomes 171, 220, 276, 302

M
Magnesium Nanoparticle 177
miRNA 25, 52
Model genome 2, 88, 269
Mouse 7, 31, 86, 213, 294
Multifunctional Nanoparticles 219, 244, 320

N
Nanobiotechnology 288
Nanocarriers 182, 238, 272, 297
Nanoemulsions 180
Nanomedicine 199, 267, 332
Nanoparticles and vaccine adjuvant 181, 330
Nanoparticles A Study of Urinary Tract Infections 183
Nanotechnology 169, 184, 287
Nanotechnology-based vaccines and immunostimulatory adjuvant 179

Natural compounds 101, 123, 142
Nitric oxide-releasing nanoparticles 178
Nitroxide radicals 200, 214

O

Oncogenic pathways 111
Oral Delivery of nanomedicines 180, 319, 329
Oxidative stress 39, 83, 199, 298

P

Passive targeting 224, 273
pH 64, 175, 204, 330
Poly(ethylene glycol) 201, 222
Polymeric micelles 182, 202, 293
Polymeric Nanoparticles 169, 183, 229, 336
Polymeric NPs 173, 327
Polymers 173, 200, 230, 318
PPARγ 62, 77, 135

R

Reactive oxygen species 39, 72, 142, 178, 299
Redox nanoparticles 200, 205
Ribozymes 25, 31
ROS 39, 72, 142, 199, 299

S

Silver Nanoparticle 176, 297
siRNA 25, 39, 231, 303
Solid lipid (SL) NPs 172
Surfactant-based nanoemulsions 182
Synergism of Antibiotics with Zinc Oxide 183

T

Targeted Drug Delivery 191, 221
Targeted Vaccine Delivery 180, 315, 331
Theranostic Nanosystems 235, 251
Titanium Nanoparticle 177
Toxicological concerns and solutions 287
Tumor Imaging 241, 290, 317
Tumor Targeting Carrier System 221, 244, 303, 328
Tumor Targeting 221, 274, 328
Tumor 6, 40, 89, 120, 131

Z

Zinc Nanoparticle 178